AF307897

Ergebnisse der Mathematik und ihrer Grenzgebiete

Band 83

Herausgegeben von P. R. Halmos   P. J. Hilton
R. Remmert   B. Szőkefalvi-Nagy

Unter Mitwirkung von L. V. Ahlfors   R. Baer
F. L. Bauer   A. Dold   J. L. Doob   S. Eilenberg
K. W. Gruenberg   M. Kneser   G. H. Müller
M. M. Postnikov   B. Segre   E. Sperner

Geschäftsführender Herausgeber: P. J. Hilton

Russell C. Walker

The Stone-Čech Compactification

Springer-Verlag
Berlin Heidelberg New York 1974

Russell C. Walker

Department of Mathematics, Carnegie-Mellon University, Pittsburgh

AMS Subject Classification (1970): 54–02, 54D35, 54D40

ISBN-13: 978-3-642-61937-3   e-ISBN-13: 978-3-642-61935-9
DOI: 10.1007/978-3-642-61935-9

Preface

Recent research has produced a large number of results concerning the Stone-Čech compactification or involving it in a central manner. The goal of this volume is to make many of these results easily accessible by collecting them in a single source together with the necessary introductory material.

The author's interest in this area had its origin in his fascination with the classic text *Rings of Continuous Functions* by Leonard Gillman and Meyer Jerison. This excellent synthesis of algebra and topology appeared in 1960 and did much to draw attention to the Stone-Čech compactification βX as a tool to investigate the relationships between a space X and the rings $C(X)$ and $C^*(X)$ of real-valued continuous functions. Although in the approach taken here βX is viewed as the object of study rather than as a tool, the influence of *Rings of Continuous Functions* is clearly evident.

Three introductory chapters make the book essentially self-contained and the exposition suitable for the student who has completed a first course in topology at the graduate level. The development of the Stone-Čech compactification and the more specialized topological prerequisites are presented in the first chapter. The necessary material on Boolean algebras, including the Stone Representation Theorem, is developed in Chapter 2. A very basic introduction to category theory is presented in the beginning of Chapter 10 and the remainder of the chapter is an introduction to the methods of categorical topology as it relates to the Stone-Čech compactification.

Chapter 3 is transitional in nature. Many of the familar results concerning $\beta \mathbb{N}$, the Stone-Čech compactification of the natural numbers, are collected here together with many newer results which are appearing in a book for the first time. Several of the ideas and arguments developed here for $\beta \mathbb{N}$ are later adapted to more general settings.

The remaining chapters each treat a particular topic in the theory of Stone-Čech compactifications. Chapters 4, 5, and 9 consider the non-homogeneity, cellularity, and connectedness properties of growths, respectively. Chapter 7 marks a return to the subject of $\beta \mathbb{N}$, concentrating on properties of the growth $\beta\mathbb{N}\setminus\mathbb{N}$. Stone-Čech compactifications of products are the subject of Chapter 8 while mappings of βX to $\beta X\setminus X$ are discussed in Chapter 6.

The more specialized chapters are largely independent, making each of them accessible as a part of a short seminar or reading course. The following chart indicates the major pattern of dependence among the chapters:

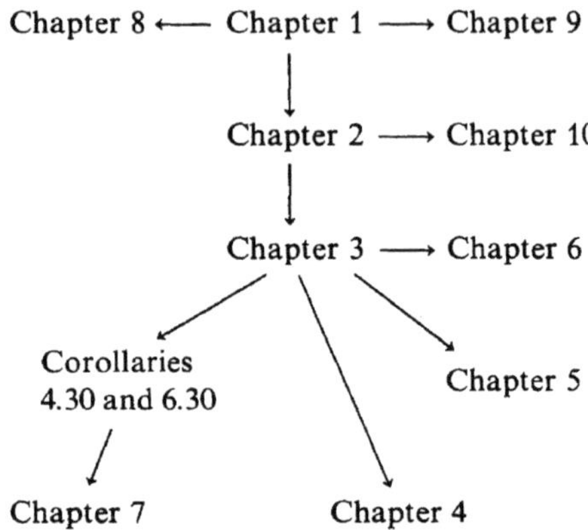

For the reader who is mainly interested in $\beta\mathbb{N}$ and $\beta\mathbb{N}\setminus\mathbb{N}$, Corollaries 4.30 and 6.30 comprise the material needed between Chapters 3 and 7. In addition, the construction used in Corollary 6.16 is needed in the proof of Theorem 7.4. In Chapter 6, the results 6.19—6.21 rely on Theorem 5.12. In Chapter 9, Lemma 9.22 and Theorem 9.23 rely on Theorem 3.41. Otherwise, the only dependence among the chapters consists of references to basic lemmas to which the reader may turn directly.

A list of exercises follows each chapter to supplement the text and to help reinforce basic ideas. The extensive bibliography encompasses both the papers and books cited as well as related papers. Reference to the bibliography is by author and year, with the occasional letter following a year distinguishing multiple papers appearing in that year. Where available, the appropriate reference to Mathematical Reviews is included in the bibliography. This together with the inclusion of references within the text should grant the reader an easy access to the literature. An author index and a detailed subject index are provided to make the book more useful as a reference.

Pittsburgh, Spring 1974

Russell C. Walker

Acknowledgements

A preliminary version of this book was written under the supervision of Stanley P. Franklin as my Doctor of Arts dissertation at Carnegie-Mellon University. I would like to thank him for his encouragement and his many suggestions which played an essential role in shaping the final product.

Much of the original manuscript and most of the revisions were read by Greg Naber. His constructive criticism and many suggestions have been invaluable and I am most grateful for his help.

I was originally introduced to the study of compactifications by Richard Alò, and our many conversations have been very helpful. His influence is most clear in the first chapter and I would particularly like to thank him for his comments relating to those topics.

During the writing of this book I have spoken or corresponded with many of the mathematicians whose work is included. The comments and suggestions which I have received have been most helpful and I would like to take this opportunity to express my appreciation.

Finally, I would like to thank Carnegie-Mellon University for financial support during the writing of the book and Nancy Colmer for her excellent typing of the manuscript.

Table of Contents

Chapter 1. Development of the Stone-Čech Compactification

1.1. A *compactification* of a topological space X is a compact space K together with an embedding $e: X \to K$ with $e[X]$ dense in K. We will usually identify X with $e[X]$ and consider X as a subspace of K. Our main topic is a very special type of compactification—one in which X is embedded in such a way that every bounded, real-valued continuous function on X will extend continuously to the compactification. Such a compactification of X will be called a *Stone-Čech compactification* and will be denoted by βX. In this chapter, several constructions of βX will be examined. We will find that βX is a useful device to study relationships between topological characteristics of X and the algebraic structure of the real-valued continuous functions defined on X and that many topological properties of X can be translated into properties of βX. The discussion will proceed in a rough approximation of historical order, although many anachronisms have been included to smooth the way and at the same time to introduce material that will be useful later.

Much of the first chapter will be a review of material developed in the classic L. Gillman and M. Jerison text, *Rings of Continuous Functions*, which will henceforth be referred to by [GJ]. The development will be almost entirely self-contained. Only a very few results which rely on the more algebraic approach of [GJ] will be stated without proof. The basic reference for results in general topology will be J. Dugundji's text, *Topology*, which will be referred to by [D].

1.2. For the most part, notation will be as in [GJ]. The ring of all real-valued continuous functions defined on a space X will be denoted by $C(X)$ and the subring consisting of all bounded members of $C(X)$ will be denoted by $C^*(X)$. We will say that a subspace S of X is *C-embedded* (resp. *C*-embedded*) in X if every member f of $C(S)$ (resp. $C^*(S)$) extends to a member g of $C(X)$ (resp. $C^*(X)$). The following diagram illustrates C-embedding:

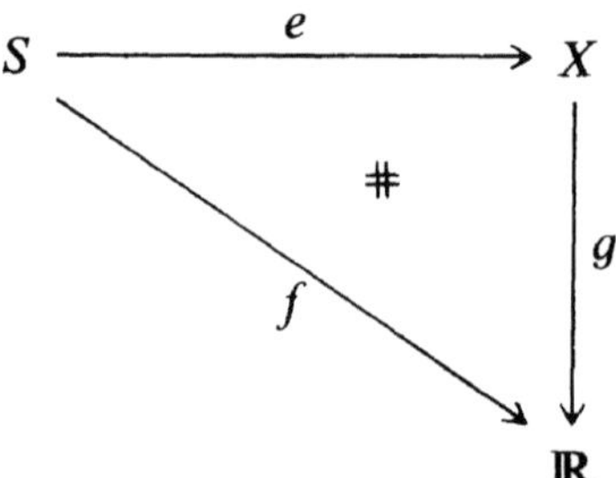

The symbol # indicates that the composition of g with the embedding e is equal to the mapping f. In such an instance, we will say that the diagram is *commutative*.

The set of points of X where a member f of $C(X)$ is equal to zero is called the *zero-set* of f and will be denoted by $\mathbf{Z}(f)$. We will frequently say that f *vanishes at* x to mean $f(x)=0$. The complement of the zero-set $\mathbf{Z}(f)$ is called a *cozero-set* and is denoted by $\mathbf{Cz}(f)$. The collections of all zero-sets and all cozero-sets of X will be denoted by $\mathbf{Z}[X]$ and $\mathbf{CZ}[X]$, respectively. The ring and lattice operations in the rings $C(X)$ and $C^*(X)$ will be defined pointwise and it will be convenient to be familiar with such relationships as $\mathbf{Z}(|f|+|g|)=\mathbf{Z}(f)\cap\mathbf{Z}(g)$ and $\{x\in X : f(x)\geqslant 1\}=\mathbf{Z}((f-1)\wedge 0)$. Note that in the second equation, the symbol "**1**" is used to represent the function which is constantly equal to 1. The symbol f^{-1} will be reserved to denote the reciprocal of a member f of $C(X)$ if the reciprocal is defined, i.e. if f does not vanish anywhere on X. The term *mapping* will always refer to a continuous function. The composition of two functions f and g will be denoted by $g\circ f$ and inverse images under a function f will be written $f^{\leftarrow}(S)$. The closure of a subspace S of X will be denoted by $\mathrm{cl}_X S$ and the subscript will be omitted if no confusion can result. Two subsets A and B of a space X are said to be *completely separated in* X if there exists a mapping f in $C(X)$ such that $f(a)=0$ for all a in A and $f(b)=1$ for all b in B. One can easily see that two sets are completely separated if and only if they are contained in disjoint zero-sets.

The concept of C^*-embedding is important because as we have observed, the Stone-Čech compactification of a space X is a compact Hausdorff space containing X as a dense C^*-embedded subspace. The following theorem is [GJ, 1.17] and will be the main tool used to show that a subspace is C^*-embedded. The argument is a modification of one used by Urysohn in 1925 B to show that a closed subset of a normal space is C^*-embedded.

Urysohn's Extension Theorem. *A subspace S of a space X is C^*-embedded in X if and only if any two completely separated sets in S are completely separated in X.*

Proof. If S is C^*-embedded in X and A and B are completely separated in S, then the extension to X of the mapping which separates them will separate them in X.

Now assume that f_1 belongs to $C^*(S)$. Then $|f_1| \leqslant \mathbf{m}$ for some integer m. For convenience, put

$$r_n = (m/2)(2/3)^n$$

for every n in $\mathbb{N}$. Then $|f_1| \leqslant 3\mathbf{r}_1$. Proceeding inductively, suppose that we have obtained an element f_n of $C^*(S)$ such that $|f_n| \leqslant 3\mathbf{r}_n$. Define

$$A_n = \{s \in S : f_n(s) \leqslant -r_n\} \quad \text{and} \quad B_n = \{s \in S : f_n(s) \geqslant r_n\}\,.$$

A_n and B_n are completely separated in S and are therefore completely separated in X by hypothesis. Hence, there exists a mapping g_n in $C^*(X)$ such that g_n is constantly equal to $-r_n$ on A_n and to r_n on B_n and $|g_n| \leqslant \mathbf{r}_n$. Now define

$$f_{n+1} = f_n - g_n|S\,.$$

Since $|f_n(s) - g_n(s)| \leqslant 2r_n$ for every s in S, it is clear that $|f_{n+1}| \leqslant 3\mathbf{r}_{n+1}$, and the induction step is complete. The Weierstrass M-test shows that the sequence $\left\{\sum_{i=1}^{n} g_i\right\}$ converges uniformly to a continuous function on X. Since

$$g_1 + \cdots + g_n|S = (f_1 - f_2) + \cdots + (f_n - f_{n+1})$$
$$= f_1 - f_{n+1}$$

and $f_{n+1}(s)$ converges to zero for every s in S, the limit of the sequence extends f_1. $\square$

1.3. The next result is $[\text{GJ}, 1.18]$ and indicates when a C^*-embedded subspace will also be C-embedded.

Theorem. *A C^*-embedded subspace is C-embedded if and only if it is completely separated from every zero-set disjoint from it.*

Proof. Let S be C-embedded in X. If a zero-set $\mathbf{Z}(h)$ misses S, define f in $C(S)$ by $f(s) = 1/h(s)$. If g is the extension of f to all of X, then gh completely separates $\mathbf{Z}(h)$ from S.

Now assume that S is a C^*-embedded subspace which is completely separated from every zero-set which misses it. Let f belong to $C(S)$.

Then the composition $\arctan \circ f$ belongs to $C^*(S)$ and has an extension to a mapping g in $C(X)$. The zero-set

$$Z = \{x \in X : |g(x)| \geqslant \pi/2\}$$

misses S so that there exists h mapping X into the closed unit interval I, such that h is constantly equal to 1 on S and to 0 on Z. Then gh agrees with $\arctan \circ f$ on S and satisfies $|(gh)(x)| < \pi/2$ for every x in X. Hence, $\tan \circ (gh)$ is defined, continuous, and real-valued on X and is an extension of f to all of X. $\quad \Box$

Completely Regular Spaces

1.4. The investigation of properties of a topological space through an embedding into a compact space is clearly limited to subspaces of compact spaces. Thus, it was a major step when Tychonoff characterized this class of spaces in 1930. A space X is *completely regular* if every closed subspace F of X is completely separated from any point x not in F and if each point is closed. This new class of spaces proved to be precisely the class that can be studied through the method of compactification.

Theorem (Tychonoff). *The completely regular spaces are precisely those spaces which can be embedded in a product of copies of the closed unit interval I.*

1.5. The proof of the preceding theorem will be immediate from the following more general result and from there it will be an easy step to the simplest construction of the Stone-Čech compactification. A family $\mathscr{F}$ of functions on X is said to *distinguish points* if for each pair of distinct points x and y, there exists an f in $\mathscr{F}$ such that $f(x)$ is not equal to $f(y)$. $\mathscr{F}$ is said to *distinguish points and closed sets* if for each closed set F in X and each point x not in F there exists f in $\mathscr{F}$ such that $f(x)$ misses $\operatorname{cl} f[F]$. The following result is adapted from J. L. Kelley's text, *General Topology*.

Embedding Lemma. *Let $\mathscr{F}$ be a family of mappings such that each member f of $\mathscr{F}$ maps the space X to a space Y_f. Then:*

(a) The evaluation mapping $e : X \to \bigtimes Y_f$ defined by $e(x)_f = f(x)$ for all points x of X is continuous.

(b) The mapping e is an open mapping onto $e[X]$ if $\mathscr{F}$ distinguishes points and closed sets.

(c) The mapping e is one-to-one if and only if $\mathscr{F}$ distinguishes points.

(d) The mapping e is an embedding if $\mathscr{F}$ distinguishes points and $\mathscr{F}$ distinguishes points and closed sets.

Proof. (a) The function e is continuous since its composition with each projection is continuous, i.e. $\pi_f \circ e = f$.

(b) If U is an open set in X and x is in U, choose f in $\mathscr{F}$ such that $f(x) \notin \mathrm{cl}\, f[X \backslash U]$. Then the set of all z in $e[X]$ such that $z_f \notin \mathrm{cl}\, f[X \backslash U]$ is a neighborhood of $e(x)$ and is contained in $f[U]$. Therefore, $f[U]$ is open in $e[X]$.

(c) is clear and (d) follows from (b) and (c). $\Box$

1.6. Tychonoff's characterization of complete regularity shows that no larger class can be studied by the method of compactification. Complete regularity plays a similar role if one attempts to study the topological properties of a space through algebraic or lattice properties of rings of continuous functions. The following theorem was obtained independently by E. Čech and M. H. Stone in 1937 and shows that completely regular spaces are not distinguishable from more general spaces through the algebraic properties of rings of continuous functions.

Theorem. *For any topological space X, there exists a completely regular space ρX which is a continuous image of X such that any real-valued mapping from X factors through ρX.*

Proof. The situation described in the theorem is illustrated in the following diagram:

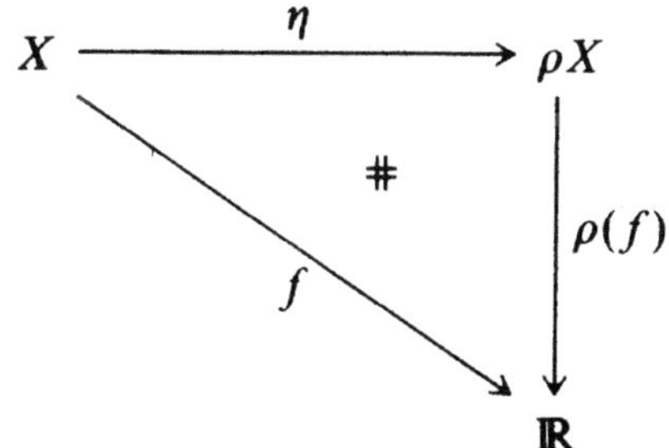

Define two points x and y of X to be equivalent if $f(x) = f(y)$ for all f in $C(X)$. This relation partitions X into equivalence classes. Let ρX denote the set of equivalence classes and let $\eta : X \to \rho X$ assign to each point of X its equivalence class. Since every f in $C(X)$ is constant on each equivalence class, we can define $\rho(f): \rho X \to \mathbb{R}$ by $\rho(f)(\eta(x)) = f(x)$. It is immediate that this definition of $\rho(f)$ makes the diagram commute. Now provide ρX with the smallest topology such that each $\rho(f)$ is continuous. Then the closed sets of ρX are of the form

$$F = \bigcap \rho(f_\alpha)^{\leftarrow}(F_\alpha)$$

where each F_α is closed in $\mathbb{R}$. With this topology, ρX is Hausdorff since points of X which are not separated by some member of $C(X)$ are identified in ρX. If F is closed in ρX and y is not in F, then there exists α such that y is not in $\rho(f_\alpha)^\leftarrow(F_\alpha)$. The point $\rho(f_\alpha)(y)$ is completely separated from F_α by some real-valued mapping g on $\mathbb{R}$ and $g \circ \rho(f_\alpha)$ completely separates y from F, making ρX completely regular. We will use the form of the closed sets of ρX to show that η is continuous. If F is a closed subset of ρX, then

$$\eta^\leftarrow(F) = \eta^\leftarrow(\bigcap \rho(f_\alpha)^\leftarrow(F_\alpha)) = \bigcap f_\alpha^\leftarrow(F_\alpha)$$

since the diagram is commutative. Because each f_α is continuous, $\eta^\leftarrow(F)$ is the intersection of closed sets and is therefore closed. $\square$

The importance of this theorem is that the correspondence $f \mapsto \rho(f)$ preserves both the ring and lattice structures of $C(X)$ and is an algebraic and lattice isomorphism between $C(X)$ and $C(\rho X)$. Thus, algebraic and lattice properties of $C(X)$ which are valid for an arbitrary space X also hold for $C(\rho X)$. [GJ, Chapter 3] provides a more detailed discussion of this aspect of ρX.

1.7. The proof of Theorem 1.6 reveals the form of the closed sets in a completely regular space. If f is a mapping of X to Y and g belongs to $C(Y)$, then $\mathbf{Z}(g \circ f) = f^\leftarrow(\mathbf{Z}(g))$, so that the inverse image of a zero-set is again a zero-set. Since any closed set in $\mathbb{R}$ is a zero-set, we have shown that any closed set in ρX is the intersection of zero-sets. We shall say that a family $\mathfrak{Z}$ of subsets of X is a *base for the closed sets* of X if any closed set is the intersection of members of $\mathfrak{Z}$. Hence, the family of zero-sets is a base for the closed sets of ρX. This property can easily be seen to characterize the class of completely regular spaces.

Proposition. *A space is completely regular if and only if the family of zero-sets of the space is a base for the closed sets (or equivalently, the family of cozero-sets is a base for the open sets).*

1.8. By utilizing Tychonoff's characterization of complete regularity, we can show that any mapping of a space X into a completely regular space will factor through ρX.

Corollary. *If f is a mapping of the space X into a completely regular space Y, then there is a mapping $\rho(f)$ of ρX into Y such that the diagram commutes:*

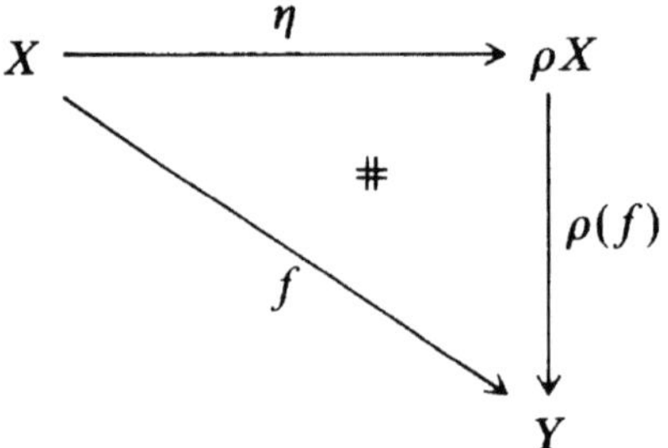

Proof. For each g in $C(Y)$, let $\mathbb{R}_g$ be a copy of the real line and let e be the evaluation map embedding Y into their product, $\times \mathbb{R}_g$. Theorem 1.6 gives a mapping $\rho(g \circ f)$ of ρX into R_g such that $g \circ f = \rho(g \circ f) \circ \eta$ since the composition $g \circ f$ maps X into R_g. To show that f factors through ρX, define a mapping h of ρX to $\times \mathbb{R}_g$ by

$$h(z)_g = \rho(g \circ f)(z)$$

for all z in ρX. Then h is continuous and $h[\rho X]$ is contained in $e[Y]$ since for each projection map π_g,

$$(\pi_g \circ h)(z) = \rho(g \circ f)(z)$$

so that

$$(\pi_g \circ h)[\rho X] \subset g[Y].$$

Since e is an embedding and $h[X]$ is contained in $e[Y]$, putting $\rho(f) = e^{\leftarrow} \circ h$ gives the required factorization of f through ρX. The following diagram describes the situation:

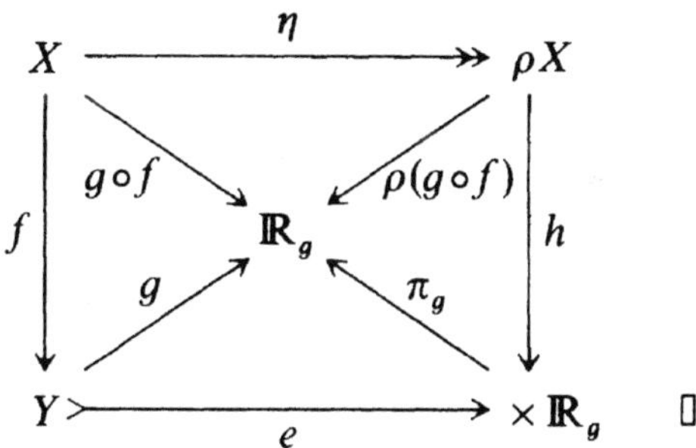

To summarize the importance of the class of completely regular spaces, Theorem 1.4 shows that no larger class can be studied by means of embeddings into compact Hausdorff spaces. Theorem 1.6 establishes that no additional information can be gained by investigating algebraic properties of rings of continuous functions for any larger class of spaces. Finally, Proposition 1.7 exhibits a relationship between the topology of a completely regular space and the real-valued mappings defined on the space which will prove to be very useful. For these reasons, unless otherwise noted,

All spaces mentioned will be presumed to be completely regular.

βX and the Extension of Mappings

1.9. The year 1937 was an important one in the developing theory of topology and its relations to algebra. M.H. Stone and E. Čech each published important papers which provided independent proofs of the existence of the compactification βX. H. Cartan introduced the notions of filter and ultrafilter in a fundamental paper which led to a new theory of convergence by generalizing both sequences and neighborhoods of the diagonal.

Stone's paper treated the relations of algebra and topology through applications of Boolean rings. The most important result in this theory is the representation of Boolean algebras utilizing totally disconnected compact Hausdorff spaces. We will consider this topic in Chapter 2. We will not describe Stone's development of the compactification, but will treat the outgrowths of his work found in the papers of others.

Čech demonstrated the existence of the compactification βX in his paper and used it to investigate properties of X by embedding X into βX. We will use modifications of Čech's methods to obtain the results of Stone. The following theorem can be interpreted algebraically as showing that $C^*(X)$ and $C^*(\beta X)$ are isomorphic.

Theorem (Stone-Čech). *Every completely regular space X has a Hausdorff compactification βX in which it is C^*-embedded.*

Proof. For every f in $C^*(X)$, let I_f denote the range of f. Since f is bounded, $\operatorname{cl} I_f$ is compact. By taking $\mathscr{F} = C^*(X)$ in the Embedding Lemma, 1.5, X can be embedded into $\times \operatorname{cl} I_f$ by means of the evaluation mapping $e(x)_f = f(x)$. Put $\beta X = \operatorname{cl}(e[X])$. Then for f in $C^*(X)$, the extension $\beta(f): \beta X \to \operatorname{cl} I_f$ is the restriction to βX of the projection π_f. The situation is described in the following diagram:

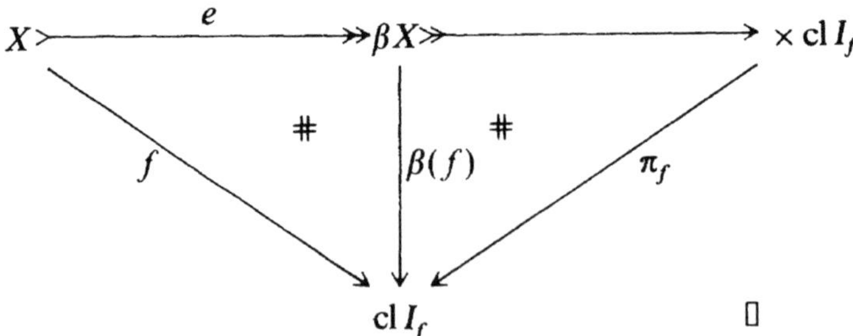

Note that some new notation has been introduced in the preceding diagram. A "tail" on an arrow will denote an embedding and a "double tail" will denote a closed embedding. A double headed arrow will indicate a mapping with dense range.

1.10. From the above construction and the observation that a mapping from a compact space to a Hausdorff space is closed, we obtain the following result.

Corollary. *If X is a compact space, βX is homeomorphic to X.*

1.11. Stone showed that not only would members of $C^*(X)$ extend to βX, but that any mapping of X into a compact space would extend to βX. We will prove Stone's result by combining the two previous results with Theorem 1.4. The similarity between the following proof and the proof of Corollary 1.8 will be investigated in Chapter 10 when ρX and βX are viewed in a common categorical context.

Theorem (M. H. Stone). *Every completely regular space X has a compactification βX such that any mapping of X to a compact space K will extend uniquely to βX.*

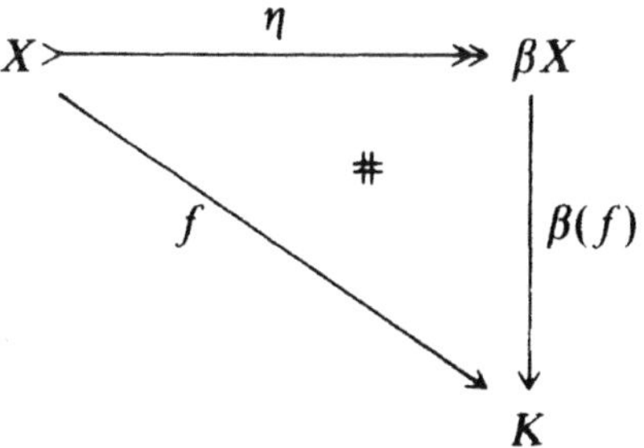

Proof. For each map g in $C^*(K)$, let I_g denote the range of g and let e be the evaluation map embedding K into their product, $\times I_g$. Since $g \circ f$ maps X to I_g, Theorem 1.9 provides an extension $\beta(g \circ f)$ of $g \circ f$ to βX. To show that f extends to βX, define a mapping h of βX to $\times I_g$ by

$$h(p)_g = \beta(g \circ f)(p)$$

for all p in βX. Then h is continuous since its composition with each projection is continuous, i.e. $(\pi_g \circ h)(p) = \beta(g \circ f)(p)$ for all g in $C^*(K)$.

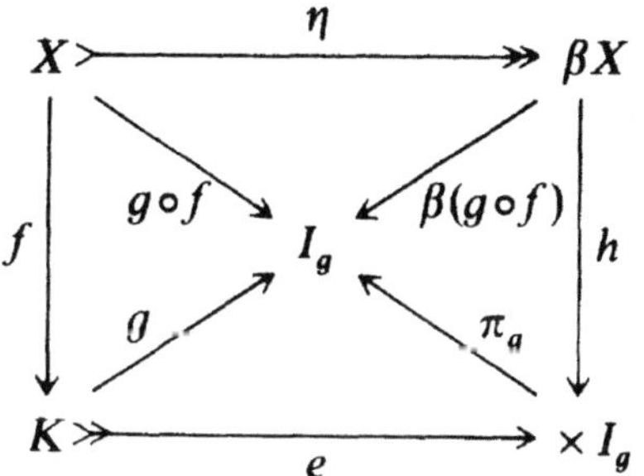

Since e is an embedding, it remains to show that the image of βX under h is contained in the image of K under e. Since for each g in $C^*(K)$, $\beta(g \circ f)[X]$ is contained in $g[K]$,

$$h[\beta X] = h[\operatorname{cl}_{\beta X} X] \subset \operatorname{cl}(e[K]) = \operatorname{cl} K = K$$

because K is compact.

The uniqueness of the extension h is immediate since any two extensions of f must agree on the dense subspace X of βX.   □

1.12. The above proof is a modification of the technique used by Čech to obtain the following special case of the theorem.

Corollary. *Any compactification of X is a continuous image of βX under a mapping which leaves points of X fixed.*

1.13. A partial ordering can be induced on the set of compactifications of a space X in the following way: If K_1 and K_2 are two compactifications of X, define $K_1 \leqslant K_2$ whenever there exists a mapping g of K_2 onto K_1 which leaves the points of X fixed. Then Corollary 1.12 shows that βX is a maximal element in the set of compactifications of X. The maximality of βX enables us to show that the Stone-Čech compactification of a space is essentially unique.

Corollary. *Any compactification of X to which every mapping of X to a compact space has an extension is homeomorphic to βX under a homeomorphism which leaves points of X fixed.*

Proof. If K is a compactification of X satisfying the stated factorization property, then the embedding of X into βX has an extension f to K and similarly the embedding of X into K has an extension g to βX. Since the restriction of $f \circ g$ to X is the identity on X, $f \circ g$ is the identity on βX since X is dense in βX. Similarly, $g \circ f$ is the identity on K. Thus, f and g are homeomorphisms leaving points of X fixed and $f = g^{\leftarrow}$.   □

1.14. Čech gave an additional characterization of βX which is important as a forerunner of the construction of βX via zero-sets as described in [GJ, Chapter 6]. We will consider a variant of the [GJ] approach beginning in Section 1.34.

Theorem (Čech). *βX is that compactification of a space X in which completely separated subsets of X have disjoint closures.*

Proof. If two subsets of X are completely separated by a mapping f in $C^*(X)$, then the extension $\beta(f)$ of f to βX completely separates the closures of the sets. Thus, βX satisfies the condition of the theorem.

Let K be a compactification of X in which completely separated subsets of X have disjoint closures. By Corollary 1.12, there exists a mapping h of βX onto K which leaves points of X fixed. Because βX is compact, h is a closed mapping and it is sufficient to show that h is one-to-one. Let p and q be distinct points of βX and let f in $C(\beta X)$ be such that $f(p)=0$ and $f(q)=1$. The sets $A_0=\{x\in X:f(x)\leqslant 1/3\}$ and $A_1=\{x\in X:f(x)\geqslant 2/3\}$ are completely separated in X and thus have disjoint closures in K. Since $h(p)$ is in $\mathrm{cl}_K A_0$ and $h(q)$ is in $\mathrm{cl}_K A_1$, we must have $h(p)\neq h(q)$. Thus, h is a closed continuous bijection and is therefore a homeomorphism between K and βX. $\quad\square$

1.15. By observing that in a normal space any two disjoint closed sets are completely separated and applying the theorem, Čech obtained the

Corollary. *When X is a normal space, βX is that compactification of X in which disjoint closed subsets of X have disjoint closures.*

3-Filters and 3-Ultrafilters

1.16. We now turn to the developments which follow from Cartan's ultrafilter concept and from the set-theoretic content of M. H. Stone's work. The following ideas are adapted from H. Wallman's 1938 paper and from P. Samuel's 1948 paper and are formulated while looking ahead in directions to be taken by E. Hewitt in 1948 and O. Frink in 1964. The material here and in subsequent sections devoted to spaces of 3-ultrafilters is based on class notes from R. A. Alò concerning material contained in his book, *Normal Topological Spaces*, written with H. L. Shapiro, and in their papers.

A family 3 of subsets of a space X is called a *ring of sets* if it is closed under finite intersections and unions. A subfamily $\mathscr{F}$ of non-empty members of a ring 3 is called a 3-*filter* if $\mathscr{F}$ satisfies the following conditions:

(1) $\mathscr{F}$ is closed under finite intersections.
(2) A member of 3 containing a member of $\mathscr{F}$ is in $\mathscr{F}$.

1.17. A 3-filter $\mathscr{F}$ is a 3-*ultrafilter* if it is not properly contained in any other 3-filter. The following characterization of 3-ultrafilters is easily verified.

Proposition. *A 3-filter $\mathscr{U}$ is a 3-ultrafilter if and only if any member of 3 which meets every member of $\mathscr{U}$ is in $\mathscr{U}$.*

1.18. In his fundamental 1937 paper, H. Cartan considered 3-filters and 3-ultrafilters where 3 is the family of all subsets of a set X. In such a case, the reference to $\mathfrak{Z}$ is usually surpressed and the terms filter and ultrafilter are used. An example of such a filter is the collection of all neighborhoods of a point x in X which is denoted by $\mathscr{B}(x)$ and called the *neighborhood filter of* x. The filter concept was introduced in order to study convergence. A filter $\mathscr{F}$ is said to *converge* to a point x if $\mathscr{F}$ contains $\mathscr{B}(x)$. A filter $\mathscr{F}$ is said to *cluster* at a point x if every member of $\mathscr{F}$ meets every member of $\mathscr{B}(x)$, i.e. if x belongs to $\bigcap\{\operatorname{cl} Z : Z \in \mathscr{F}\}$.

Observe that the definition of convergence of a filter is valid in an arbitrary topological space. Convergence of $\mathfrak{Z}$-filters for rings other than the power set is usually defined only for classes of spaces in which each point has a neighborhood base contained in the ring. The proper choice of $\mathfrak{Z}$ for the class of completely regular spaces will be considered in Section 1.27.

1.19. In addition to describing convergence, collections of $\mathfrak{Z}$-filters, usually $\mathfrak{Z}$-ultrafilters, have been used to construct topological spaces. Let $\omega_X(\mathfrak{Z})$ denote the collection of all $\mathfrak{Z}$-ultrafilters on X. When no confusion can result, we shall simply write $\omega(\mathfrak{Z})$. For Z in $\mathfrak{Z}$, let Z^ω denote the members of $\omega(\mathfrak{Z})$ which contain Z. Taking $\{Z^\omega : Z \in \mathfrak{Z}\}$ as a base for the closed sets imposes a topology on $\omega(\mathfrak{Z})$ which is frequently useful in the formation of compactifications.

(a) M.H. Stone introduced this topology with $\mathfrak{Z}$ the family of all open-and-closed subsets of X, which we shall call *clopen sets*. He showed that $\omega(\mathfrak{Z})$ is compact and that if X has a base for the open sets composed of clopen sets, then X can be embedded into $\omega(\mathfrak{Z})$.

(b) H. Wallman in 1938 considered the case where $\mathfrak{Z}$ is the family of all closed sets of a T_1-space X. Wallman showed that under these hypotheses, $\omega(\mathfrak{Z})$ is compact and contains X as a dense subspace, but that $\omega(\mathfrak{Z})$ is not necessarily Hausdorff. He showed that $\omega(\mathfrak{Z})$ would be Hausdorff precisely when X is normal. In fact, we will later show that when X is a normal space and is embedded in $\omega(\mathfrak{Z})$, then Z^ω is the closure of the set Z in $\omega(\mathfrak{Z})$ and $\omega(\mathfrak{Z})$ is βX. The definition of Z^ω shows that disjoint closed sets of X have disjoint closures in $\omega(\mathfrak{Z})$. Coupling this with Čech's characterization of βX for normal X contained in Corollary 1.15 gives $\omega(\mathfrak{Z}) = \beta X$. It was this result which was to motivate Frink's work as we shall see later. Wallman utilized his compactification to show that it is not possible to distinguish between T_1-spaces and compact spaces by means of homology theory.

(c) The space $\omega(\mathfrak{Z})$ is compact if $\mathfrak{Z}$ is the power set of X, but X cannot be embedded into $\omega(\mathfrak{Z})$ except when X is discrete. If X is discrete, the discussion of the Wallman compactification above shows that $\omega(\mathfrak{Z}) = \beta X$. This is the interpretation of βX which will be most useful in the case of discrete spaces in later chapters.

βX and Maximal Ideal Spaces

1.20. In his 1937 paper, M.H. Stone showed that if Y is a compact space, then the maximal ideals of $C^*(Y)$ are in one-to-one correspondence with the points of Y and that the maximal ideal corresponding to a point of Y is the set of mappings which vanish at that point. Stone also introduced a topology for the set of maximal ideals of a Boolean ring, i.e. a ring in which $x^2 = x$ for every x in the ring. In this section, we will consider the relationships between a space X and the rings $C^*(X)$ and $C(X)$ which were developed from Stone's work by I. Gelfand and A. Kolmogoroff in their 1939 paper.

We first consider the topology introduced by Stone for the set of maximal ideals of a Boolean ring and show that the topology can be used in a wider class of rings. If A is a commutative ring with unity, let $\mathcal{M}(A)$ denote the collection of all maximal ideals of A. For a subset $\mathcal{H}$ of $\mathcal{M}(A)$, the *kernel* of $\mathcal{H}$ is defined to be $\bigcap \mathcal{H}$ and for any ideal I of A, define the *hull* of I to be

$$\{ M \in \mathcal{M}(A) : I \subset M \} \, .$$

The *Stone topology* on the set of maximal ideals $\mathcal{M}(A)$ is then obtained by defining the closure of any subset $\mathcal{H}$ of $\mathcal{M}(A)$ to be the hull of the kernel of $\mathcal{H}$, i.e.

$$\mathrm{cl}\, \mathcal{H} = \{ M \in \mathcal{M}(A) : \bigcap \mathcal{H} \subset M \} \, .$$

For this reason, the Stone topology is often called the *hull-kernel topology*. With this topology, the space $\mathcal{M}(A)$ is called the *structure space* of A.

1.21. For each element a of A, let $\mathscr{C}(a)$ denote the set of maximal ideals containing a. Each set $\mathscr{C}(a)$ is closed since it is the hull of the ideal consisting of all multiples of a. The family $\{ \mathscr{C}(a) : a \in A \}$ is a base for the closed sets of $\mathcal{M}(A)$ since

$$\mathrm{cl}\, \mathcal{H} = \bigcap \{ \mathscr{C}(a) : a \in \bigcap \mathcal{H} \}$$

when $\mathcal{H}$ is a subset of $\mathcal{M}(A)$. The following basic result follows the outline of [GJ, ex. 7M].

Proposition. *If A is a commutative ring with unity, $\mathcal{M}(A)$ with the Stone topology is a compact Hausdorff space if and only if for each pair M and M' of distinct maximal ideals there exist a not in M and a' not in M' such that $a a'$ belongs to every maximal ideal of A.*

Proof. If A satisfies the stated hypothesis and M and M' are distinct maximal ideals, then

$$U = \mathscr{M}(A) \backslash \mathscr{C}(a) \quad \text{and} \quad U' = \mathscr{M}(A) \backslash \mathscr{C}(a')$$

are neighborhoods of M and M', respectively. To show that U and U' are disjoint, we write:

$$U \cap U' = (\mathscr{M}(A) \backslash \mathscr{C}(a)) \cap (\mathscr{M}(A) \backslash \mathscr{C}(a'))$$

$$= \mathscr{M}(A) \backslash (\mathscr{C}(a) \cup \mathscr{C}(a'))$$

$$= \mathscr{M}(A) \backslash \mathscr{C}(a a').$$

The last equality holds because every maximal ideal is prime. Hence, since $a a'$ belongs to every maximal ideal, $\mathscr{C}(a a') = \mathscr{M}(A)$ implying that $U \cap U' = \emptyset$ and $\mathscr{M}(A)$ is Hausdorff.

Conversely, if $\mathscr{M}(A)$ is Hausdorff any pair of distinct maximal ideals M and M' must be separated by disjoint basic open sets U and U' as above. But

$$\emptyset = U \cap U' = \mathscr{M}(A) \backslash (\mathscr{C}(a) \cup \mathscr{C}(a'))$$

implies that $\mathscr{C}(a) \cup \mathscr{C}(a')$ is all of $\mathscr{M}(A)$ so that $a a'$ belongs to every maximal ideal of A.

It remains to demonstrate the compactness of $\mathscr{M}(A)$. Let $\{F_\alpha\}$ be a family of closed sets. We will show that if $\{F_\alpha\}$ has empty intersection, then some finite subfamily has empty intersection. Since each F_α is an intersection of basic closed sets, it is sufficient to assume that each set in the family is a basic set, i.e. that for every α there is some a_α in A such that $F_\alpha = \mathscr{C}(a_\alpha)$. The result will follow from determining when the intersection of a basic family $\{\mathscr{C}(a_\alpha)\}$ will be empty in terms of the elements $\{a_\alpha\}$. We will show that $\bigcap \mathscr{C}(a_\alpha) = \emptyset$ exactly when the subset $\{a_\alpha\}$ of A generates A. If $\bigcap \mathscr{C}(a_\alpha) = \emptyset$, then for every M in $\mathscr{M}(A)$ there is some a_α not belonging to M. Thus, the only ideal containing $\{a_\alpha\}$ is the ring A itself so that $\{a_\alpha\}$ generates A. The converse is clear since each of the steps is reversible.

Now if $\bigcap \mathscr{C}(a_\alpha) = \emptyset$, $\{a_\alpha\}$ generates A and there exist members r_i of A such that the identity element of A can be written $1 = r_1 a_{\alpha_1} + \cdots + r_n a_{\alpha_n}$ for some finite family $\{a_{\alpha_i}\}$. But then the ideal generated by $\{a_{\alpha_i}\}$ contains the identity and hence is all of A. Thus, the finite subfamily $\{a_{\alpha_i}\}$ generates A so that $\bigcap \mathscr{C}(a_{\alpha_i}) = \emptyset$. Hence, $\mathscr{M}(A)$ is compact. $\quad\Box$

1.22. The next step in relating $C^*(X)$ to X is to characterize the maximal ideals of $C^*(X)$. One need only consider a compact space since $C^*(X)$ and $C^*(\beta X)$ are isomorphic.

Proposition (M. H. Stone). *If Y is a compact space, the maximal ideals of $C^*(Y)$ are in one-to-one correspondence with the points of Y and are given by*

$$M^{*p} = \{ f \in C^*(Y) : f(p) = 0 \}$$

for p a point of Y.

Proof. Each M^{*p} is clearly an ideal. Since distinct points of Y are separated by a member of $C^*(Y)$, M^{*p} and M^{*q} are distinct whenever p is not equal to q. To complete the proof it is sufficient to show that any proper ideal I is contained in M^{*p} for some p. Suppose on the contrary that for every point p of Y, there is a member f_p of I such that $f_p(p) \neq 0$. Then there is some neighborhood $U(p)$ of p on which f_p is never equal to zero. Since Y is compact, the covering $\{U(p)\}$ of Y has a finite subcover, $\{U(p_i)\}$. Then the mapping defined by $g = f_{p_1}^2 + \cdots + f_{p_n}^2$ belongs to I and is never zero so that its reciprocal g^{-1} belongs to $C^*(Y)$. Thus, every h in $C^*(Y)$ can be written as $h = g \cdot g^{-1} \cdot h$ and therefore belongs to I so that I is not a proper ideal. This is a contradiction. $\quad\square$

1.23. In order to obtain the following main theorem, it is sufficient to show that the correspondence between the points of a compact space Y and the maximal ideals of $C^*(Y)$ is a homeomorphism. Let $\mathcal{M}^*(Y)$ denote $\mathcal{M}(C^*(Y))$ and $\mathcal{M}(Y)$ denote $\mathcal{M}(C(Y))$.

Theorem. *A compact space Y is homeomorphic with the maximal ideal space $\mathcal{M}^*(Y)$.*

Proof. If p and q are distinct points of Y, there exist f and g in $C^*(Y)$ such that $f(p) = g(q) = 1$ and $fg = 0$. Hence, f does not belong to M^{*p} and g does not belong to M^{*q} although their product fg belongs to every maximal ideal of $C^*(Y)$. Proposition 1.21 thus shows that $\mathcal{M}^*(Y)$ is a compact Hausdorff space.

Denote the bijection of Y with $\mathcal{M}^*(Y)$ which sends a point p of Y to the ideal M^{*p} by τ^*. To show that τ^* is a homeomorphism it is sufficient to show that for S a subset of Y, $\mathrm{cl}(\tau^*[S]) = \tau^*[\mathrm{cl}\,S]$. If p is in cl S, then every member of $C^*(Y)$ which vanishes on all of S also vanishes at p. But then by the definition of closure in a space of maximal ideals,

$$\mathrm{cl}(\tau^*[S]) = \bigcap \{ \mathscr{C}(f) : f[S] = \{0\} \} \, ,$$

and this shows that $\tau^*(p) = M^{*p}$ belongs to $\mathrm{cl}(\tau^*[S])$. On the other hand, if p is not in cl S, then there is a member of $C^*(Y)$ which vanishes on all of S but not at p. But this shows that $\tau^*(p) = M^{*p}$ fails to belong to $\mathrm{cl}(\tau^*[S])$. $\quad\square$

1.24. The following corollary is immediate from the observation that $C^*(X)$ is isomorphic to $C^*(\beta X)$.

Corollary. *βX is homeomorphic with the maximal ideal space $\mathcal{M}^*(X)$.*

1.25. If $C^*(X)$ and $C^*(Y)$ are isomorphic for compact spaces X and Y, then $\mathcal{M}^*(X)$ and $\mathcal{M}^*(Y)$ are homeomorphic. The following result expresses the fact that the ring of bounded real-valued mappings on a compact space determines the space to within homeomorphism.

Corollary (M. H. Stone). *If X and Y are compact spaces, then X and Y are homeomorphic if and only if $C^*(X)$ and $C^*(Y)$ are isomorphic.*

1.26. Gelfand and Kolmogoroff also showed that βX is homeomorphic to $\mathcal{M}(X)$. The proof is more complex than that just given in the case of the maximal ideal space $\mathcal{M}^*(X)$. The difficulty involved is to characterize the maximal ideals of $C(X)$ in terms of the points of βX. Once this has been achieved, it will remain to exhibit the homeomorphism of βX with $\mathcal{M}(X)$, and this step will be similar to that for $\mathcal{M}^*(X)$. In the case of $C^*(X)$, we can utilize the ring isomorphism $f \mapsto \beta(f)$ of $C^*(X)$ and $C^*(\beta X)$ to characterize the maximal ideals of $C^*(X)$ in terms of βX.

Proposition. *The maximal ideals of $C^*(X)$ are in one-to-one correspondence with the points of βX and are given by*

$$M^{*p} = \{ f \in C^*(X) : \beta(f)(p) = 0 \}$$

where p is a point of βX.

In terms of zero-sets, this tells us that M^{*p} is the collection of all members of $C^*(X)$ such that p belongs to $\mathbf{Z}(\beta(f))$. The maximal ideals of $C(X)$ can also be characterized in terms of zero-sets, but the process is more complicated and will require the development of the notion of $\mathfrak{Z}$-filters in such a way as to relate the topological structure of X to the algebraic structure of $C(X)$.

1.27. We first consider a description of convergence in completely regular spaces. In a completely regular space X, if U is an open set containing the point x, there exists a map f in $C(X)$ such that $f(x) = 1$ and $f[X \setminus U] = \{0\}$. Then we have that the set Z defined by

$$Z \equiv \{ x : f(x) \geqslant 1/2 \} = \mathbf{Z}((f - 1/2) \wedge 0)$$

is contained in U and is both a zero-set and a neighborhood of x. Thus, in order to describe convergence in a completely regular space, it is sufficient to consider zero-set neighborhoods.

Further, if $\mathbf{Z}(f_1)$ and $\mathbf{Z}(f_2)$ are zero-sets, then $\mathbf{Z}(f_1 f_2) = \mathbf{Z}(f_1) \cup \mathbf{Z}(f_2)$ and $\mathbf{Z}(f_1^2 + f_2^2) = \mathbf{Z}(f_1) \cap \mathbf{Z}(f_2)$ so that $\mathbf{Z}[X]$ is a ring of sets and we can consider $\mathfrak{Z}$-filters and $\mathfrak{Z}$-ultrafilters with $\mathfrak{Z} = \mathbf{Z}[X]$. We shall refer to the ring of zero-sets often and $\mathfrak{Z}$-filters where $\mathbf{Z}[X]$ is the ring will be called *z-filters*. For each point x of X, let $\mathcal{O}(x)$ denote the neighborhoods of x which are also zero-sets. Then if $\mathcal{B}(x)$ is the neighborhood filter of x, $\mathcal{O}(x) = \mathcal{B}(x) \cap \mathbf{Z}[X]$. We have shown above that $\mathcal{O}(x)$ contains enough neighborhoods of x to describe convergence. $\mathcal{O}(x)$ is clearly a z-filter. A z-filter $\mathcal{F}$ *converges* to x if $\mathcal{O}(x)$ is contained in $\mathcal{F}$ and $\mathcal{F}$ *clusters* at x if x belongs to $\bigcap \mathcal{F}$. The following property makes it convenient to discuss convergence in terms of z-ultrafilters as opposed to z-filters.

1.28. Proposition. *A z-ultrafilter converges to any cluster point.*

Proof. If $\mathcal{U}$ is a z-ultrafilter and a point x belongs to $\bigcap \mathcal{U}$, then every member of $\mathcal{O}(x)$ meets every member of $\mathcal{U}$. But then $\mathcal{O}(x)$ is contained in $\mathcal{U}$ by Proposition 1.17. $\square$

1.29. We now relate z-filters and z-ultrafilters to the ring $C(X)$. The relationships between z-filters and ideals of $C(X)$ were first explored by E. Hewitt in his fundamental 1948 paper on rings of continuous functions. Consider the function

$$\mathbf{Z} : C(X) \to \mathbf{Z}[X]$$

which sends each mapping in $C(X)$ to its zero-set. The following result is [GJ, 2.3] and shows that the image of an ideal under Z is a z-filter and that the pre-image of a z-filter is an ideal.

Proposition. (a) *If $\mathscr{I}$ is a proper ideal in $C(X)$, then $\mathbf{Z}[\mathscr{I}] = \{\mathbf{Z}(f) : f \in \mathscr{I}\}$ is a z-filter on X.*

(b) *If $\mathscr{F}$ is a z-filter on X, then $\mathbf{Z}^{\leftarrow}[\mathscr{F}] = \{f \in C(X) : \mathbf{Z}(f) \in \mathscr{F}\}$ is an ideal in $C(X)$.*

Proof. (a) Since a proper ideal can contain no unit and the units of $C(X)$ are those maps which have void zero-sets, all members of $\mathbf{Z}[\mathscr{I}]$ are non-empty. If f_1 and f_2 are in $\mathscr{I}$, then $f_1^2 + f_2^2$ is in $\mathscr{I}$ and since $\mathbf{Z}(f_1) \cap \mathbf{Z}(f_2) = \mathbf{Z}(f_1^2 + f_2^2)$, $\mathbf{Z}[\mathscr{I}]$ is closed under finite intersections. Let $\mathbf{Z}(f)$ be in $\mathbf{Z}[\mathscr{I}]$ for f in $\mathscr{I}$. If $\mathbf{Z}(g)$ contains $\mathbf{Z}(f)$, then

$$\mathbf{Z}(g) = \mathbf{Z}(f) \cup \mathbf{Z}(g) = \mathbf{Z}(fg)$$

is in $\mathbf{Z}[\mathscr{I}]$. Thus, $\mathbf{Z}[\mathscr{I}]$ is a z-filter.

(b) Let $\mathscr{I} = \mathbf{Z}^{\leftarrow}[\mathscr{F}]$. Since the empty set is not in $\mathscr{F}$, $\mathscr{I}$ does not contain a unit. If f and g are in $\mathscr{I}$,

$$\mathbf{Z}(f - g) \supset \mathbf{Z}(f) \cap \mathbf{Z}(g)$$

and $f - g$ is in $\mathscr{I}$ since $\mathscr{F}$ is closed under supersets in $\mathbf{Z}[X]$ and finite intersections. Thus, $\mathscr{J}$ is an additive subgroup. If f is in $\mathscr{J}$ and g is in $C(X)$, then $\mathbf{Z}(fg) \supset \mathbf{Z}(f)$ and fg is in $\mathscr{J}$ since $\mathscr{F}$ is closed under supersets in $\mathbf{Z}[X]$.   ☐

1.30. Since $\mathbf{Z}$ preserves containment, it is clear that the proposition yields a one-to-one correspondence between the z-ultrafilters on X and the maximal ideals of $C(X)$. The previous proposition together with the characterization of $\mathfrak{Z}$-ultrafilters given in Proposition 1.17 allows the identification of those members of $C(X)$ which belong to a maximal ideal M: f is in M if $\mathbf{Z}(f)$ meets the zero-set of every member of M. We can now establish the main result of Gelfand and Kolmogoroff's 1939 paper.

Theorem (Gelfand and Kolmogoroff). *The maximal ideals of $C(X)$ are in one-to-one correspondence with the points of βX and are given by*

$$M^p = \{ f \in C(X) : p \in \mathrm{cl}_{\beta X} \mathbf{Z}(f) \}$$

for p in βX.

Proof. We show first that each M^p is a maximal ideal by showing that $\mathbf{Z}[M^p]$ is a z-ultrafilter. It is clear that $\mathbf{Z}[M^p]$ is closed under supersets in $\mathbf{Z}[X]$ and that $\mathbf{Z}[M^p]$ does not contain the empty set. Since disjoint zero-sets of X are completely separated, they would have disjoint closures in βX. Thus, since p is in the closure of every $\mathbf{Z}$ in $\mathbf{Z}[M^p]$, no two members of $\mathbf{Z}[M^p]$ can be disjoint by Theorem 1.14. To show that $\mathbf{Z}[M^p]$ is maximal, suppose that a zero-set Z meets every member of $\mathbf{Z}[M^p]$. Then if p is not in $\mathrm{cl}_{\beta X} Z$, there exists a zero-set neighborhood Z' of p in βX which misses Z. But then $Z' \cap X$ is in $\mathbf{Z}[M^p]$ and misses Z, which is a contradiction. Thus, M^p is a maximal ideal.

It remains to show that every maximal ideal is of the form M^p for some p in βX. If M is maximal, then $\{\mathrm{cl}_{\beta X} Z : Z \in \mathbf{Z}[M]\}$ is a family of closed sets with the finite intersection property in a compact space. Thus, there exists p in $\bigcap \{\mathrm{cl}_{\beta X} Z : Z \in \mathbf{Z}[M]\}$ so that $\mathbf{Z}[M]$ clusters at p. Then $\mathbf{Z}[M]$ converges to p and must converge to p alone since βX is Hausdorff. Thus, $M = M^p$.   ☐

1.31. Corollary. *βX is homeomorphic with the maximal ideal space $\mathscr{M}(X)$.*

Proof. The theorem establishes a one-to-one correspondence τ between βX and $\mathscr{M}(X)$. The proof that τ is a homeomorphism is similar to the proof that τ^* is a homeomorphism in Corollary 1.23.   ☐

The proofs here are based on a 1954 paper of Gillman, Henriksen, and Jerison in which Theorem 1.30 is discussed with its applications. The characterization of maximal ideals is also treated in [GJ, Chapter 7].

1.32. The proof of Theorem 1.30 exhibits a one-to-one correspondence between the points of βX and the maximal ideals of $C(X)$. We also have a one-to-one correspondence between the maximal ideals of $C(X)$ and the z-ultrafilters on X. Following [GJ], if p belongs to βX, we will denote the z-ultrafilter corresponding with the maximal ideal M^p by A^p. Then Proposition 1.29 shows that

$$A^p = \mathbf{Z}^{\leftarrow}[M^p] = \{\mathbf{Z}(f): f \in M^p\} .$$

From Theorem 1.30, it is evident that if p belongs to X, then $\bigcap A^p = \{p\}$, and that if p belongs to $\beta X \setminus X$, then $\bigcap A^p = \emptyset$. In the first case, we shall say that A^p is *fixed* and in the second, that A^p is *free*. The same terms will also be applied to the ideals M^p as well as to any z-filter or ideal. Thus we can view the construction of βX as adding a point p to X for each free maximal ideal M^p in such a way that the resulting space is compact and that A^p converges to p.

1.33. Gelfand and Kolmogoroff observed that if $\mathbb{N}$ is the countable discrete space, then $C^*(\mathbb{N})$ and $C^*(\beta \mathbb{N})$ are isomorphic, but that $C(\mathbb{N})$ and $C(\beta \mathbb{N})$ are not isomorphic. Thus, $C(X)$ is a more sensitive invariant than $C^*(X)$ for distinguishing between topological spaces. We shall see in Theorem 1.56 that E. Hewitt developed this idea by introducing the class of real-compact spaces which play a role with respect to $C(X)$ analogous to that played by the class of compact spaces with respect to $C^*(X)$ in Corollary 1.25.

Spaces of $\mathfrak{Z}$-Ultrafilters

1.34. We have seen two different ways to represent the Stone-Čech compactification. We first showed that βX could be obtained as the closure of a copy of X embedded in a product of intervals and we later saw that βX is homeomorphic to the spaces of maximal ideals of the rings $C^*(X)$ and $C(X)$. In the next several sections we will consider the Stone-Čech compactification in the context of spaces of $\mathfrak{Z}$-ultrafilters.

In 1.19b, it was mentioned that H. Wallman showed that for a normal space X, βX is homeomorphic to the space of $\mathfrak{Z}$-ultrafilters on X when $\mathfrak{Z}$ is the ring of closed subsets of X. In Chapter 6 of [GJ], βX is described as the space of all z-ultrafilters on X by considering the ring of zero-sets of X.

1.35. Motivated largely by these two familiar examples of rings of sets— the zero-sets of a completely regular space and the closed sets of a normal space, in 1964 O. Frink introduced the concept of a normal base. A collection of closed subsets of a T_1-space X is called a *normal base* for X if $\mathfrak{Z}$ satisfies the following conditions:

(1) $\mathfrak{Z}$ is a ring of sets.

(2) $\mathfrak{Z}$ is *disjunctive*, i.e. if a closed subset of X does not contain a point of X, then there exists a member of $\mathfrak{Z}$ containing the point and missing the closed set.

(3) $\mathfrak{Z}$ is a base for the closed sets of X, i.e. every closed set is an intersection of members of $\mathfrak{Z}$.

(4) $\mathfrak{Z}$ is *normal*, i.e. disjoint members of $\mathfrak{Z}$ are contained in disjoint complements of members of $\mathfrak{Z}$.

After we obtain the necessary preliminary results, we will impose the topology used by Wallman on the space of $\mathfrak{Z}$-ultrafilters when $\mathfrak{Z}$ is a normal base and show that the resulting space is a Hausdorff compactification.

1.36. The following result can be obtained by a straightforward application of Zorn's Lemma and will be used to show that $\omega(\mathfrak{Z})$ is compact.

Proposition. *Every $\mathfrak{Z}$-filter is contained in a $\mathfrak{Z}$-ultrafilter if $\mathfrak{Z}$ is any ring of sets.*

1.37. The next result will allow X to be embedded into $\omega(\mathfrak{Z})$. If x is a point in X, let $\varphi(x) = \{Z \in \mathfrak{Z} : x \in \mathfrak{Z}\}$.

Proposition. *Each $\varphi(x)$ is a $\mathfrak{Z}$-ultrafilter if $\mathfrak{Z}$ is a disjunctive ring of sets.*

Proof. Suppose that Z belongs to $\mathfrak{Z}$ and that x is not in Z. Since Z is closed and $\mathfrak{Z}$ is disjunctive, there exists Z' containing x and missing Z. Then Z does not belong to $\varphi(x)$ and every element of $\mathfrak{Z}$ which meets every member of $\varphi(x)$ contains x. Hence, Proposition 1.17 shows that $\varphi(x)$ is a $\mathfrak{Z}$-ultrafilter. $\square$

Thus, we have obtained a function

$$\varphi : X \to \omega(\mathfrak{Z})$$

by associating to each point x of X the $\mathfrak{Z}$-ultrafilter $\varphi(x)$ consisting of all members of $\mathfrak{Z}$ containing x. Because X is a T_1-space and $\mathfrak{Z}$ is a base for the closed sets of X, the function φ is one-to-one. We will identify X with its image $\varphi[X]$ and regard X as a subset of $\omega(\mathfrak{Z})$. We now show that $\omega(\mathfrak{Z})$ with the topology defined in Section 1.19 is a compact Hausdorff space and that the function φ is an embedding. Recall that a base for the closed sets of $\omega(\mathfrak{Z})$ is given by

$$\{Z^\omega : Z \in \mathfrak{Z}\}$$

where $Z^\omega = \{\mathcal{U} \in \omega(\mathfrak{Z}) : Z \in \mathcal{U}\}$. Since $\mathfrak{Z}$ is a base for the closed sets of X and $Z^\omega \cap X = Z$, φ is an embedding.

1.38. The closure of any subset of $\omega(\mathfrak{Z})$ is the intersection of all the basic closed sets containing the given set. The description of the closures of sets in $\mathfrak{Z}$ is useful in relating X to $\omega(\mathfrak{Z})$. The following result is from the 1968A paper of R.A. Alò and H.L. Shapiro.

Proposition. *If Z is in $\mathfrak{Z}$, then Z^ω is the closure of Z in $\omega(\mathfrak{Z})$.*

Proof. Since $Z \subset Z^\omega$, it is clear that $\mathrm{cl}\,Z \subset Z^\omega$. Now suppose that Z_0^ω is a basic set containing Z. Then $Z_0 = Z_0^\omega \cap X \supset Z$ so that $Z^\omega \subset Z_0^\omega$. Thus, $Z^\omega \subset \mathrm{cl}\,Z$. $\square$

Since two members of $\mathfrak{Z}$ both belong to a $\mathfrak{Z}$-ultrafilter exactly when their intersection belongs to the $\mathfrak{Z}$-ultrafilter, it is clear that if Z_1 and Z_2 are in $\mathfrak{Z}$,

(a)
$$(Z_1 \cap Z_2)^\omega = Z_1^\omega \cap Z_2^\omega .$$

Similarly, we have

(b)
$$(Z_1 \cup Z_2)^\omega = Z_1^\omega \cup Z_2^\omega .$$

The basic open sets of $\omega(\mathfrak{Z})$ can be identified by taking complements of the basic closed sets:

$$\omega(\mathfrak{Z}) \setminus Z^\omega = \{\mathcal{U} \in \omega(\mathfrak{Z}) : Z \notin \mathcal{U}\}$$
$$= \{\mathcal{U} \in \omega(\mathfrak{Z}) : Z' \subset X \setminus Z \text{ for some } Z' \in \mathcal{U}\} .$$

For $U = X \setminus Z$, denote the basic open set obtained from U by

$${}^\omega U = \{\mathcal{U} \in \omega(\mathfrak{Z}) : Z' \subset U \text{ for some } Z' \in \mathcal{U}\} .$$

1.39. We can now establish Frink's basic result.

Theorem. *If $\mathfrak{Z}$ is a normal base for a T_1-space X, then $\omega(\mathfrak{Z})$ is a compact Hausdorff space and φ is a dense embedding of X into $\omega(\mathfrak{Z})$.*

Proof. (a) $\omega(\mathfrak{Z})$ is Hausdorff:

For $\mathcal{U}$ and $\mathcal{V}$ distinct points of $\omega(\mathfrak{Z})$, Proposition 1.17 shows that there exist Z_1 in $\mathcal{U}$ and Z_2 in $\mathcal{V}$ such that $Z_1 \cap Z_2 = \emptyset$. Now since $\mathfrak{Z}$ is normal, there exist A_1 and A_2 in $\mathfrak{Z}$ such that $Z_i \subset X \setminus A_i$ for both values of i and $(X \setminus A_1) \cap (X \setminus A_2) = \emptyset$. But then ${}^\omega(X \setminus A_1)$ and ${}^\omega(X \setminus A_2)$ are disjoint basic neighborhoods of $\mathcal{U}$ and $\mathcal{V}$, respectively.
(b) $\omega(\mathfrak{Z})$ is compact:

It is sufficient to show that any family $\mathcal{A}^\omega$ of basic closed sets with the finite intersection property has non-empty intersection Let $\mathcal{A} = \{Z \in \mathfrak{Z} : Z^\omega \in \mathcal{A}^\omega\}$. Then $\mathcal{A}$ has the finite intersection property and is contained in a $\mathfrak{Z}$-ultrafilter $\mathcal{U}$. But since $\mathcal{U} \supset \mathcal{A}$, $\mathcal{U}$ is in Z^ω for each Z in $\mathcal{A}$, and $\mathcal{U}$ belongs to $\bigcap \mathcal{A}^\omega$.

Since we have seen earlier that φ is an embedding, all that remains is to show that:

(c) $\varphi[X]$ is dense in $\omega(\mathfrak{Z})$:

Let ${}^\omega U$ be a non-empty basic open set of $\omega(\mathfrak{Z})$. Then there exists $\mathscr{U}$ in ${}^\omega U$. If Z is a member of $\mathscr{U}$ such that Z is a subset of U, then the image under φ of a point of Z belongs to ${}^\omega U \cap \varphi[X]$. □

With X regarded as a subspace of $\omega(\mathfrak{Z})$, it is clear from the definition of closure that a $\mathfrak{Z}$-ultrafilter $\mathscr{U}$ belongs to $Z^\omega = \mathrm{cl}\, Z$ for any Z in $\mathscr{U}$. Further, since distinct $\mathfrak{Z}$-ultrafilters must contain disjoint $\mathfrak{Z}$-sets and $Z_1^\omega \cap Z_2^\omega = (Z_1 \cap Z_2)^\omega$, $\mathscr{U}$ is actually the intersection of such closures, i.e.

$$\{\mathscr{U}\} = \bigcap \{Z^\omega : Z \in \mathscr{U}\}\,.$$

Hence, the point $\mathscr{U}$ is the limit of the $\mathfrak{Z}$-ultrafilter $\mathscr{U}$, and $\mathscr{U}$ is the only $\mathfrak{Z}$-ultrafilter on X which can have the point $\mathscr{U}$ as its limit.

The ring of closed subsets of a regular T_1-space which fails to be normal will satisfy all the requirements of a normal base except normality. In such a case, $\omega(\mathfrak{Z})$ will not be Hausdorff. Thus, in order to guarantee that the space of $\mathfrak{Z}$-ultrafilters will be Hausdorff, it is necessary to add the additional condition that the ring $\mathfrak{Z}$ be a normal family, a condition which abstracts a crucial property of the family of closed subsets of a normal space. Also note the similarity between the normality condition and the condition of Proposition 1.21 for maximal ideal spaces, particularly as it is applied in the proof of Theorem 1.23.

1.40. Observe that when X is a normal space and $\mathfrak{Z}$ is the normal base of all closed subsets of X, the definition of closure of members of $\mathfrak{Z}$ and 1.38(a) make the following corollary immediate from Corollary 1.15.

Corollary. *If $\mathfrak{Z}$ is the family of all closed subsets of a normal space X, then $\omega(\mathfrak{Z})$ is βX.*

1.41. The Alexandroff one point compactification αX of a locally compact space X can also be obtained in this way by choosing as the normal base the zero-sets of those mappings in $C(X)$ which are constant on the complement of a compact set. (See Exercise 1 F.)

Many compactifications can be obtained by the methods of Wallman and Frink. However, it is not known if every compactification can be achieved in this way. In most cases where a compactification has been shown to be a Wallman-type compactification, the normal base has been a subcollection of the family of zero-sets. In 1968 B, R. A. Alò and H. L. Shapiro showed that any compact space which has a normal base consisting of regular closed sets will be a Wallman-type compactification of each of its dense subspaces. In particular, they showed that this is the case for every product of closed intervals of $\mathbb{R}$. A set is said to be *regular closed* if it is the closure of its interior.

1.42. In [GJ, Chapter 6], the Stone-Čech compactification is obtained as a Wallman-type compactification. A similar approach will be taken here. We will first show that the zero-sets of X form a normal base and then show that the resulting compactification is βX.

Proposition. *The collection* $\mathbf{Z}[X]$ *of zero-sets of a completely regular space* X *is a normal base which is closed under countable intersections, and* $\omega(\mathbf{Z}[X])$ *is* βX.

Proof. We have already seen that $\mathbf{Z}[X]$ is closed under finite unions and is a base for the closed sets of X. Let $\{\mathbf{Z}(f_n):n\geqslant 1\}$ be countable family of zero-sets and put $g_n=|f_n|\wedge 1/2^n$. The Weierstrass M-test shows that the limit of the sequence $\{g_1+\cdots+g_n:n\geqslant 1\}$ is a continuous function g and it is evident that the zero-set of g is $\bigcap\mathbf{Z}(f_n)$. It is easy to see that $\mathbf{Z}[X]$ is disjunctive by noting that the zero-set neighborhoods form a base for the neighborhoods of a point. Let Z_1 and Z_2 be disjoint zero-sets. Then there is a real-valued mapping f which completely separates Z_1 and Z_2. The zero-sets $\{x:f(x)\leqslant 1/2\}$ and $\{x:f(x)\geqslant 1/2\}$ have disjoint complements which show that $\mathbf{Z}[X]$ is a normal family.

Any two completely separated subsets of X are contained in disjoint zero-sets and 1.38(a) shows that disjoint zero-sets have disjoint closures in $\omega(\mathbf{Z}[X])$. Thus, completely separated sets of X have disjoint closures in $\omega(\mathbf{Z}[X])$ and Theorem 1.14 shows that $\omega(\mathbf{Z}[X])$ is βX. $\square$

1.43. Since the completely regular spaces are precisely the subspaces of compact Hausdorff spaces by Tychonoff's Theorem 1.4, Proposition 1.42 and Theorem 1.39 combine to yield another characterization of the class of completely regular spaces.

Corollary (Frink). *A* T_1-*space is completely regular if and only if it admits a normal base.*

Note that this characterization is an internal one in the sense that it contains no reference to the real-valued mappings on the space. Frink's motivation for considering normal bases was to obtain such a characterization.

1.44. We have seen that βX can be obtained as a space of z-ultrafilters in such a way that each point of $\beta X \setminus X$ is the limit of a unique free z-ultrafilter. To relate this property to the characterizations of βX in terms of extensions of mappings, it is necessary to consider the behavior of z-filters under mappings.

If f is a mapping of X to Y and $\mathscr{F}$ a z-filter on X, the family of sets $f[\mathscr{F}]=\{f[Z]:Z\in\mathscr{F}\}$ need not be a z-filter. This can easily be seen by choosing f to be the identity from $\mathbb{R}$ with the discrete topology to $\mathbb{R}$ with the usual topology. However, a subfamily of $f[\mathscr{F}]$ is sufficient to reflect the most interesting properties of $\mathscr{F}$. Define

$$f^{\#}\mathscr{F} \equiv \{Z \in \mathbf{Z}[Y] : f^{\leftarrow}(Z) \in \mathscr{F}\}.$$

It is clear that $f^{\#}\mathscr{F}$ is a z-filter. Since the zero-set neighborhoods of a point form a base for the neighborhoods of the point, it can easily be seen that if a z-filter $\mathscr{F}$ converges to x in X, then $f^{\#}\mathscr{F}$ converges to $f(x)$ in Y. Further discussion of the function $f^{\#}$ can be found in [GJ, 4.12, 4.13, 10.17].

In Proposition 1.28, we saw that convergence can be described in terms of z-ultrafilters. It would be convenient if $f^{\#}$ preserved z-ultra-filters but this is not the case. Following [GJ, ex. 4H], let X be $[0,1]$ with the discrete topology and let f from X to the closed unit interval I be the identity mapping. Since $I = [0,1]$ is compact and any z-filter is a family of closed sets with the finite intersection property, every z-filter on I is fixed. Thus, if $\mathscr{U}$ is any free z-ultrafilter on X, $f^{\#}\mathscr{U}$ is fixed. But any fixed z-ultrafilter on I contains its limit point as a member, and $f^{\#}\mathscr{U}$ can contain no finite subset set since $\mathscr{U}$ is free. Thus, $f^{\#}\mathscr{U}$ is not a z-ultrafilter.

However, $f^{\#}$ does preserve the prime z-filters, and we shall see that this family includes the z-ultrafilters and shares many of the characteristics of z-ultrafilters. A z-filter $\mathscr{F}$ is *prime* if the union of two zero-sets belongs to $\mathscr{F}$ only if one of the zero-sets belongs to $\mathscr{F}$.

Let $\mathscr{F}$ be a prime z-filter on X and suppose that $Z_1 \cup Z_2$ is in $f^{\#}\mathscr{F}$. Then $f^{\leftarrow}(Z_1 \cup Z_2) = f^{\leftarrow}(Z_1) \cup f^{\leftarrow}(Z_2)$ is in $\mathscr{F}$ as is either $f^{\leftarrow}(Z_1)$ or $f^{\leftarrow}(Z_2)$. But then the corresponding Z_i is in $f^{\#}\mathscr{F}$ and we have demonstrated the following useful property of prime z-filters.

Proposition. *If f is a mapping of X to Y and $\mathscr{F}$ is a prime z-filter on X, then $f^{\#}\mathscr{F}$ is a prime z-filter on Y.*

1.45. Additional properties of the class of prime z-filters are contained in the next proposition.

Proposition. (a) *Any z-ultrafilter is prime.*

(b) *If $\mathscr{F}$ is a prime z-filter on X, the following are equivalent for a point x in X:*

(1) *x is a cluster point of $\mathscr{F}$.*

(2) *$\mathscr{F}$ converges to x.*

(3) $\bigcap \mathscr{F} = \{x\}$.

Proof. (a) Let $Z_1 \cup Z_2$ be in the z-ultrafilter $\mathscr{U}$. Then if neither Z_1 nor Z_2 is in $\mathscr{U}$, there exist A_1 and A_2 in $\mathscr{U}$ such that $A_1 \cap Z_1 = \emptyset$ and

$A_2 \cap Z_2 = \emptyset$. But then $(Z_1 \cup Z_2) \cap (A_1 \cap A_2) = \emptyset$, and $Z_1 \cup Z_2$ could not be in $\mathcal{U}$, which is a contradiction.

(b) Let the prime z-filter $\mathcal{F}$ cluster at x and let V be a zero-set neighborhood of x. By the complete regularity of X, V contains a neighborhood of x of the form $X \setminus Z$ for Z a zero-set. Since $V \cup Z = X$ is in $\mathcal{F}$, either V or Z is in $\mathcal{F}$. But Z misses x and therefore cannot be in $\mathcal{F}$. Thus, V is in $\mathcal{F}$ and $\mathcal{F}$ converges to x. This shows that (1) implies (2) and the other implications are clear. $\square$

Prime z-filters are discussed in [GJ, Chapter 2 and Section 3.17] which is the source of the previous Proposition.

Characterizations of βX

1.46. We now summarize the characteristic properties of the Stone-Čech compactification by displaying their equivalence in a somewhat encyclopedic theorem similar to Theorem 6.5 of [GJ].

Theorem. *Every completely regular space X has a unique compactification βX which has the following equivalent properties:*

   (1) *X is C^*-embedded in βX.*
   (2) *Every mapping of X into a compact space extends uniquely to βX.*
   (3) *Every point of βX is the limit of a unique z-ultrafilter on X.*
   (4) *If Z_1 and Z_2 are zero-sets in X, then*

$$\mathrm{cl}_{\beta X} Z_1 \cap \mathrm{cl}_{\beta X} Z_2 = \mathrm{cl}_{\beta X}(Z_1 \cap Z_2).$$

   (5) *Disjoint zero-sets in X have disjoint closures in βX.*
   (6) *Completely separated sets in X have disjoint closures in βX.*
   (7) *βX is maximal in the partially ordered set of compactifications of X.*

Proof. The existence of a compactification satisfying one or more of the listed properties has already been demonstrated. That βX satisfies (1), (2), (6), and (7) seems to be clearest by considering βX through embedding X into a product of intervals as in Theorem 1.9. Condition (3) seems to stand out most clearly through the isomorphism of βX with the space of maximal ideals $\mathcal{M}(X)$ established in Theorem 1.31. Conditions (4) and (5) are most transparent by applying Theorem 1.39 to the normal base $\mathbf{Z}[X]$.

The uniqueness of a compactification satisfying (2) was shown in Corollary 1.13. It remains to show that the seven properties are equivalent.

(3)$\Rightarrow$(2): Let A^p be the unique z-ultrafilter converging to the point p of βX. If f is a mapping of X into a compact Hausdorff space Y, $f^{\#} A^p$

is a family of closed sets with the finite intersection property and therefore $\bigcap f^{\#} A^{p}$ contains a point y of Y. By Propositions 1.44 and 1.45, $f^{\#} A^{p}$ is a prime z-filter converging to y and $\bigcap f^{\#} A^{p} = \{y\}$. This defines a function $\beta(f)$ from βX to Y. If p is in X, p belongs to $\bigcap A^{p}$ and $y = \beta(f)(p)$ is in $\bigcap f^{\#} A^{p}$ so that $\beta(f)$ is an extension of f.

To show that $\beta(f)$ is continuous, let F be a zero-set neighborhood of $\beta(f)(p)$ and let F' be a zero-set in Y such that $Y \backslash F'$ is contained in F and is a neighborhood of $\beta(f)(p)$. Then $F \cup F' = Y$ and if Z and Z' are the inverse images of F and F' under f, we have that $Z \cup Z' = X$. Thus, $\operatorname{cl}_{\beta X} Z \cup \operatorname{cl}_{\beta X} Z' = \beta X$. Since $\beta(f)(p)$ is not in F', p is not in $\operatorname{cl}_{\beta X} Z'$ so that $\beta X \backslash \operatorname{cl}_{\beta X} Z'$ is a neighborhood of p. Every point in this neighborhood belongs to $\operatorname{cl}_{\beta X} Z$, so that by the definition of $\beta(f)$, $\beta(f)[\beta X \backslash \operatorname{cl}_{\beta X} Z'] \subset F$, and $\beta(f)$ is continuous. The uniqueness of the extension is immediate since any two extensions must agree on the dense subspace X.

$(2) \Rightarrow (1)$ is clear since I is compact.

$(1) \Rightarrow (6)$ follows from Urysohn's Extension Theorem, Theorem 1.2 which actually shows that the two are equivalent.

$(6) \Rightarrow (5)$ is immediate since disjoint zero-sets are completely separated.

$(5) \Rightarrow (4)$: It is clear that $\operatorname{cl}_{\beta X}(Z_1 \cap Z_2) \subset \operatorname{cl}_{\beta X} Z_1 \cap \operatorname{cl}_{\beta X} Z_2$. On the other hand, if $p \in \operatorname{cl}_{\beta X} Z_1 \cap \operatorname{cl}_{\beta X} Z_2$, then for every zero-set neighborhood V of p in βX, $p \in \operatorname{cl}_{\beta X}(Z_1 \cap V)$ and $p \in \operatorname{cl}_{\beta X}(Z_2 \cap V)$. Then (5) implies that $V \cap Z_1$ and $V \cap Z_2$ cannot be disjoint so that $V \cap (Z_1 \cap Z_2) \neq \emptyset$ and $p \in \operatorname{cl}_{\beta X}(Z_1 \cap Z_2)$.

$(4) \Rightarrow (3)$: Since X is dense in βX, the trace on X of the zero-set neighborhoods of a point p in βX is a z-filter $\mathscr{F}$ on X. By Proposition 1.36, $\mathscr{F}$ is contained in a z-ultrafilter $\mathscr{U}$ and $\mathscr{U}$ converges to p. But distinct z-ultrafilters must contain disjoint zero-sets by Proposition 1.17, and (4) shows that any pair of disjoint zero-sets must have disjoint closures in βX. Hence, exactly one z-ultrafilter converges to p.

$(2) \Rightarrow (7)$ is shown in Corollary 1.12.

$(7) \Rightarrow (1)$: As in Theorem 1.9, use the evaluation mapping e to embed X into a product of unit intervals indexed by $C^{*}(X)$. Then if K is the closure of X in the product, K is a compactification of X and thus is a continuous image of the maximal compactification βX under a mapping

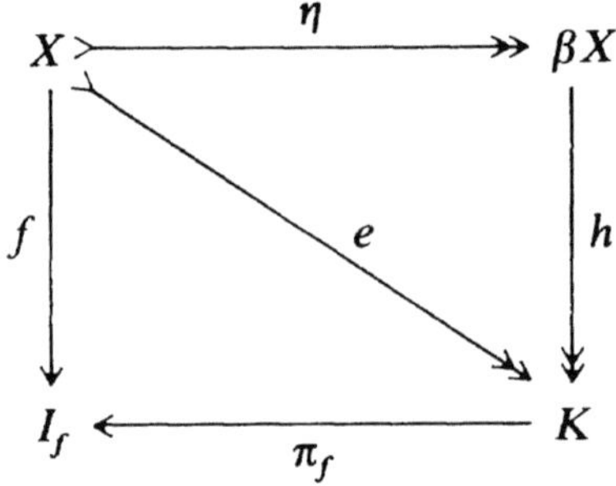

h which leaves points of X fixed. A member f of $C^*(X)$ extends to K by composing e with the projection π_f, and $\pi_f \circ h$ extends f to βX. $\Box$

Note that in the proof, the compactness of βX has been used only once and that was in condition (7). Thus, *if we replace βX with any space T which contains X as a dense subspace, conditions (1)—(6) remain equivalent.*

The main steps in the development of the Stone-Čech compactification which led up to the preceding theorem are outlined below. The arrows indicate major sources of influence.

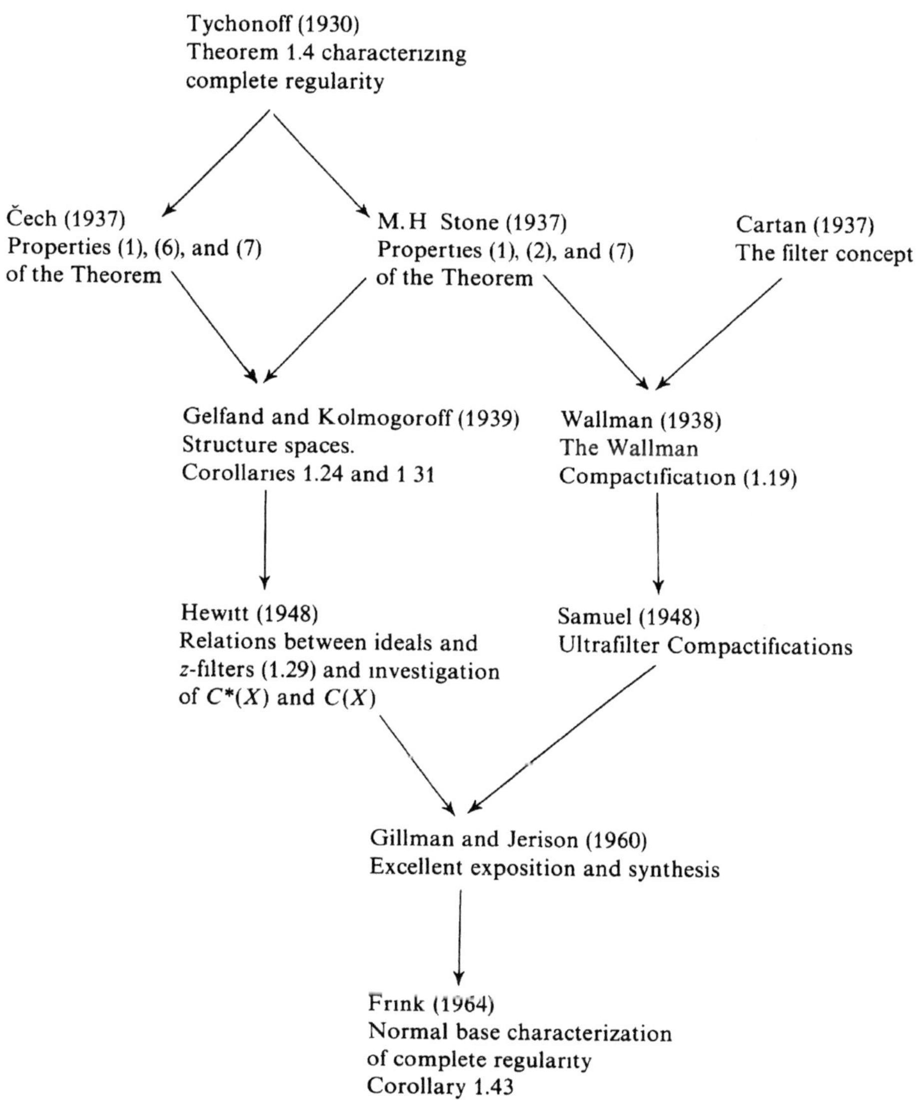

1.47. We now consider a few of the most useful properties of the Stone-Čech compactification. First we have a technical result which is taken from [GJ, 3.11].

Proposition. *Any two disjoint closed sets, one of which is compact, are completely separated. Hence, every compact subspace is C^*-embedded.*

Proof. Let K and F be disjoint closed subsets of X with K compact. For each x in K, choose a zero-set neighborhood Z_x of x and a zero-set Z_x' containing F and missing Z_x. The cover $\{Z_x\}$ of K has a finite subcover $\{Z_{x_i}\}$. Then $\bigcup Z_{x_i}$ and $\bigcap Z_{x_i}'$ are disjoint zero-sets containing K and F respectively. Hence, K and F are completely separated.

To prove the second statement, we show that a compact subspace S of X is C^*-embedded by applying Theorem 1.2. Any two subsets of S which are completely separated in S have disjoint compact closures and thus are completely separated in X by the first statement. $\square$

1.48. Proposition. *A subspace S of X is C^*-embedded in X if and only if $\beta S = \mathrm{cl}_{\beta X} S$.*

Proof. If S is C^*-embedded in X, S is clearly C^*-embedded in βX and therefore $\mathrm{cl}_{\beta X} S$ is a compactification of S in which S is C^*-embedded. Conversely, $\mathrm{cl}_{\beta X} S$ is compact and therefore is C^*-embedded in βX. It follows that S is C^*-embedded in X since S is C^*-embedded in $\mathrm{cl}_{\beta X} S$ by assumption. $\square$

The proposition will be most useful when S is a closed, C^*-embedded copy of the countable discrete space $\mathbb{N}$.

1.49. The following proposition will be useful in the creation of examples. A subspace T of βX which contains X is clearly dense in βX. Further, such a subspace T is C^*-embedded in βX since a mapping in $C^*(T)$ can first be restricted to X and then extended to βX. Thus, we have verified the

Proposition. *If T is a subspace of βX containing X, then βT is βX.*

1.50. The work of Wallman and Frink make it clear that the process of forming βX is one of "fixing" the free z-ultrafilters of X by attaching a point p for each free z-ultrafilter A^p in such a way that A^p converges to p. Because of this interpretation of βX, we will often identify the points of $\beta X \setminus X$ with the free z-ultrafilters of X. Since any z-filter is contained in a z-ultrafilter and it is clear that a free z-filter cannot be contained in a fixed z-ultrafilter, we have verified the following characterizations of compactness:

Proposition. *The following are equivalent:*
   (1) *X is compact.*

(2) *Every z-ultrafilter on X converges to a point of X.*

(3) *Every z-filter on X is fixed.*

(4) *Every ideal in $C(X)$ is fixed.*

We see then that an ideal M^p or a z-ultrafilter A^p is fixed for p in X and free for p in $\beta X\backslash X$. The subspace $\beta X\backslash X$ will be referred to as the *growth* of X and will be denoted by X^*.

Note that if Z belongs to a z-filter $\mathscr{F}$, then the trace of the elements of $\mathscr{F}$ on Z is a family of closed sets of Z having the finite intersection property. Hence, if Z is compact, the trace must have non-empty intersection and $\mathscr{F}$ must then be fixed. Thus, we have shown that *compact zero-sets can belong only to fixed z-ultrafilters* and that *no free z-filter can contain a compact zero-set.*

1.51. Since a space X is C^*-embedded in βX, one is led to expect that there is a simple and perhaps useful relationship between the zero-sets of X and those of βX. It would be tempting to conjecture that a zero-set of βX is just the closure of a zero-set of X. However, this is not the case. Consider the mapping f on $\mathbb{R}$ defined by $f(x)=1/(1+|x|)$. It is clear that f never vanishes on $\mathbb{R}$, but since f approaches zero on every non-compact subset of $\mathbb{R}$, $\beta(f)(p)=0$ for all p in $\beta\mathbb{R}\backslash\mathbb{R}$. Thus, $\beta\mathbb{R}\backslash\mathbb{R}$ is the zero-set of $\beta(f)$ but is not the closure of any zero-set of $\mathbb{R}$.

The actual relationship between the two families of zero-sets is somewhat more complicated, but will still prove to be useful. For the proof of the following proposition, it is helpful to recall that the z-ultrafilters on X are of the form

$$A^p = \{Z\in \mathbf{Z}[X] : p\in \mathrm{cl}_{\beta X} Z\}$$

for p in βX. Thus, the closure of a zero-set is given by

$$\mathrm{cl}_{\beta X} Z = \{p\in \beta X : Z\in A^p\}\ .$$

Proposition. *The zero-sets of βX are countable intersections of closures in βX of zero-sets of X.*

Proof. If Z is in $\mathbf{Z}[\beta X]$, then $Z=\mathbf{Z}(\beta(f))$ for some f in $C^*(X)$. Thus,

$$\mathbf{Z}(\beta(f)) = \bigcap_{n=1}^{\infty} \{p\in \beta X : |\beta(f)(p)| \leqslant 1/n\}\ .$$

But then we can write that

$$\mathbf{Z}(\beta(f)) = \bigcap_{n=1}^{\infty} \mathrm{cl}_{\beta X}\{x\in X : |f(x)| \leqslant 1/n\}$$

and the result follows since $\{x\in X : |f(x)| \leqslant 1/n\}$ is a zero-set for each n. $\quad\square$

Observe that from the definition of closure, the proof also shows that the zero-sets of X involved in computing $\mathbf{Z}(\beta(f))$ belong to A^p if p is in $\mathbf{Z}(\beta(f))$.

1.52. If f is a mapping of X to Y, then the composition of f and the embedding of Y into βY extends to a mapping $\beta(f)$ of βX into βY. Thus, we have the following commutative diagram:

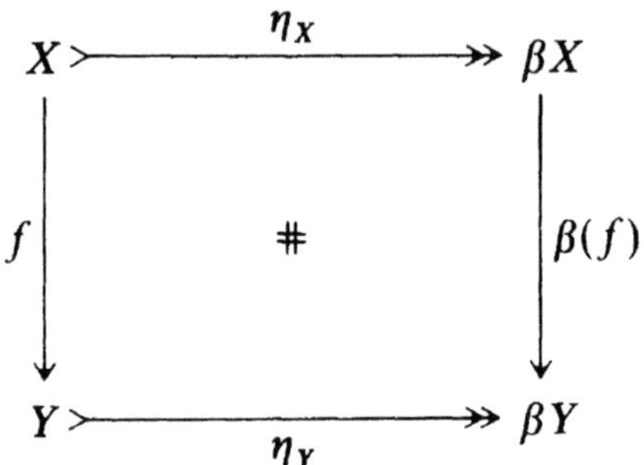

We shall see in Chapter 10 that the existence of this diagram is a major step in placing the Stone-Čech compactification in its categorical context.

Generalizations of Compactness

1.53. The classes of realcompact spaces, pseudocompact spaces, and locally compact spaces each contain the class of compact spaces and will be defined later. Our main interest in these classes of spaces will stem from the investigation of the interactions between a space X and its "growth" $\beta X \setminus X$. The subspace $\beta X \setminus X$ will be denoted by X^*. We will give a criterion to recognize a space of each class in terms of the growth of its Stone-Čech compactification.

Any member f of $C(X)$ is a mapping of X into $\alpha \mathbb{R} = \mathbb{R} \cup \{\infty\}$, the one point compactification of $\mathbb{R}$, and thus has an extension f^α which maps βX into $\alpha \mathbb{R}$.

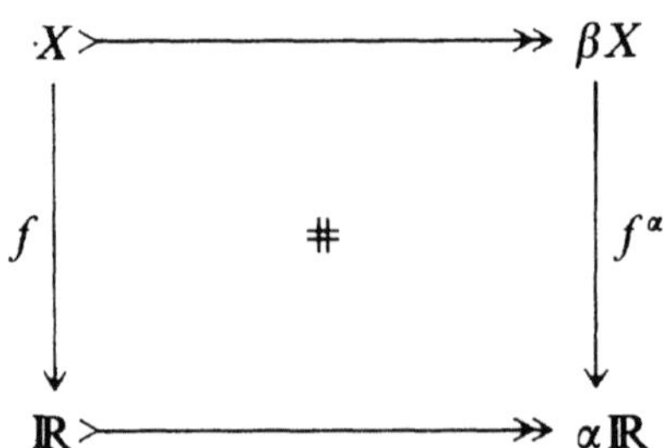

If f is unbounded, there will be a point in the growth of X at which f^α will take the value ∞. For each map f in $C(X)$, define

$$\upsilon_f X \equiv \beta X \setminus \{ p \in \beta X : f^\alpha(p) = \infty \} \ .$$

Thus, $\upsilon_f X$ is the set of points of βX at which f^α is finite, and we will call $\upsilon_f X$ the set of *real points* of f. Let υX be the subspace of βX consisting of points which are real points for every f in $C(X)$, i.e.

$$\upsilon X = \bigcap \{ \upsilon_f X : f \in C(X) \} \ .$$

A space X is said to be *realcompact* if $X = \upsilon X$, i.e. if the only points which are real points for every f in $C(X)$ are the points of X itself. It is immediate that any compact space is realcompact and that the subspace υX of βX is the largest subspace of βX to which every member of $C(X)$ can be extended without any extension taking on the value ∞. If the extension of f to υX is denoted by $\upsilon(f)$, then the correspondence $f \mapsto \upsilon(f)$ is an isomorphism of $C(X)$ with $C(\upsilon X)$.

Our main use of realcompactness will be to recognize the subspace υX of βX as a "dividing line" between two different kinds of points of βX. The distinction between points of υX and points of $\beta X \setminus \upsilon X$ can be recognized through the corresponding z-ultrafilters. A z-ultrafilter A^p is called a *real z-ultrafilter* if the intersection of any countable subfamily of A^p is non-empty. The ideal M^p will be called real if A^p is a real z-ultrafilter. We can now use the representation of the zero-sets of βX obtained in Proposition 1.51 to characterize realcompactness in terms of zero-sets and also in terms of z-ultrafilters.

Theorem (Hewitt). *The following are equivalent for any space X:*
   (1) X is realcompact.
   (2) Every point of $\beta X \setminus X$ is contained in a zero-set of βX which misses X.
   (3) Every real z-ultrafilter on X is fixed.

Proof. (1)$\Leftrightarrow$(2): If $X = \upsilon X$, then for every point p in $\beta X \setminus X$ there a mapping f in $C(X)$ such that $f > 0$ and $f^\alpha(p) = \infty$. But then if g is the reciprocal of f, $p \in \mathbf{Z}(\beta(g)) \subset \beta X \setminus X$. The converse is obtained by reversing the steps.

   (2)$\Rightarrow$(3): Suppose that p is in $\beta X \setminus X$. We will show that the corresponding free z-ultrafilter A^p cannot be real. By (2), there is a zero-set Z such that $p \in Z \subset \beta X \setminus X$. Proposition 1.51 implies that $Z = \bigcap \mathrm{cl}_{\beta X} Z_n$. But since this intersection is contained in the growth, $\bigcap Z_n = \emptyset$ and A^p is not a real z-ultrafilter.

   (3)$\Rightarrow$(2): For p in $\beta X \setminus X$, A^p is not real. Hence, there exists a countable family $\{ \mathbf{Z}(f_n) \}$ contained in A^p such that $\bigcap \mathbf{Z}(f_n) = \emptyset$ and f_n is in

$C^*(X)$ for every n. Then the zero-set $\mathbf{Z}(\beta(f_n))$ of the extension of f_n contains p. Then $\bigcap \mathbf{Z}(\beta(f_n))$ is a zero-set containing p and missing X. $\quad\square$

1.54. Examples. The argument given in Section 1.51 to show that $\beta\mathbb{R}\setminus\mathbb{R}$ is a zero-set of $\beta\mathbb{R}$ also shows that $\mathbb{R}$ *is realcompact*. A similar argument will show that *the discrete space* $\mathbb{N}$ *of natural numbers is realcompact. The space of rationals,* $\mathbb{Q}$, *is also realcompact* as will be shown in Exercise 1D.

1.55. Let f be a mapping of X to Y. We saw in Section 1.52 that there is an extension $\beta(f)$ of f such that the following diagram commutes:

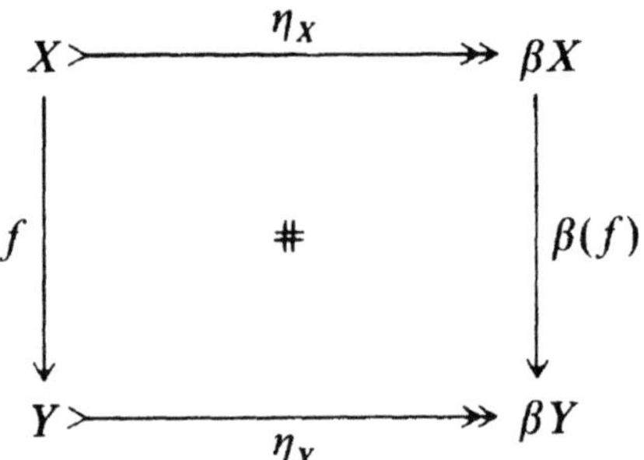

We will establish an analogous result for υX by showing that the image of the restriction $\beta(f)|\upsilon X$ is a subspace of υY. The result will follow from the definition of υX. If g belongs to $C(Y)$, then there is an extension g^α of g which maps βY to $\alpha\mathbb{R}$. Similarly, $g\circ f$ belongs to $C(X)$ and therefore has an extension $(g\circ f)^\alpha$ of βX into $\alpha\mathbb{R}$. Since the two mappings $(g\circ f)^\alpha$ and $g^\alpha\circ\beta(f)$ both agree with $g\circ f$ on X, they are equal and we have the following commutative diagram:

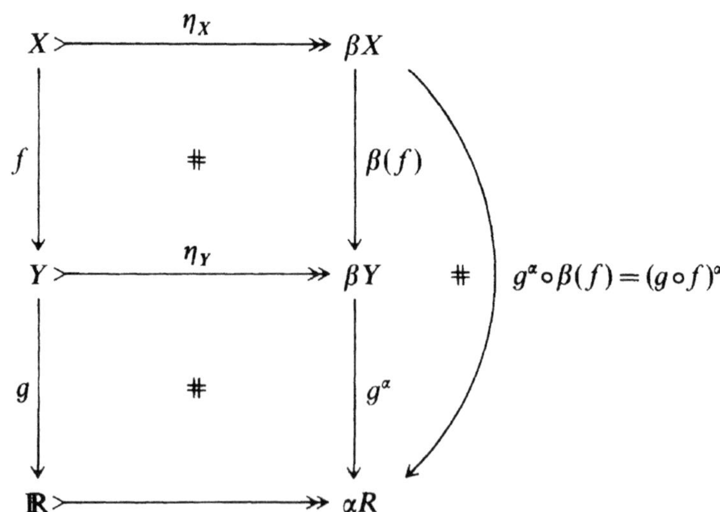

Now if q belongs to vX, we must show that $\beta(f)(q)$ is in vY. From the definition of vX, we see that q is a real point of $g \circ f$ for every g in $C(Y)$. Thus, for every g in $C(Y)$, $(g^\alpha \circ \beta(f))(q) = (g \circ f)^\alpha(q) \neq \infty$. Hence, we have shown that

$$\beta(f)(q) \in vY = \bigcap \{v_g Y : g \in C(Y)\} \ .$$

By defining $v(f)$ to be the restriction $\beta(f)|vX$, we have obtained the following commutative diagram:

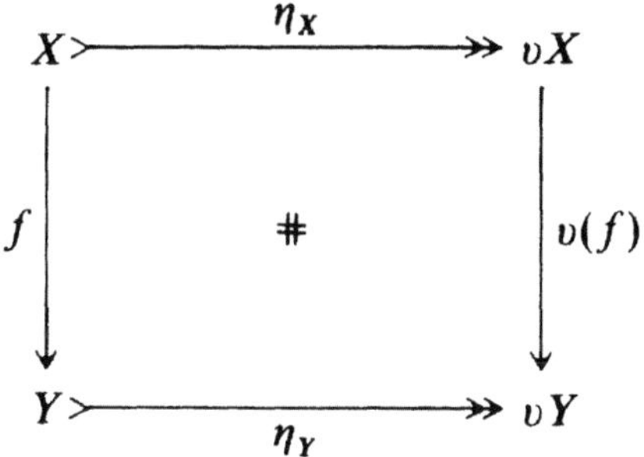

If we consider the special case where the space Y in the preceding diagram is realcompact, then we have verified the following result:

Proposition. *Every mapping of the space X into a realcompact space Y will extend uniquely to vX.*

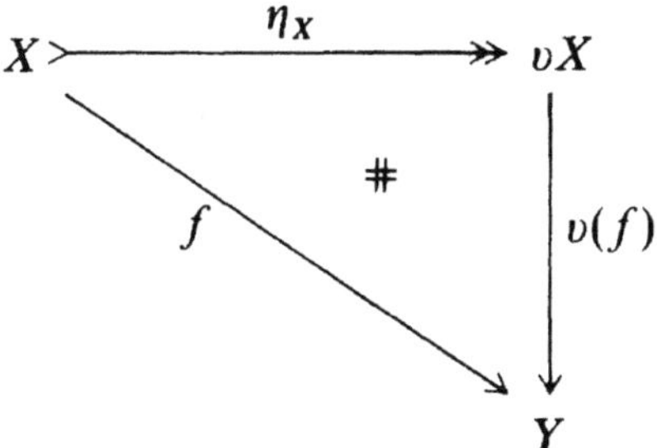

Note that this result is analogous to Theorem 1.11 for βX and to Corollary 1.8 for ρX.

The space vX is called the *Hewitt-Nachbin realcompactification* of X since E. Hewitt and L. Nachbin gave independent constructions of the extension vX. Hewitt's discussion of the class of realcompact spaces appears in his 1948 paper. Nachbin's construction of vX is based on uniformities and was not published. [GJ, Chapter 15] discusses Nachbin's work and Shirota's important contribution which came in 1952 and showed that barring measurable cardinals, the class of realcompact spaces is the same as the class of spaces which admit a complete uniform structure.

1.56. In Corollary 1.25, we saw that $C^*(X)$ is an algebraic invariant for the class of compact spaces. Since υX is a subspace of βX and hence of the space of ideals $\mathcal{M}(X)$, we have the analogous theorem for $C(X)$.

Theorem (Hewitt). *If X and Y are realcompact spaces, X is homeomorphic to Y if and only if $C(X)$ and $C(Y)$ are isomorphic.*

The theorem follows from the fact that the property of being a real maximal ideal is preserved under the ring isomorphism of $C(X)$ with $C(Y)$.

The brief treatment of realcompactness given here is a very narrow one designed only to facilitate the study of βX by utilizing the subspace υX. The material included is based on [GJ, Chapter 8] which offers a much more complete discussion of realcompactness. The most complete treatment of realcompactness is given in the thesis of M. Weir.

1.57. A space X is *pseudocompact* if every real-valued mapping on X is bounded, i.e. if $C^*(X)$ is equal to $C(X)$.

Theorem (Hewitt). *The following are equivalent for any space X:*
  (1) *X is pseudocompact.*
  (2) *Every non-empty zero-set of βX meets X.*
  (3) *Every z-ultrafilter on X is real.*
  (4) *Every z-filter on X has the countable intersection property.*

Proof. (1)$\Rightarrow$(2): If $\mathbf{Z}(\beta(f))$ is contained in $\beta X \backslash X$, f does not vanish on X and is not bounded away from zero on X. Thus, f^{-1} is an unbounded member of $C(X)$.

(2)$\Rightarrow$(1): If there is an unbounded member of $C(X)$, then there is an unbounded member f of $C(X)$ which does not vanish on X, and the zero-set of the extension of f^{-1} is non-empty and misses X.

(2)$\Leftrightarrow$(3): The zero-sets containing a point p of the growth of X are of the form $\bigcap \mathrm{cl}_{\beta X} Z_n$ with $\{Z_n\} \subset A^p$. Since such a zero-set can miss X exactly when $\bigcap Z_n = \emptyset$, it is clear that A^p is a real z-ultrafilter precisely when every zero-set containing p meets X.

(3)$\Leftrightarrow$(4) is clear. $\square$

1.58. Since every z-ultrafilter is real when X is pseudocompact and every real z-ultrafilter is fixed for X realcompact, we have verified the following characterization of compactness. The second statement in the corollary follows since υX is the subspace of βX corresponding to the real z-ultrafilters on X.

Corollary. *A space is compact if and only if it is realcompact and pseudocompact. Further, $\upsilon X = \beta X$ if and only if X is pseudocompact.*

1.59. A space is *locally compact* if each point of the space has a basis of compact neighborhoods. The locally compact spaces are precisely those spaces which can be compactified by the addition of a single point. (See Exercise 1F.) Since such a compactification αX is the image of βX, $\beta X \setminus X$ is a closed subspace of βX since it is the inverse image of the added point. We have shown the necessity in the following characterization.

Theorem. *A space is locally compact if and only if it is an open subspace of its Stone-Čech compactification.*

Sufficiency follows from the fact that an open subspace of a compact space is locally compact.

F-Spaces and *P*-Spaces

1.60. In addition to the generalizations of compactness just discussed, the class of *F*-spaces will be of frequent interest in the investigation of Stone-Čech compactifications. This class of spaces was introduced in 1956 B by L. Gillman and M. Henriksen as the class of spaces for which $C(X)$ is a ring in which every finitely generated ideal is a principal ideal. This is also the definition which is used in the discussion of *F*-spaces contained in [GJ, Chapter 14] from which the material in this section is drawn. Since the algebraic content is not emphasized here, we will adopt one of the characterizations of the class of *F*-spaces given by Gillman and Henriksen as our definition: An *F-space* is a space in which every cozero-set is C^*-embedded. The next theorem offers the characterizations of the class which will be most useful in the present context. If f belongs to $C(X)$, let $\mathrm{pos}(f)$ and $\mathrm{neg}(f)$ denote the set of points on which f is positive and negative, respectively.

Theorem. *The following are equivalent for any space X:*
   (1) *X is an F-space.*
   (2) *Disjoint cozero-sets of X are completely separated.*
   (3) *For f in $C(X)$, $\mathrm{pos}(f)$ and $\mathrm{neg}(f)$ are completely separated.*
   (4) *βX is an F-space.*

Proof. (1)$\Rightarrow$(2): Let U and V be disjoint cozero-sets of X. Then the function which is constantly equal to 0 on U and to 1 on V is continuous on the cozero-set $U \cup V$ and its extension separates U from V.

   (2)$\rightarrow$(3): This is immediate since $\mathrm{pos}(f) = \mathbf{Cz}(f \vee \mathbf{0})$ and $\mathrm{neg}(f) = \mathbf{Cz}(f \wedge \mathbf{0})$.

   (3)$\Rightarrow$(4): Let $\mathbf{Cz}(h)$ be a cozero-set in βX and let A and B be completely separated subsets of $\mathbf{Cz}(h)$. Then there exists k in $C^*(\mathbf{Cz}(h))$ which is positive on A and negative on B. Define f on X by

$$f(x) = \begin{cases} 0 & \text{on } X \backslash \mathbf{C}\mathbf{z}(h)\,, \\ k(x)|h(x)| & \text{on } X \cap \mathbf{C}\mathbf{z}(h)\,. \end{cases}$$

Since k is bounded, f is continuous on X and $A \cap X$ is contained in $\mathrm{pos}(f)$ and $B \cap X$ is contained in $\mathrm{neg}(f)$. Since $\mathrm{pos}(f)$ and $\mathrm{neg}(f)$ are completely separated in X, their closures are disjoint in βX and contain A and B, respectively. Thus, A and B are completely separated in βX showing that $\mathbf{Cz}(h)$ is C^*-embedded in βX by Theorem 1.2 and therefore βX is an F-space.

(4)$\Rightarrow$(1): This follows from the isomorphism of $C(\beta X)$ with $C^*(X)$ and is left to the algebraic approach of [GJ, 14.25]. □

1.61. Corollary. *Any cozero-set or C^*-embedded subspace of an F-space is also an F-space.*

1.62. Examples of F-spaces include the growths of certain Stone-Čech compactifications. A *σ-compact space* is one which is the union of countably many compact subspaces.

Proposition (Gillman and Henriksen). *The growth of a σ-compact locally compact space is an F-space.*

The proof of the preceding result is particularly suited to the more algebraic techniques developed in detail in [GJ] and will not be included here. The proof given in [GJ, 14.27] shows that X^* satisfies (3) of Theorem 1.60 and is based on an algebraic treatment of the extension of a member of $C(X)$ to a mapping of βX into $\alpha \mathbb{R}$, the one-point compactification of $\mathbb{R}$.

1.63. Since a zero-set $\mathbf{Z}(f)$ can be written

$$\mathbf{Z}(f) = \bigcap \{x \in X : |f(x)| < 1/n\}\,,$$

any zero- set is a G_δ, i.e. an intersection of countably many open sets. The complement of a G_δ is a union of countably many closed sets and is called an F_σ.

If Z is a zero-set of βX contained in $X^* = \beta X \backslash X$, then $\beta X \backslash Z$ is an open F_σ in the compact space βX and is therefore locally compact and σ-compact. By Proposition 1.49, $\beta(\beta X \backslash Z) = \beta X$, and we have verified the

Corollary. *Any zero-set of βX which does not meet X is an F-space.*

1.64. Another relationship between Stone-Čech compactifications and the class of F-spaces is that every infinite compact F-space contains a copy of $\beta \mathbb{N}$.

Proposition. *Every countable subspace of an F-space is C*-embedded and hence, every infinite compact F-space contains a copy of $\beta\mathbb{N}$.*

Proof. Let X be an F-space and let N be a countable subspace of X. Let $\{x_n\}$ and $\{y_n\}$ be subsets of N which are completely separated in N. We will show that $\{x_n\}$ and $\{y_n\}$ are completely separated in X in order to apply Urysohn's Extension Theorem, 1.2. Since neither sequence meets the closure of the other, $\{x_n\}$ and $\{y_n\}$ are contained in disjoint cozero-sets, U and V, of X: Choose closed neighborhoods of each x_n and y_n by alternating the choices between the two sequences and each time choosing the neighborhood to miss the union of the closure of the other sequence with the neighborhoods previously chosen for the other sequence. Since each neighborhood contains a cozero-set neighborhood, U and V can each be chosen to be the union of countably many cozero-sets. Since the cozero-sets U and V are completely separated in X, $\{x_n\}$ and $\{y_n\}$ are completely separated in X and N is C^*-embedded in X by Theorem 1.2.

Since any infinite Hausdorff space contains a countable discrete subspace $\mathbb{N}$ and the closure of such a subspace in a compact F-space X is a compactification in which $\mathbb{N}$ is C^*-embedded, an infinite compact F-space contains a copy of $\beta\mathbb{N}$. $\square$

1.65. Gillman and Henriksen also introduced the class of P-spaces. The class was first defined in their 1954 paper and was also discussed in 1956 B.

A point of a topological space is called a *P-point* if every G_δ containing the point is a neighborhood of the point. Recall that every zero-set is a G_δ. Further, it is easy to see that in a completely regular space, every G_δ containing a point contains a zero-set which also contains the point. Hence, a point is a P-point if and only if every zero-set containing the point is a neighborhood of the point. A space is called a *P-space* if every point of the space is a P-point. Thus, a P-space is one in which every G_δ, or equivalently every zero-set, is open. Since every cozero-set in a P-space is clopen, it is immediate that every P-space is an F-space. But while Proposition 1.62 indicates that there are many compact F-spaces, the additional open sets in a P-space make it hard for a P-space to be compact. In fact,

Proposition. *Every pseudocompact P-space is finite.*

Proof. Since the complement of a countable set is a G_δ, every sequence in a P-space is closed, discrete, and is also C^*-embedded by the preceding proposition since a P-space is an F-space. Since a zero-set is clopen, every sequence in a P-space is completely separated from any disjoint zero-set and thus is C-embedded by Theorem 1.3. Hence, an infinite

P-space contains a C-embedded copy of $\mathbb{N}$ and fails to be pseudo-compact. ☐

1.66. Just as a P-space can rarely be compact, the product of two F-spaces cannot often be an F-space. The following result is contained in the 1960 paper of P. C. Curtis, Jr. and several related results are contained in the thesis of N. Hindman and the 1969 paper of Comfort, Hindman, and Negrepontis.

Proposition. *If $X \times Y$ is an F-space, then either X or Y is a P-space.*

Proof. If X and Y both fail to be P-spaces, then there exist points x_0 in X and y_0 in Y together with non-negative mappings f in $C(X)$ and g in $C(Y)$ such that $f(x_0) = g(y_0) = 0$ and x_0 belongs to $\mathrm{cl}(\mathrm{pos}(f))$ and y_0 to $\mathrm{cl}(\mathrm{pos}(g))$. Define h in $C(X \times Y)$ by $h(x, y) = f(x) - g(y)$. Then the point (x_0, y_0) belongs to the closures of both $\mathrm{pos}(h)$ and $\mathrm{neg}(h)$ which contradicts (3) of Theorem 1.60. ☐

1.67. As an application of the preceding proposition, consider the following example which lays to rest the tempting, but false, conjecture that $\beta X \times \beta Y$ is $\beta(X \times Y)$. (In Chapter 8 we will investigate the problems inherent in describing the Stone-Čech compactifications of product spaces.)

Example. Proposition 1.62 shows that the growth of $\mathbb{R} \times \mathbb{R}$ is an F-space. In $\beta\mathbb{R} \times \beta\mathbb{R}$, the complement of $\mathbb{R} \times \mathbb{R}$ is $(\beta\mathbb{R} \times \mathbb{R}^*) \cup (\mathbb{R}^* \times \beta\mathbb{R})$. If this subspace were an F-space, then the C^*-embedded subspace $\beta\mathbb{R} \times \mathbb{R}^*$ would also be an F-space. But neither $\beta\mathbb{R}$ nor $\mathbb{R}^*$ is a P-space by Proposition 1.65. Hence, $\beta(\mathbb{R} \times \mathbb{R})$ *is not* $\beta\mathbb{R} \times \beta\mathbb{R}$.

P-points and P-spaces are discussed in [GJ, ex. 4 JKL, 14.29] and the class of P-spaces is investigated in detail in the thesis and paper of A. K. Misra.

Other Approaches to βX

1.68. We have seen three approaches to the construction of βX: as an embedding of X into a product of intervals indexed by $C^*(X)$, as the structure space of $C^*(X)$ or $C(X)$, and as the space of z-ultrafilters on X. A fourth approach is through uniformities and had its beginning in 1937 when A. Weil obtained βX via a construction based on uniform structures. The following use of uniformities is discussed in [GJ, Chapter 15]. If for each f in $C^*(X)$ we define $\psi_f : X \times X \to \mathbb{R}$ by

$$\psi_f(x, y) = |f(x) - f(y)|,$$

then ψ_f is a pseudometric on X. Every member of $C^*(X)$ is uniformly continuous with respect to the uniformity generated by all such pseudo-metrics as f runs through $C^*(X)$ and the uniformity generated is pre-compact. Thus, the completion of X with respect to this uniformity is a compact space containing X as a dense, C^*-embedded subspace and hence is βX. The realcompactification vX is obtained in the same manner by allowing f to run through all of $C(X)$.

Another characterization of βX in terms of uniformities was given by R. Alò in 1969. Call a uniformity on X *admissible* if it generates the topology of X and *continuous* if it generates a topology weaker than that of X. Then βX is that compactification of X to which every admissible precompact uniformity has a continuous extension.

M.H. Stone's original construction in 1937 was based on a combination of algebraic properties of $C^*(X)$ with a ring of subsets of X. His construction was simplified by I. Gelfand and G. Shilov in 1941. In 1941, S. Kakutani gave a construction of βX based on Banach lattices.

P. Alexandroff, in 1939, modified the Wallman technique of compactification described in Section 1.19 to obtain a compactification (not Hausdorff, of course) of an arbitrary regular space. J. Flachsmeyer, in 1961, obtained βX and other compactifications by using inverse limit spaces. P. Zenor, in 1970, constructed both βX and vX as an inverse limit of systems of metric spaces.

In the first volume of their treatise, *Linear Operators*, N. Dunford and J. Schwartz demonstrate the existence of the Stone-Čech compactification from the Stone-Weierstrass Theorem for algebras of complex-valued functions.

Speaking very loosely, a proximity on a set is a relation on the power set of the set in which related sets are thought of as being "close". Each proximity will induce a topology on the set by defining the closure of a set to be the set of singletons which are "close" to the set. Proximities were introduced by V.A. Efremovic in his 1951 and 1952 papers. The class of completely regular spaces can be characterized as that class of spaces in which the topology is induced by a proximity. There is a one-to-one correspondence between the proximities which induce the topology of a space and the set of compactifications of a space. The relationship between proximities and compactifications was considered by Yu M. Smirnov in his 1952 and 1967 papers. Basic discussions of proximities are contained in the texts of R. Engelking and S. Willard.

Finally, in 1971, R. Alò and L. Sennott characterized βX as that compactification of X to which every mapping of X into a Fréchet space such that the image of X is totally bounded will extend. Here the term *Fréchet space* is not used in the usual topological sense, but refers to a topological vector space in which the topology is induced by a complete metric.

Exercises

1A. *Topologically complete spaces*

In 1937, Čech termed a space X *topologically complete* if X is a G_δ in some compactification. His motivation was the fact that a metrizable space is completely metrizable if and only if it is topologically complete.

1. A space is topologically complete if and only if it is a G_δ in its Stone-Čech compactification.
2. Let X be a topologically complete space contained in a space T. Then X is a G_δ in its closure in T. [$\mathrm{cl}_{\beta T} X$ is a compactification of X and βX is maximal.]

References: Čech, 1937. For a proof of the stated characterization of complete metric spaces, see Nagata, 1968.

1B. *σ-compact and locally compact spaces*

1. A space X is *σ-compact* if it is the union of at most countably many compact subspaces. A space X is σ-compact if and only if X^* is a G_δ in βX.
2. Every G_δ containing a compact subspace contains a zero-set containing the compact subspace.
3. Every compact G_δ is a zero-set.
4. A space X is σ-compact and locally compact if and only if X^* is a zero-set in βX.
5. A σ-compact and locally compact space is realcompact.

1C. *Realcompact spaces*

A subspace F of X is *real-closed* in X if every point not in F is contained in a G_δ which misses F. The *real-closure* of a subset S of X is the smallest real-closed subset containing S.

1. The real-closure of a subspace is contained in its closure.
2. υX is the real-closure of X in βX.
3. X is realcompact if and only if it is real-closed in some compactification.
4. A real-closed subspace of a realcompact space is realcompact.
5. Every subspace of a first countable realcompact space is realcompact, hence every subspace of $\mathbb{R}$ is realcompact.

Reference: Real-closure has also been called Q-closure. Mrówka, 1957.

1D. *More realcompact Spaces*

A *Lindelöf space* is one in which every open cover has a countable subcover. A *second countable space* is one which has a countable base for the open sets.

1. Every second countable space is Lindelöf.
2. A space is Lindelöf if and only if every family of closed sets with the countable intersection property has non-empty intersection.
3. Every Lindelöf space is realcompact.
4. Every subspace of $\mathbb{R}^n$ is realcompact if n is a positive integer.
5. The rationals, $\mathbb{Q}$, and the irrationals, $\mathbb{P}$, are realcompact.
6. Every countable space is realcompact.
7. A separable metric space is second countable. Hence, every separable metric space is realcompact.

Reference: [GJ, 8.2]. The various spaces were originally shown to be realcompact in Hewitt, 1948.

1E. *The maximal Compactification*

1. Any space X admits at most $2^{2^{|X|}}$ filters.
2. Any space X admits a *set* $\{K_\alpha\}$ of compactifications.
3. Any space X admits a maximal compactification. [Embed X as the diagonal in $\underset{\alpha}{\times} K_\alpha$.]

Reference: This construction of βX was suggested to the author by B. Banaschewski.

1F. *One-point compactification*

1. An open subspace of a compact space is locally compact.
2. A space X has a Hausdorff compactification αX whose growth is a single point if and only if X is locally compact. [Let neighborhoods of the added point be the point together with the complement of a compact subspace of X.]
3. If X is locally compact and Z is a compact subspace contained in an open set U, then there exists an open set V such that $Z \subset V \subset \mathrm{cl}\, V \subset U$ and $\mathrm{cl}\, V$ is compact.

In the remainder of the problem, let X be a non-compact, locally compact space. Let $\mathfrak{Z}$ be the zero-sets of those mappings in $C^*(X)$ which are constant on the complement of a compact set.

4. $\mathfrak{Z}$ is a disjunctive ring and is a base for the closed subsets of X.
5. If $\mathbf{Z}(f)$ belongs to $\mathfrak{Z}$, then either $\mathbf{Z}(f)$ or $\mathrm{cl}(\mathbf{Cz}(f))$ is compact.
6. If Z_1 and Z_2 in $\mathfrak{Z}$ are both non-compact, then $Z_1 \cap Z_2 \neq \emptyset$.
7. $\mathfrak{Z}$ is a normal base. [(3) and Proposition 1.47.]
8. $\omega(\mathfrak{Z})$ is αX, the one-point compactification of X.

References: The one-point compactification appears in Alexandroff and Urysohn, 1929. αX was shown to be a Wallman-type compactification by Brooks, 1967. In 1952, Fan and Gottesman obtained a similar result using the family of open sets such that either the closure of the set or the complement of the set is compact.

1G. *Completely regular ultrafilters*

A collection of sets $\mathscr{B}$ contained in a filter $\mathscr{F}$ is a *base* for $\mathscr{F}$ if the intersection of two sets in $\mathscr{F}$ contains a member of $\mathscr{B}$. A filter $\mathscr{F}$ on X is said to be a *completely regular filter* if $\mathscr{F}$ has a base $\mathscr{B}$ of open sets such that for each set A in $\mathscr{B}$, there exists B in $\mathscr{B}$ such that B is contained in A and B is completely separated from $X \backslash A$.

1. Any completely regular filter is contained in a maximal completely regular filter. [Zorn's Lemma.]

2. A completely regular filter $\mathscr{F}$ is maximal if and only if for each pair of open sets A and B with B contained in A and B completely separated from $X \backslash A$, either A belongs to $\mathscr{F}$ or A does not belong to $\mathscr{F}$ and B misses some set of $\mathscr{F}$. [If $f : X \to I$ completely separates B from $X \backslash A$, consider the filter generated by $\mathscr{F}$ and $\{f^{\leftarrow}([0,a)) : 0 < a < 1\}$.]

3. If p belongs to $\beta X \backslash X$, then the trace of the neighborhoods of p on X is a maximal completely regular filter. [$^{\omega}U$, 1.38.]

4. For x in X, $\mathscr{B}(x)$, the neighborhood filter of x is a maximal completely regular filter.

5. Distinct maximal completely regular ultrafilters on X cluster at distinct points of βX. [Distinct maximal completely regular filters contain completely separated sets.]

6. There is a one-to-one correspondence between maximal completely regular filters on X and the points of βX.

References: Adapted from N. Bourbaki, 1966, Exercise IX 1.8. This approach to the Stone-Čech compactification appears in Alexandroff, 1939, and is also discussed in Banachewski, 1964.

Chapter 2. Boolean Algebras

2.1. This chapter is intended mainly as an introduction for the reader who is not already familar with Boolean algebras. The central result is Stone's Representation Theorem which shows that any Boolean algebra can be identified with the family of clopen subsets of some totally disconnected compact space. The reader who is familar with the relationships between Boolean algebras and totally disconnected compact spaces will probably wish to omit this chapter and perhaps return to specific topics within the chapter as they are used later. The material on separability will be used in connection with $\beta \mathbb{N} \backslash \mathbb{N}$ in Chapters 3 and 7. The homomorphism described in Proposition 2.16 will be used in Chapter 3 to relate $\mathbb{N}$ to $\beta \mathbb{N} \backslash \mathbb{N}$. The only occasion in this chapter where the Stone-Čech compactification plays an important part is in Examples 2.14 and 2.15 which are based on properties of $\beta \mathbb{Q}$. Propositions 2.3 and 2.5 together with the Stone Representation Theorem will be needed in Chapter 10 in the context of projectives.

Many of the routine verifications are left to the exercises following the chapter. The reader who would like a more detailed account of Boolean algebras should consult the books of Dwinger, Halmos, and Sikorski.

2.2. A *partially ordered set* is a set X together with a binary relation $\leqslant$ defined on X which satisfies the following conditions for arbitrary elements x, y, and z of X:

(1) $\leqslant$ is *reflexive*: $x \leqslant x$ for all x.
(2) $\leqslant$ is *transitive*: $x \leqslant y$ and $y \leqslant z$ implies that $x \leqslant z$.
(3) $\leqslant$ is *antisymmetric*: $x \leqslant y$ and $y \leqslant x$ implies that $x = y$.

An *upper bound* for a subset A of X is an element x such that $a \leqslant x$ for all a in A, and a *lower bound* for A is an element z such that $z \leqslant a$ for all a in A. A *least upper bound* for the set A is an upper bound x such that for any other upper bound y, $x \leqslant y$. A *greatest lower bound* is defined similarly. The least upper bound of a family $\{x_\alpha\}$ is written $\bigvee_\alpha x_\alpha$ and the greatest lower bound is written $\bigwedge_\alpha x_\alpha$. The terms *join* and *supremum* (resp. *meet* and *infimum*) are often used instead of least upper bound (resp. greatest lower bound).

A *lattice* is a partially ordered set in which every pair of elements has a supremum and an infimum. A lattice is said to be *complete* if every set has a supremum and an infimum.

Examples. The rings $C^*(X)$ and $C(X)$ for any space X are lattices with the partial order defined by $f \leqslant g$ if $f(x) \leqslant g(x)$ for every point x of X.

A normal base as defined in Section 1.35 is a lattice where the partial order is defined by set inclusion and meets and joins are given by intersections and unions, respectively. The normal base of zero-sets of a space X is a lattice in which every countable set has a greatest lower bound.

The family of all subsets of a set is a lattice with the relation of containment and the operations of union and intersection.

A lattice is called *distributive* if the operations meet and join satisfy the two identities:

$$x \wedge (y \vee z) = (x \wedge y) \vee (x \wedge z)$$

and

$$x \vee (y \wedge z) = (x \vee y) \wedge (x \vee z) .$$

The two identities can be shown to be equivalent, which is a useful fact when working with examples since it simplifies verification of the properties of a distributive lattice.

A lattice is called *complemented* if it contains distinct elements 0 and 1 such that $0 \leqslant x \leqslant 1$ for all elements x and to each element x is assigned an element x' such that $x \vee x' = 1$ and $x \wedge x' = 0$. The element x' is called the *complement* of x. The elements 0 and 1 are often referred to as the zero element and the unit, respectively. A *Boolean algebra* is a complemented distributive lattice. A Boolean algebra is said to be *complete* if it is complete as a lattice. Observe that in order to show that a Boolean algebra is complete, it suffices to show only that every family has a supremum because the existence of infimia follows from the existence of suprema by taking complements.

2.3. Of the examples listed in the preceding section, only the family of all subsets of a set is a Boolean algebra, and that example is complete. We now consider a second example of a complete Boolean algebra—the family of all regular closed subsets of a topological space. A subset F of a space X is called a *regular closed set* if $F = \mathrm{cl}(\mathrm{int}\, F)$. Similarly, a subset G of X is said to be a *regular open set* if $G = \mathrm{int}(\mathrm{cl}\, G)$.

Proposition. *The family $R(X)$ of regular closed subsets of a space X is a complete Boolean algebra with the following operations:*

(1) $A \leqslant B$ *if and only if* $A \subseteq B$.

(2) $\bigvee_\alpha A_\alpha = \operatorname{cl}(\bigcup_\alpha \operatorname{int} A_\alpha)$.

(3) $\bigwedge_\alpha A_\alpha = \operatorname{cl}(\operatorname{int}(\bigcap_\alpha A_\alpha))$.

(4) $A' = \operatorname{cl}(X \setminus A)$.

Proof. Let $\{A_\alpha\}$ be any collection of regular closed subsets of X.

Set $B = \operatorname{cl}(\bigcup_\alpha \operatorname{int} A_\alpha)$. Then B is the closure of an open set and is clearly in $R(X)$. Since B contains $\operatorname{cl}(\operatorname{int} A_\alpha) = A_\alpha$ for each α, B is an upper bound of $\{A_\alpha\}$. Suppose that E is another upper bound. Then $\operatorname{int} A_\alpha$ is contained in $\operatorname{int} E$ for every α and therefore

$$B = \operatorname{cl}(\bigcup_\alpha \operatorname{int} A_\alpha) \subseteq \operatorname{cl}(\operatorname{int} E) = E .$$

Hence, B is the least upper bound of $\{A_\alpha\}$.

Set $F = \operatorname{cl}(\operatorname{int}(\bigcap_\alpha A_\alpha))$. Since $\operatorname{int}(\bigcap_\alpha A_\alpha)$ is contained in $\operatorname{int} A_\alpha$ for all α, F is contained in A_α for every α and F is a lower bound for $\{A_\alpha\}$. Now suppose that E is another lower bound. Then E is contained in $\bigcap A_\alpha$ so that $\operatorname{int} E$ is contained in $\operatorname{int}(\bigcap_\alpha A_\alpha)$. Therefore,

$$E = \operatorname{cl}(\operatorname{int} E) \subseteq \operatorname{cl}(\operatorname{int}(\bigcap_\alpha A_\alpha)) = F$$

and F is the greatest lower bound.

Take the empty set and X to be the zero element and the unit element, respectively. For any A in $R(X)$, put $A' = \operatorname{cl}(X \setminus A)$. Then $A \wedge A'$ is the closure of the interior of the boundary of A and is therefore empty. Similarly, $A \vee A'$ is the closure of a dense set and is therefore all of X.

It remains to show that the distributive law holds. Let A, B, and C belong to $R(X)$. Note first that

$$\operatorname{cl}(\operatorname{int} B \cup \operatorname{int} C) = B \vee C = \operatorname{cl}(\operatorname{int}(B \vee C)) .$$

Now recall that in any topological space, if two sets S and T have the same closure and G is an open set, then

$$\operatorname{cl}(G \cap S) = \operatorname{cl}(G \cap T) .$$

Applying these two formulas, we obtain the distributive law:

$$A \wedge (B \vee C) = \operatorname{cl}(\operatorname{int} A \cap \operatorname{int}(B \vee C))$$

$$= \operatorname{cl}(\operatorname{int} A \cap (\operatorname{int} B \cup \operatorname{int} C))$$

$$= \operatorname{cl}((\operatorname{int} A \cap \operatorname{int} B) \cup (\operatorname{int} A \cap \operatorname{int} C))$$

$$= \operatorname{cl}(\operatorname{int} A \cap \operatorname{int} B) \cup \operatorname{cl}(\operatorname{int} A \cap \operatorname{int} C)$$

$$= (A \wedge B) \cup (A \wedge C)$$

$$= (A \wedge B) \vee (A \wedge C). \quad \square$$

By interchanging the roles of the closure and interior operators, one can show that the Boolean algebra of regular open sets of a space is also a complete Boolean algebra.

2.4. Denote the family of clopen subsets of a space X by $CO(X)$. Since the complement of a clopen set is clopen, it is easy to see that the clopen subsets form a Boolean algebra with the operations of set-theoretic complementation, union, and intersection. However, if X is connected, the only clopen sets are X itself and the empty set. We now consider three classes of highly non-connected spaces which are important because a space X belonging to any one of the classes possesses a sufficiently rich supply of clopen subsets to make $CO(X)$ an interesting Boolean algebra.

A space is said to be *totally disconnected* if the only connected subsets are the singletons. In his 1937 paper, M.H. Stone showed that any distributive lattice can be represented as the family of clopen subsets of some totally disconnected space. A space is said to be *zero-dimensional* if it has a base of clopen sets. From the definition, we can expect to be able to relate $CO(X)$ to the space X if X is zero dimensional. Note that this is not the same definition adopted by [GJ]. Their definition is equivalent to what we will call strongly zero-dimensional in Section 3.34 and implies that the space has a base of clopen sets. Proposition 3.35 will show that the two concepts are equivalent in Lindelöf spaces but Example 3.39 will show that they are not identical.

It is easy to see that a zero-dimensional space is totally disconnected. If a space is also compact, the converse is true.

Proposition. *A compact space is zero-dimensional if and only if it is totally disconnected.*

Proof. Let X be a compact, totally disconnected space and let U be an open neighborhood of a point x of X. The component of a point in a compact space is the intersection of the clopen sets containing the point [GJ, 16.15]. Now because X is totally disconnected, we can write $\{x\} = \bigcap \{U_\alpha : \alpha \in \mathscr{A}\}$ where $\{U_\alpha : \alpha \in \mathscr{A}\}$ is the family of clopen sets containing x. For every point y not equal to x, there is some U_α which misses y. Hence, $\{U\} \cup \{X \backslash U_\alpha : \alpha \in \mathscr{A}\}$ is an open cover of X. Because X is compact, there is a finite subcover $\{U\} \cup \{X \backslash U_{\alpha_i} : i = 1, \ldots, n\}$. But then $X \backslash U$ is contained in $\bigcup (X \backslash U_{\alpha_i})$ so that $\bigcap U_{\alpha_i}$ is a clopen neighborhood of x contained in U. Thus, X has a base of clopen sets and is zero-dimensional. $\square$

2.5. A space is said to be *extremally disconnected* if the closure of every open subspace is open. This class of spaces was introduced by M. H. Stone in his 1937A paper in which he showed its importance by proving the following

Proposition. *The Boolean algebra of clopen subsets of a zero-dimensional space is complete if and only if the space is extremally disconnected.*

Proof. Let X be a zero-dimensional space. We first show that $CO(X)$ is complete if X is extremally disconnected. Let $\{A_\alpha\}$ be a family of clopen subsets of X. Then the set $B = \mathrm{cl}(\bigcup A_\alpha)$ is an upper bound for $\{A_\alpha\}$ since B is open if X is extremally disconnected. Since any clopen set containing every A_α also contains $\bigcup A_\alpha$, B is the least upper bound for $\{A_\alpha\}$.

Now assume that $CO(X)$ is complete and let U be an open subset of X. Let $\{A_\alpha\}$ be the family of clopen subsets of X contained in U and let V be the least upper bound of $\{A_\alpha\}$. The proof is completed by showing that $\mathrm{cl}\, U = V$ and is therefore open. Since X is zero-dimensional, $U = \bigcup A_\alpha$ and therefore

$$\mathrm{cl}\, U = \mathrm{cl}(\bigcup A_\alpha) \subseteq \mathrm{cl}\, V = V.$$

To show that V is contained in $\mathrm{cl}\, U$, consider the set $V \setminus \mathrm{cl}\, U$. This is an open set and if it is non-empty it contains a non-empty clopen set S since X is zero-dimensional. Then $V \setminus S$ is clopen and contains U and is therefore an upper bound for $\{A_\alpha\}$. But since V is the least upper bound, S must be empty and V is contained in $\mathrm{cl}\, U$. $\Box$

The Stone Representation Theorem

2.6. An *ideal* in a Boolean algebra L is a subset $\mathscr{I}$ of L satisfying:

    (1) $a \vee b$ belongs to $\mathscr{I}$ whenever both a and b belong to $\mathscr{I}$.

    (2) If b belongs to $\mathscr{I}$ and $a \leqslant b$, then a belongs to $\mathscr{I}$.

It is clear that an ideal is not all of L only if the unit element is not in the ideal. Such an ideal is called a *proper ideal*. For any subset A of L, there is a smallest ideal containing A, called the *ideal generated by A*. The members of the ideal generated by A are all members of L which are dominated by the supremum of some finite subset of A. A *principal ideal* is an ideal generated by a single element of L, and the principal ideal generated by a is denoted by L_a.

A *filter* in L is a subset $\mathscr{F}$ of L satisfying:

    (1) $a \wedge b$ belongs to $\mathscr{F}$ whenever both a and b belong to $\mathscr{F}$.

(2) If b belongs to $\mathscr{F}$ and $b \leqslant a$, then a belongs to $\mathscr{F}$.

A filter is proper, i.e. not all of L, exactly when it does not contain the zero element.

The relationship between ideals and filters in a Boolean algebra is apparent by comparing their definitions and observing that a Boolean algebra satisfies the *de Morgan Laws*:

(1) $$a \vee b = (a' \wedge b')'$$

and

(2) $$a \wedge b = (a' \vee b')' .$$

Thus, the complements of members of an ideal form a filter and conversely. For this reason, a filter is frequently called a *dual ideal* and a *principal dual ideal* is the set of all elements greater than or equal to a single element. The principal dual ideal generated by an element a is denoted by L^a.

A filter or ideal is called *maximal* if the only filter or ideal properly containing it is L itself.

Proposition. *A proper filter (resp. ideal) is maximal if and only if for every element a of L, either a or a' belongs to the filter (resp. ideal).*

Proof. If a proper ideal $\mathscr{I}$ satisfying the stated condition is properly contained in an ideal $\mathscr{J}$, then there exists an element a in $\mathscr{J}$ but not in $\mathscr{I}$. Then a' belongs to $\mathscr{I}$ so that a and a' belong to $\mathscr{J}$. Thus, $a \vee a' = 1$ belongs to $\mathscr{J}$ and $\mathscr{J}$ is all of L.

Now suppose that $\mathscr{I}$ is maximal and that a is not in $\mathscr{I}$. Then the ideal generated by $\mathscr{I}$ and $\{a\}$ contains $\mathscr{I}$ properly and is therefore all of L. Hence, 1 is the supremum of a and finitely many elements of $\mathscr{I}$. But this implies that a' is dominated by the supremum of finitely many elements of $\mathscr{I}$ and therefore belongs to $\mathscr{I}$. $\square$

2.7. A *field of sets* is a family of subsets of a set X which is closed under finite unions, finite intersections, and complementation. An example is the family of clopen subsets of any topological space. It is clear that any field is a Boolean algebra. The converse of this statement is known as the Stone Representation Theorem which states that every Boolean algebra can be represented as the field of clopen subsets of some totally disconnected compact space. A field of subsets of X is called a *reduced field* if for every pair of distinct points of X, there is a member of the field containing one of the points but not the other. The field of clopen sets of a zero-dimensional space is easily seen to be reduced.

A filter (resp. ideal) in a field is said to be *determined by a point* if it is the set of all members of the field containing the (resp. not containing) the point. It is clear that such a filter or ideal is necessarily maximal. A *perfect field* is a field in which every maximal ideal, or equivalently every maximal filter, is determined by a point. Because any filter of clopen subsets of a compact space has non-empty intersection, the clopen sets of a compact space form a perfect field. Every maximal filter is determined by each point belonging to the intersection of the filter. Because an infinite discrete space has maximal filters of clopen sets with empty intersection, such a field is not perfect. It is, however, reduced since a discrete space is zero-dimensional. On the other hand, the field of clopen sets of a compact connected space having more than two points is perfect, but not reduced.

The following proposition shows that every field which is both perfect and reduced is the family of clopen subsets of a totally disconnected compact space.

Proposition. *If $\mathscr{E}$ is a perfect reduced field of subsets of a set X, then a topology can be defined on X such that the space X is compact and totally disconnected and $\mathscr{E}$ is the Boolean algebra of clopen subsets of X.*

Proof. Take $\mathscr{E}$ to be the basis for a topology on X, i.e. a set G contained in X is open if and only if G is a union of members of $\mathscr{E}$. Since every set of $\mathscr{E}$ is open in this topology and $\mathscr{E}$ is a field, the sets of $\mathscr{E}$ are also closed. Since $\mathscr{E}$ is a reduced field, for any two points of X, there is a clopen set containing one and missing the other so that X is totally disconnected.

To show that X is compact, we show that any open covering $\mathscr{U}$ of X has a finite subcover. We can assume that $\mathscr{U}$ is made up of members of $\mathscr{E}$. Let $\mathscr{I}$ be the ideal of $\mathscr{E}$ generated by $\mathscr{U}$. If X is not the union of finitely many elements of $\mathscr{U}$, then X does not belong to $\mathscr{I}$ and $\mathscr{I}$ is proper. Choose a member E of $\mathscr{E}$ which is not contained in any of the sets of $\mathscr{I}$. Then the union of any chain of ideals containing $\mathscr{I}$ but not containing E is a proper ideal. Hence, Zorn's Lemma shows that $\mathscr{I}$ is contained in a maximal ideal. Because $\mathscr{E}$ is perfect, there is some point of X which does not belong to any member of the maximal ideal, and hence, does not belong to any member of $\mathscr{U}$. Thus, $\mathscr{U}$ cannot be a covering of X and the assumption that $\mathscr{U}$ does not admit a finite subcover leads to a contradiction.

It remains to prove that every clopen subset U of X belongs to $\mathscr{E}$. The clopen set U is the union of a family $\{V_\alpha\}$ of members of $\mathscr{E}$. But then $\{V_\alpha\} \cup \{X \backslash U\}$ is an open cover of X and therefore has a finite subcover. Hence, U is the union of finitely many sets belonging to $\mathscr{E}$ so that U belongs to $\mathscr{E}$. $\square$

2.8. A *Boolean algebra homomorphism h* from a Boolean algebra L

to a Boolean algebra M is a function which preserves the Boolean operations, i.e.

(1) $h(a \wedge b) = h(a) \wedge h(b)$,
(2) $h(a \vee b) = h(a) \vee h(b)$,
(3) $h(a') = (h(a))'$.

Note that in the presence of (3), de Morgan's Laws show that (1) and (2) are equivalent. Thus, to verify that a function is a Boolean algebra homomorphism, it is sufficient to demonstrate that it satisfies (3) and either (1) or (2). A Boolean algebra homomorphism is called a *monomorphism* if it is one-to-one. An *isomorphism* is a monomorphism which is onto.

Proposition. *A Boolean algebra homomorphism h is a monomorphism if and only if $h(a) = 0$ implies that a is 0.*

Proof. Necessity is clear. To show sufficiency, assume that $h(a) = h(b)$. Then $a \wedge b' = 0$ since $h(a \wedge b') = h(a) \wedge (h(b))' = 0$. Similarly, $a' \wedge b = 0$. But this implies that $a \leqslant b$ and $b \leqslant a$, and therefore $a = b$. □

2.9. The following proposition shows that the set of all maximal filters of a Boolean algebra admits a reduced field of subsets which is a homomorphic image of the Boolean algebra. This fact combined with Proposition 2.7 will yield a proof of the Stone Representation Theorem. The proofs are based on those given in Sikorski's book.

Proposition. *If $S(L)$ is the set of maximal filters of a Boolean algebra L and if, for every element a of L, $h(a)$ denotes the set of maximal filters containing a, then the family $\mathscr{E} = \{h(a) : a \in L\}$ is a reduced field of subsets of $S(L)$ and h is a homomorphism onto $\mathscr{E}$. If in addition, every non-zero element of L belongs to some maximal filter, then h is an isomorphism.*

Proof. Let $\mathscr{F}$ be a maximal filter of L. Then by definition of h,

$$a \in \mathscr{F} \quad \text{if and only if} \quad \mathscr{F} \in h(a).$$

Since $a \wedge b$ is in $\mathscr{F}$ if and only if a is in $\mathscr{F}$ and b is in $\mathscr{F}$,

$$h(a \wedge b) = h(a) \cap h(b).$$

Because $S(L)$ is made up of maximal filters,

$$h(a') = S(L) \setminus h(a).$$

Thus, h is a homomorphism and $\mathscr{E} = \{h(a) : a \in L\}$ is a field of subsets of $S(L)$ since it is the image of L under h.

If $\mathscr{F}_1$ and $\mathscr{F}_2$ are distinct maximal filters of L, then there is an element a of L belonging to $\mathscr{F}_1$ but not $\mathscr{F}_2$. Hence, $\mathscr{F}_1$ belongs to $h(a)$ and $\mathscr{F}_2$ does not, thus showing that $\mathscr{E}$ is a reduced field.

Finally, if every non-zero element of L belongs to some maximal filter, $h(a) = \emptyset$ implies that $a = 0$, and h is therefore an isomorphism. $\square$

2.10. The Stone Representation Theorem. *Every Boolean algebra is isomorphic to the Boolean algebra of clopen sets of a compact totally disconnected space.*

Proof. Let L be a Boolean algebra and let $S(L)$, $\mathscr{E}$, and h be as in the preceding proposition. If a in L is not 0, then the principal filter generated by a is proper, and therefore can be extended to a maximal filter. It follows from the preceding proposition that h is an isomorphism of L onto the reduced field $\mathscr{E}$.

We show that $\mathscr{E}$ is a perfect field so that we can apply Proposition 2.7. Let $\mathscr{G}$ be a maximal filter of $\mathscr{E}$. Then the set $\mathscr{F}$ of all elements a of L such that $h(a)$ belongs to $\mathscr{G}$ is a maximal filter of L by isomorphism, so that $\mathscr{F}$ is also a point of $S(L)$. If b belongs to L, we have

$$h(b) \in \mathscr{G} \quad \text{if and only if} \quad b \in \mathscr{F}$$

and as in the proof of the preceding proposition,

$$h(b) \in \mathscr{G} \quad \text{if and only if} \quad \mathscr{F} \in h(b).$$

Hence, the filter $\mathscr{G}$ is determined by the point $\mathscr{F}$ of $S(L)$.

The proof is completed by appealing to Proposition 2.7 to show that $\mathscr{E}$ is the Boolean algebra of clopen subsets of $S(L)$ for the topology generated by $\mathscr{E}$ and that this topology is compact and totally disconnected. $\square$

The set $S(L)$ with the topology described in Proposition 2.7 is called the *Stone space* of L. This topology was introduced in 1937 by M.H. Stone in his development of the Representation Theorem. As we saw in Chapter 1, the same topology has been used to construct a variety of compactifications of topological spaces. Note that if D is an infinite discrete space, the clopen sets and the zero-sets of D coincide and the Stone space of the algebra of clopen sets of D is βD.

If L and M are Boolean algebras, there is a natural one-to-one correspondence between homomorphisms of L into M and mappings of $S(M)$ into $S(L)$. We will first show that a homomorphism f of L into M induces a homomorphism g of $CO(S(L))$ into $CO(S(M))$. Let

$h_L: L \to CO(S(L))$ and $h_M: M \to CO(S(M))$ be the isomorphisms provided by the Stone Representation Theorem and for U in $CO(S(L))$, define

$$g(U) = h_M \circ f \circ h_L^{\leftarrow}(U).$$

Then g is a homomorphism and we have the following diagram:

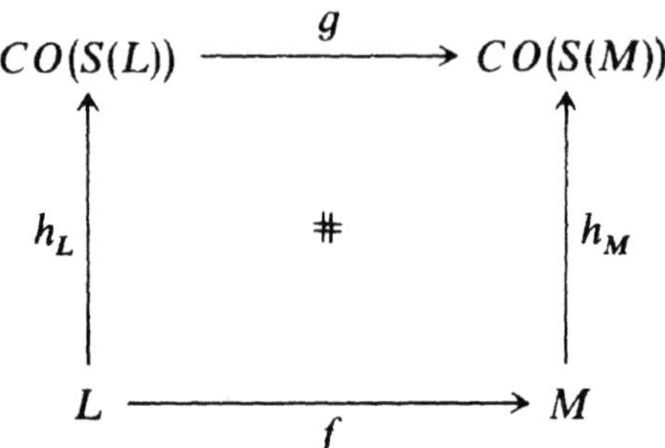

Now we will use g to define a mapping $\varphi: S(M) \to S(L)$. For every x in $S(M)$, let $\mathscr{F}(x)$ denote the maximal filter of $CO(S(M))$ determined by x. Then it is easily seen that the family

$$g^{\leftarrow}\mathscr{F}(x) \equiv \{U \in CO(S(L)): g[U] \in \mathscr{F}(x)\}$$

is a maximal filter in $CO(S(L))$, and is therefore determined by a point y of $S(L)$. Define $\varphi(x) = y$. The function φ is well-defined because distinct points of $S(L)$ determine different maximal filters since $S(L)$ is zero-dimensional. If U is a clopen subset of $S(L)$, the definition of φ shows that

(*) $\varphi(x) \in U$ if and only if $x \in g[U]$.

Thus, $\varphi^{\leftarrow}(U) = g[U]$ is clopen because g is a homomorphism. Hence, φ is continuous since $S(L)$ has a base of clopen sets. Note that the process of obtaining the mapping φ from the original homomorphism f has "reversed the arrows":

$$L \xrightarrow{\;f\;} M\,,$$

$$S(L) \xleftarrow{\;\varphi\;} S(M)\,.$$

The inverse procedure to obtain a homomorphism sending L to M from a mapping of $S(M)$ into $S(L)$ is less complicated. The definition of the homomorphism is actually dictated by (*) above. If $\varphi: S(M) \to S(L)$

is a mapping, then $\varphi^{\leftarrow}$ preserves the Boolean operations and sends clopen sets to clopen sets. Thus, the definition

$$g[U] = \varphi^{\leftarrow}(U)$$

defines a homomorphism of $CO(S(L))$ into $CO(S(M))$. The homomorphism g then gives a homomorphism f of L into M by defining $f = h_M^{\leftarrow} \circ g \circ h_L$. Further, it is evident that if f is an isomorphism, then φ is a homeomorphism, and conversely.

Two Examples

2.11. By an *automorphism* of a topological space we shall mean a homeomorphism of the space onto itself. Similarly, an automorphism of a Boolean algebra is an isomorphism of the algebra with itself. A topological space is said to be *rigid* if the identity map is the only automorphism of the space. In 1948, G. Birkhoff asked if there existed an infinite Boolean algebra whose only automorphism was the identity. The Stone Representation Theorem and the preceding discussion of duality between homomorphisms and mappings show that an affirmative answer to Birkhoff's question is equivalent to demonstrating the existence of a rigid totally disconnected compact space. In 1951, M. Katětov provided an affirmative answer by exhibiting a rigid space whose Stone-Čech compactification is both rigid and totally disconnected. Two preliminary results are required before examining Katětov's example. A point in a topological space is called a κ-*point* if it is the limit of a sequence of distinct points of the space. It is immediate that any homeomorphism must carry κ-points to κ-points. The next result will be used to study automorphisms of Stone-Čech compactifications.

2.12. Proposition. *No point belonging to a G_δ of βX which misses X can be a κ-point of βX.*

Proof. Assume on the contrary that a sequence $\{x_n\}$ of distinct points of βX converges to a point p and that p belongs to a G_δ which misses X. Put $Z = X \cup \{x_n\} \cup \{p\}$. Then there exists a sequence $\{G_i\}$ of open sets of the subspace Z such that $\bigcap G_i = \{p\}$. By perhaps passing to a subsequence and eliminating a point of the sequence from some of the G_i's, we can assume that for each i, $\{x_j : j > i\}$ is contained in $\bigcap\{G_j : j \leqslant i\}$ and $\{x_j : j \leqslant i\}$ misses $\bigcap\{G_j : j \leqslant i\}$. In other words, we are assuming that for each i, x_i is the only point of the sequence contained in the set $(X \backslash G_i) \cap (\bigcap\{G_j : j < i\})$.

Because any two disjoint closed sets one of which is compact are completely separated (Proposition 1.47), there exists a sequence of mappings $\{h_n\}$ in $C^*(Z)$ such that:

(a) $0 \leqslant h_n(z) \leqslant 2^{-n}$ for each z in Z,

(b) $h_n(x_i) = 0$ for $i > n$,

(c) $h_n(p) = 0$,

(d) $h_n(z) = 2^{-n}$ for each z in $Z \setminus \bigcap \{G_i : i \leqslant n\}$.

Put $g_n = h_1 + \cdots + h_n$ for each n. The Weierstrass M-test shows that the sequence g_n converges to a mapping g in $C^*(Z)$, and the construction of $\{g_n\}$ assures that:

(e) $0 \leqslant g(z) \leqslant 1$ for each z in Z,

(f) $g(p) = 0$,

(g) $g(z) > 0$ if $z \neq p$,

(h) $g(x_n) = 2^{-1} + 2^{-2} + \cdots + 2^{-n}$.

Condition (h) shows that $\{g(x_{2k})\}$ and $\{g(x_{2k+1})\}$ are disjoint closed subspaces of the half-open interval $(0,1]$. Because $(0,1]$ is normal, there exists a mapping l of $(0,1]$ into $[0,1]$ such that $l[\{g(x_{2k})\}] = \{0\}$ and $l[\{g(x_{2k+1})\}] = \{1\}$. Put $f = l \circ (g | Z \setminus \{p\})$ and consider the extension $\beta(f)$ of f to βX. Then $\{\beta(f)(x_{2k})\}$ converges to 0 and $\{\beta(f)(x_{2k+1})\}$ converges to 1. However, this is a contradiction of the continuity of $\beta(f)$ because both subsequences $\{x_{2k}\}$ and $\{x_{2k+1}\}$ converge to p. Thus, p cannot be a κ-point. $\square$

2.13. The following result is [GJ, 16.16] and will be employed in Katětov's example to show that if X is a countable zero-dimensional space, then βX is totally disconnected. A topological space is said to be a *Lindelöf space* if every open covering of the space has a countable subcovering. A *disconnection* of a space is a covering of the space by two disjoint clopen sets.

Proposition. *Disjoint closed subspaces of a zero-dimensional Lindelöf space are separated by a disconnection.*

Proof. Let X be such a space and let F and K be disjoint closed subsets of X. For each point x of X, choose a clopen neighborhood $U(x)$ of x which meets at most one of the sets F and K. Since X is Lindelöf, the covering $\{U(x)\}$ has a countable subcovering $\{U_n\}$. Put

$$V_n = U_n \setminus \bigcup \{U_i : i < n\}.$$

The V_n are thus disjoint clopen sets and cover X. Then setting

$$W = \bigcup \{V_n : V_n \cap F = \emptyset\}$$

yields a disconnection $\{W, X \setminus W\}$ where F is contained in $X \setminus W$ and K is contained in W. $\square$

2.14. We will need one further result before considering the example. In 1937, B. Pospíšil showed that the cardinality of $\beta\mathbb{N}$ is 2^c. [GJ, 9.2] provides one proof of this result and another will be given in Theorem 3.2. Since every non-degenerate interval G of $\mathbb{Q}$, the space of rationals, contains a closed C^*-embedded copy of $\mathbb{N}$, it follows from Proposition 1.48 that $G^* = \mathrm{cl}_{\beta\mathbb{Q}} G \setminus \mathbb{Q}$ contains 2^c points. We will require this result in the

Example (M. Katětov). *There exists a countable, rigid, normal space such that every point of the space is a κ-point.*

Proof. Consider the space $\mathbb{Q}$ of rationals and write $\mathbb{Q}^2 = \mathbb{Q} \times \mathbb{Q}$ and $(\beta\mathbb{Q})^2 = \beta\mathbb{Q} \times \beta\mathbb{Q}$. For every point y of $(\beta\mathbb{Q})^2$, let $\Phi(y)$ denote the set of all points of $(\beta\mathbb{Q})^2$ that are images of y under mappings of the subspace $\mathbb{Q}^2 \cup \{y\}$ into $(\beta\mathbb{Q})^2$ which have $\mathbb{Q}^2$ as the image of itself. Since there are at most c mappings of $\mathbb{Q}^2$ into itself and $\mathbb{Q}^2$ is dense in $(\beta\mathbb{Q})^2$, the cardinality of $\Phi(y)$ is at most c. Let $\{G_n\}$ denote a countable open base for the topology of $\mathbb{Q}$ and put $G_n^* = \mathrm{cl}_{\beta\mathbb{Q}} G_n \setminus G_n$ for each n. Then the cardinality of each G_n^* is 2^c.

The required space Y will be constructed by adding countably many points of $(\beta\mathbb{Q})^2$ to $\mathbb{Q}^2$. We will use Proposition 2.12 to show that any automorphism of Y must send $\mathbb{Q}^2$ into $\mathbb{Q}^2$. The fact that each $\Phi(y)$ has only c points while each G_n^* contains 2^c points will allow the added points to be chosen so that they force the automorphism to be the identity on $\mathbb{Q}^2$. Since $\mathbb{Q}^2$ is dense in Y, the automorphism must therefore also be the identity on Y.

We will first describe the points to be added to $\mathbb{Q}^2$.

(a) There exists a sequence of points (p_n, q_n) in $(\beta\mathbb{Q})^2$ such that:
  (1) $(p_n, q_n) \in G_n^* \times G_n^*$,
  (2) If (p, q) and $(\tilde{p}, \tilde{q})$ are any points of $\mathbb{Q}^2$, then for all $m, n = 1, 2, \ldots$ with $m < n$, $(p_n, \tilde{q}) \notin \Phi(p_m, q)$, $(p_n, \tilde{q}) \notin \Phi(p, q_m)$, $(\tilde{p}, q_n) \notin \Phi(p, q_m)$, $(\tilde{p}, q_n) \notin \Phi(p_m, q)$, and $(p, q_n) \notin \Phi(p_n, q)$.

Choose p_1 arbitrarily in G_1^*. Then $\bigcup_{q \in \mathbb{Q}} \Phi(p_1, q)$ contains at most c points. Since G_1^* contains 2^c points and $\mathbb{Q} \times \{r\}$ and $\mathbb{Q} \times \{s\}$ are disjoint for distinct points r and s of G_1^*, there must exist a point q_1 in G_1^* such that $\mathbb{Q} \times \{q_1\}$ misses $\bigcup_{q \in \mathbb{Q}} \Phi(p_1, q)$.

Now suppose that p_m and q_m have been chosen for all $m < n$. Then the set

$$\bigcup_{m < n} \left\{ \left(\bigcup_{q \in \mathbb{Q}} \Phi(p_m, q) \right) \cup \left(\bigcup_{q \in \mathbb{Q}} \Phi(p, q_m) \right) \right\}$$

can contain at most c points of $G_n^* \times G_n^*$, so that by a similar argument to that for q_1, there must exist a point p_n in G_n^* such that $\{p_n\} \times \mathbb{Q}$ misses the above union. Then the point q_n is chosen so that $\mathbb{Q} \times \{q_n\}$ misses the union

$$\bigcup_{m<n}\{(\bigcup_{q\in\mathbb{Q}}\Phi(p,q_m))\cup(\bigcup_{q\in\mathbb{Q}}\Phi(p_m,q))\}\cup(\bigcup_{q\in\mathbb{Q}}\Phi(p_n,q))$$

This completes the induction step.

Because no point of $\mathbb{Q}$ has a compact neighborhood in $\mathbb{Q}$, Proposition 1.59 shows that every neighborhood in $\beta\mathbb{Q}$ of a point of $\mathbb{Q}$ must meet $\beta\mathbb{Q}\setminus\mathbb{Q}$. In fact, because $\{G_n\}$ is a base for the open sets of $\mathbb{Q}$, every neighborhood in $\beta\mathbb{Q}$ of a point of $\mathbb{Q}$ must contain some G_n^*. It therefore follows that:

(b) Each of the sequences $\{p_n\}$ and $\{q_n\}$ is dense in $\beta\mathbb{Q}$.

Now put $A_n = \{p_n\} \times \mathbb{Q}$ and $B_n = \mathbb{Q} \times \{q_n\}$ for each n. Then (a) shows that:

(c) If x belongs to $A_m \cup B_m$ for some $m < n$, then $(A_n \cup B_n) \cap \Phi(x) = \emptyset$, and

(d) If x belongs to A_n, $B_n \cap \Phi(x) = \emptyset$.

Set $Y = \mathbb{Q}^2 \cup (\bigcup_n (A_n \cup B_n))$. We will show that Y is the required space. Any automorphism of Y must send a point which has a countable base to a point which also has a countable base. Thus, to show that any automorphism of Y will send points of $\mathbb{Q}^2$ to $\mathbb{Q}^2$ it will suffice to show:

(e) A point y of Y has a countable neighborhood base if and only if y belongs to $\mathbb{Q}^2$: It is clear from Proposition 2.12 that no point outside of $\mathbb{Q}^2$ can have a countable neighborhood base. Now let $\{U_n\}$ be a countable neighborhood base at a point y of $\mathbb{Q}^2$ and let V be a neighborhood of y in Y. Choose a neighborhood W of y in Y such that

$$y \in W \subset \mathrm{cl}_Y W \subset V.$$

Then there is an integer m such that U_m is contained in $W \cap \mathbb{Q}^2$. Now we have

$$y \in Y \setminus \mathrm{cl}_Y(Y \setminus \mathrm{cl}_Y U_m) \subset \mathrm{cl}_Y U_m \subset (\mathrm{cl}_Y W) \subset V,$$

so that the family $\{Y \setminus \mathrm{cl}_Y(Y \setminus \mathrm{cl}_Y U_n)\}$ is a countable neighborhood base at y in Y.

If h is an automorphism of Y, (e) shows that $h[\mathbb{Q}^2] = \mathbb{Q}^2$ and therefore that for every y in Y, $h(y)$ is in $\Phi(y)$ and y is in $\Phi(h(y))$. Further, (c) and (d) imply that $h[A_n] = A_n$ and $h[B_n] = B_n$ for every n.

Now assume that some point (p,q) of $\mathbb{Q}^2$ is not mapped to itself by h, i.e. that $h(p,q) = (\tilde{p}, \tilde{q})$ and, without loss of generality, that $q \neq \tilde{q}$. Choose a neighborhood U of q in $\beta\mathbb{Q}$ such that $\tilde{q}$ is not in $\mathrm{cl}_{\beta\mathbb{Q}} U$. Let B be the union of the sets B_n such that q_n belongs to $\mathrm{cl}_{\beta\mathbb{Q}} U$. Since $h[B_n] = B_n$, we have $h[B] = B$. Also, (p,q) belongs to $\mathrm{cl}_Y B$ since the set of all q_n is dense in $\beta\mathbb{Q}$ and therefore every neighborhood of (p,q) contains points (p, q_n) with q_n in $\mathrm{cl}_{\beta \mathbf{Q}}\, U$. Because h is continuous, $(\tilde{p}, \tilde{q}) = h(p, q)$ belongs to $\mathrm{cl}_Y B = \mathrm{cl}_Y h[B]$. But this is a contradiction since U was chosen so that $\tilde{q}$ was not in $\mathrm{cl}_{\beta \mathbf{Q}}\, U$. Hence, $h(p, q) = (p, q)$. Since $\mathbf{Q}^2$ is dense in Y, h must also be the identity on Y and Y is rigid.

It is immediate that every point of Y is a κ-point since if (p_n, q) is in Y, there is a sequence $\{r_i\}$ in $\mathbf{Q}$ converging to q and (p_n, r_i) converges to (p_n, q).

Finally, Y is Lindelöf because it is countable. But a regular Lindelöf space is paracompact, and hence, normal. [D, pp. 163, 174.] $\square$

2.15. The previously advertised rigid totally disconnected compact space is now obtained as βY where Y is the preceding example.

Example (M. Katětov). *There exists a rigid totally disconnected compact space.*

Proof. By Proposition 2.13, disjoint closed subsets of $\mathbf{Q}$ are separated by a disconnection. Then we can show that:

(a) $\beta \mathbf{Q}$ is totally disconnected: If p and q are distinct points of $\beta \mathbf{Q}$, choose disjoint closed neighborhoods U and V of p and q, respectively. Then if W is a clopen set of $\mathbf{Q}$ containing $V \cap \mathbf{Q}$ and missing $U \cap \mathbf{Q}$, $\mathrm{cl}_{\beta \mathbf{Q}}\, W$ is a clopen set of $\beta \mathbf{Q}$ containing q and missing p. Thus, no connected subspace of $\beta \mathbf{Q}$ contains more than a single point.

Hence, Proposition 2.4 shows that $\beta \mathbf{Q}$ and therefore $(\beta \mathbf{Q})^2$ and its countable subspace Y all have bases consisting of clopen sets. Therefore, by the same argument as in (a),

(b) βY is totally disconnected.

Now let h be an automorphism of βY. Since Y is countable, every point in $\beta Y \setminus Y$ is contained in a G_δ which misses Y. Thus, Proposition 2.12 together with the fact that every point of Y is a κ-point shows that:

(c) $h[Y] = Y$.

Since Y is rigid, the restriction $h|Y$ is the identity on Y, and it follows that h is the identity on βY. Hence, βY is also rigid. $\square$

2.16. The second example is a Boolean algebra homomorphism of $R(X)$ into $R(\beta X \setminus X)$. Recall that in the first chapter, we denoted $\beta X \setminus X$ by X^*. If A is a closed subspace of X, the definition

$$A^* = \mathrm{cl}_{\beta X}\, A \setminus X$$

is consistent with the notation of X^* for the growth $\beta X \setminus X$ because X is dense in βX. If we restrict our attention to the regular closed sets of X, then the assignment

$$A \mapsto A^*$$

is sometimes a homorphism of $R(X)$ into $R(X^*)$. We will use this assignment in the next chapter to study $\mathbb{N}^*$. In 1971 B, R. G. Woods has shown that this operation will be a homomorphism for any realcompact, metric, or nowhere locally compact space.

In the Boolean algebra of regular closed sets, complementation is defined by

$$A' = \mathrm{cl}_X(X \backslash A).$$

Thus, the hypotheses in the following proposition is just the requirement that $A \mapsto A^*$ preserve complementation.

Proposition. *The function that sends a subset A of X to A^* is a Boolean algebra homomorphism from $R(X)$ into $R(X^*)$ if and only if for every A in $R(X)$,*

$$[\mathrm{cl}_X(X \backslash A)]^* = \mathrm{cl}_{X^*}(X^* \backslash A^*).$$

Proof. To show sufficiency, we must first demonstrate that if A belongs to $R(X)$, then A^* belongs to $R(X^*)$:

$$\begin{aligned}
\mathrm{cl}_{X^*}(\mathrm{int}_{X^*} A^*) &= \mathrm{cl}_{X^*}[X^* \backslash \mathrm{cl}_{X^*}(X^* \backslash A^*)] \\
&= \mathrm{cl}_{X^*}[X^* \backslash [\mathrm{cl}_X(X \backslash A)]^*] \\
&= [\mathrm{cl}_X[X \backslash \mathrm{cl}_X(X \backslash A)]]^* \\
&= [\mathrm{cl}_X(\mathrm{int}_X A)]^* \\
&= A^*
\end{aligned}$$

since A is a regular closed set in X. Now since the closure operation distributes over finite unions and the join of two regular closed sets is just their union, we have

$$(A \vee B)^* = (A \cup B)^* = A^* \cup B^* = A^* \vee B^*$$

and using the hypothesis once again,

$$(A')^* = [\mathrm{cl}_X(X \backslash A)]^* = \mathrm{cl}_{X^*}(X^* \backslash A^*) = (A^*)'$$

and the function is a Boolean algebra homomorphism.

The converse is evident from the last line since a Boolean algebra homomorphism must preserve complementation. $\square$

The Completion of a Boolean Algebra

2.17. A Boolean algebra L contained in a Boolean algebra M is said to *generate* M if every element of M is a supremum of elements of L. A homomorphism of Boolean algebras is said to be *complete* if it preserves any suprema which exist. A *completion* of L is a pair (M,e) where M is a complete Boolean algebra and e is a complete monomorphism of L into M and $e[L]$ generates M. We will usually think of L as a subalgebra of M. The Stone Representation Theorem shows immediately that every Boolean algebra L has a completion since L is isomorphic with the algebra of clopen sets of $S(L)$ which can be seen from Proposition 2.3 to generate the complete Boolean algebra of regular closed subsets of $S(L)$ because $S(L)$ is zero-dimensional. We will see that not only is $R(S(L))$ a completion of L, but that in a sense it is a "best" completion in that it adds the fewest additional points to L. The discussion of completions is based on that given in the book of Halmos.

A completion (M,e) is said to be *minimal* if for any other completion (B,k), there is a complete monomorphism f from M to B such that $k = f \circ e$.

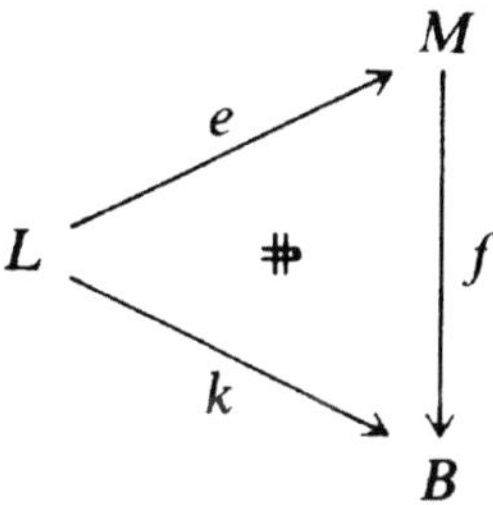

The following proposition shows that any two minimal completions are isomorphic. Observe that the proof is analogous to the proof of Corollary 1.13 concerning uniqueness of a maximal compactification. Here the key will be that complete homomorphisms which agree on a generating subalgebra must be the same, whereas in Corollary 1.13 we used that mappings which agree on a dense subspace are the same.

Proposition. *Any two minimal completions of a Boolean algebra L are isomorphic by an isomorphism which leaves points of L fixed.*

Proof. Let (B,h) and (C,k) be minimal completions of L. Then there exist complete monomorphisms f and g such that $f \circ h = k$ and $g \circ k = h$.

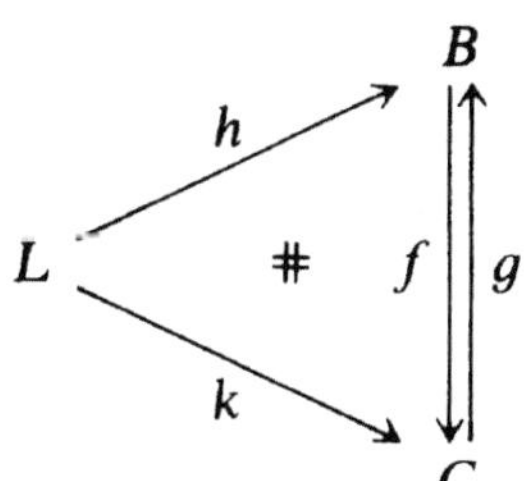

Then by substitution, $f \circ g \circ k = k$ and $g \circ f \circ h = h$. But then the two complete homomorphisms $f \circ g$ and $g \circ f$ agree with the identity of L on the generating algebra L and therefore f and g are isomorphisms and are inverses of each other. $\quad \square$

2.18. Theorem. *The minimal completion of the Boolean algebra of clopen subsets of a compact totally disconnected space is the Boolean algebra of regular closed subsets of the space.*

Proof. Let X be a compact totally disconnected space. Then $R(X)$ is complete by Proposition 2.3 and the injection e of $CO(X)$ into $R(X)$ is a homomorphism since for clopen sets the operations of $R(X)$ reduce to the ordinary set theoretic operations of $CO(X)$. We must show that if a family $\{A_\alpha\}$ has a supremum in $CO(X)$, then its supremum in $R(X)$ is the same set. If $F = \mathrm{cl}(\bigcup A_\alpha)$ is the supremum in $R(X)$, then because every clopen set is regular closed, the supremum B of $\{A_\alpha\}$ in $CO(X)$ contains F. If $B \backslash F$ is non-empty, $B \backslash F$ contains a clopen set E and every A_α is contained in $B \backslash E$, contradicting the assumption that B is the supremum in $CO(X)$. Thus, $B = F$.

To establish that $R(X)$ is a completion of $CO(X)$, it remains to show that $CO(X)$ generates $R(X)$. If F belongs to $R(X)$, let $\{A_\alpha\}$ be the family of clopen subsets of $\mathrm{int}\, F$. Then $\bigcup_\alpha A_\alpha = \mathrm{int}\, F$ and $\bigvee_\alpha A_\alpha = \mathrm{cl}(\bigcup A_\alpha) = F$, and F is a supremum of members of $CO(X)$.

We now must show that $R(X)$ is the minimal completion, i.e. if (C, k) is another completion, then there exists a complete monomorphism f from $R(X)$ to C such that

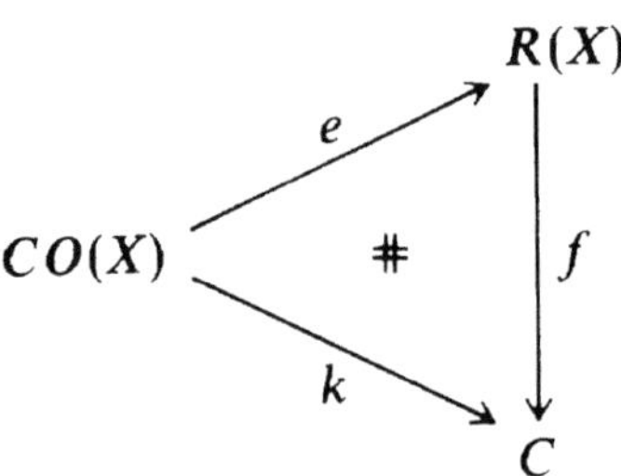

is a commutative diagram. If F belongs to $R(X)$, write $\mathrm{int}\, F$ as the union of the clopen sets which it contains, i.e. $\mathrm{int}\, F = \bigcup_\alpha A_\alpha$ and define $f(F)$ to be the supremum of $\{k(A_\alpha)\}$. It is then evident that the diagram commutes since, if F is clopen, F is not only the supremum of $\{A_\alpha\}$, but F is also equal to some A_{α_0} so that $f(F) = f(A_{\alpha_0}) = k(A_{\alpha_0})$. If $f(U) = 0$, then $f(V) = 0$ for all clopen subsets V of U and hence $k(V) = 0$. But k is one-to-one, so that each V must be the empty set. Hence, U is empty, and f is one-to-one.

To show that f preserves complementation, write the interiors of a set F in $R(X)$ and its complement F' as unions of clopen sets, i.e. $\operatorname{int} F = \bigcup_\alpha A_\alpha$ and $\operatorname{int} F' = \bigcup_\beta B_\beta$. Since any pair A_α and B_β are disjoint, it follows that $k(A_\alpha) \wedge k(B_\beta) = 0$. It therefore follows from

$$(\bigvee_\alpha A_\alpha) \wedge (\bigvee_\beta B_\beta) = \bigvee_{\alpha,\beta}(A_\alpha \wedge B_\beta)$$

that $k(F) \wedge k(F') = 0$. Since $F \vee F' = 1$ in $R(X)$, there can be no proper clopen set of X containing all the A_α's and B_β's so that $\bigvee_{\alpha,\beta}(A_\alpha \vee B_\beta) = 1$. Then $\bigvee_{\alpha,\beta}(k(A_\alpha) \vee k(B_\beta)) = 1$ and hence $f(F) \vee f(F') = 1$ so that $f(F') = (f(F))'$.

To show that f preserves all suprema, let $\{F_\alpha\}$ be a family of members of $R(X)$ and let $F = \bigvee_\alpha F_\alpha$. Since f is clearly order preserving, $\bigvee_\alpha f(F_\alpha) \leqslant f(F)$. To verify the opposite inequality, we show that for any clopen subset A of $\operatorname{int} F$, $f(A) \leqslant \bigvee_\alpha f(F_\alpha)$. Let $\{A_{\alpha,\beta}\}$ be the family of all clopen subsets of $\operatorname{int} F_\alpha$ and write

$$A = A \cap F = \bigvee_\alpha(A \cap F_\alpha) = \bigvee_\alpha(A \cap (\bigvee_\beta A_{\alpha,\beta})) .$$

Then it follows that

$$A = \bigvee_{\alpha,\beta}(A \cap A_{\alpha,\beta})$$

and hence that

$$k(A) = \bigvee_{\alpha,\beta}(k(A) \wedge k(A_{\alpha,\beta})) = \bigvee_\alpha(k(A) \wedge f(F_\alpha)) = k(A) \wedge (\bigvee_\alpha f(F_\alpha)) .$$

Hence, $k(A) \leqslant \bigvee_\alpha f(F_\alpha)$ and it follows from the definition of f that $f(F) \leqslant \bigvee_\alpha f(F_\alpha)$. $\square$

Separability in Boolean Algebras

2.19. In the remainder of the chapter we consider three conditions on a Boolean algebra which deal with the insertion of its elements into certain strictly increasing sequences. We will use the notation $a < b$ to mean that a is less than but not equal to b.

The first such condition deals with the simplest type of sequence, i.e. one involving only two elements. A Boolean algebra is said to be *dense in itself* if whenever one element is properly less than another, then a third element can be properly interposed between them, i.e. if $a < b$, then there exists c such that $a < c < b$. The next Proposition shows that a Boolean algebra is dense in itself exactly when its Stone space has no isolated points.

Proposition. *A compact zero-dimensional space Y has no isolated points if and only if the algebra $CO(Y)$ is dense in itself.*

Proof. If y is an isolated point of Y, put $A = \emptyset$ and $B = \{y\}$. Then no member of $CO(Y)$ can be interposed between A and B and $CO(Y)$ is not dense in itself.

On the other hand, if $CO(Y)$ fails to be dense in itself, there exist members A and B of $CO(Y)$ such that A is contained in B and no proper clopen subset of B is a proper superset of A. But since Y is zero-dimensional and $B\backslash A$ is clopen, $B\backslash A$ must be a singleton, and Y contains an isolated point. $\square$

2.20. We will consider two kinds of separability for infinite sequences. A Boolean algebra is said to be *Cantor separable* if no strictly increasing sequence has a least upper bound, i.e. if

$$a_1 < \cdots < a_n < \cdots < b,$$

then there exists an element c such that $a_n < c < b$ for every n. A Boolean algebra is said to be *DuBois-Reymond* separable if a strictly increasing sequence can be separated from a strictly decreasing sequence dominating the increasing one, i.e. if

$$a_1 < \cdots < a_n < \cdots < b_n < \cdots < b_1,$$

then there exists h such that $a_n < h < b_n$ for all n.

Both kinds of separability imply topological properties of the Stone space. The next result shows that if L is Cantor separable, then every non-empty G_δ in $S(L)$ has non-empty interior.

Proposition. *Every non-empty G_δ in a zero-dimensional space Y has non-empty interior if $CO(Y)$ is Cantor separable.*
Proof. Let $\bigcap U_i$ be a non-empty G_δ in Y. Choose a point y in $\bigcap U_i$ and let V_i be a clopen neighborhood of y contained in U_i. We can assume that $\{V_i\}$ is a decreasing sequence. Hence,

$$Y\backslash V_1 \subset \cdots \subset Y\backslash V_n \subset \cdots \subset Y$$

is an increasing sequence. Because $CO(Y)$ is Cantor separable, there exists a clopen set C such that $Y\backslash V_n \subset C \subset Y$ for all n. Therefore, $\emptyset \subset Y\backslash C \subset V_n$ for all n. Hence, $Y\backslash C$ is contained in $\bigcap V_n$, and $\bigcap U_n$ therefore has non-empty interior. $\square$

2.21. Next we will see that if L is DuBois-Reymond separable, then no sequence of distinct points in $S(L)$ converges, i.e. $S(L)$ contains no κ-points.

Proposition. *A zero-dimensional space Y contains no κ-points if $CO(Y)$ is DuBois-Reymond separable.*

Proof. Suppose that $\{y_n\}$ is a sequence of distinct points converging to y and that $y_n \neq y$ for any n. We will construct a neighborhood of y which misses infinitely many points of the sequence, thus contradicting convergence and showing that no point of Y can be a κ-point.

Choose U_1 to be a clopen set containing y_1 and missing y and the other points of the sequence. Choose U_2 to be a clopen set containing y_2 and missing U_1, y, and all other points of the sequence. Continuing by induction, we obtain a sequence $\{U_i\}$ of pairwise disjoint clopen sets such that y_i is in U_i for each i and y belongs to none of the sets.

For each $n \geq 1$, put $A_n = \bigcup \{U_{2i-1} : i \leq n\}$ and $B_n = Y \setminus (\bigcup \{U_{2i} : i \leq n\})$. Observe that A_n contains the first n points having odd indices and B_n excludes the first n points having even indices. Further, the sequences $\{A_n\}$ and $\{B_n\}$ satisfy:

$$A_1 \subset \cdots \subset A_n \subset \cdots \subset B_n \subset \cdots \subset B_1 .$$

DuBois-Reymond separability implies that there exists a clopen set H such that $A_n \subset H \subset B_n$ for every n. Thus, H contains all points having even indices and excludes all points having odd indices. Then y belongs to either H or $Y \setminus H$. Either set fails to contain infinitely many points of the sequence, which contradicts the assumption that it converges to y. $\square$

2.22. We now consider sufficient conditions for the algebra of clopen sets to be Cantor or DuBois-Reymond separable. In the next chapter, we will see that $\mathbb{N}^*$ satisfies the hypotheses of the following two propositions and therefore is both Cantor and DuBois-Reymond separable. The next two results are taken from R. G. Wood's 1972 A paper.

Proposition. *The Boolean algebra of clopen subsets of a totally disconnected compact space without isolated points and in which every zero-set is regular closed is Cantor separable.*

Proof. Let $\{A_n\}$ be a strictly increasing sequence of clopen subsets of a space X satisfying the stated hypotheses and let B be a clopen subset properly containing each A_n. Because B is compact, $\bigcup A_n$ is properly contained in B since $\{A_n\}$ is an open covering of its union $\bigcup A_n$ which has no finite subcovering. The set $\bigcup A_n$ is a cozero-set since it is a countable union of cozero-sets. Therefore, $B \setminus \bigcup A_n = \bigcap (B \setminus A_n)$ is a non-empty

G_δ which must contain a non-empty zero-set. Since the zero-sets are regular closed, $B \setminus \bigcup A_n$ contains a non-void open set and therefore $\text{cl}(\bigcup A_n)$ is a proper subset of B. Since X has no isolated points, there exist distinct points x and y contained in $B \setminus \text{cl}(\bigcup A_n)$. But then $\text{cl}(\bigcup A_n) \cup \{x\}$ and $(X \setminus B) \cup \{y\}$ are disjoint compact subsets of the compact totally disconnected space X. Since X has a base of clopen sets by Proposition 2.4, there is a clopen subset C of X containing the first set and missing the second. But then we have $A_n \subset C \subset B$ and both containments are proper. Hence, the Boolean algebra of clopen sets is Cantor separable. $\quad\square$

2.23. Recall that one of several equivalent defining properties of the class of *F-spaces* which was demonstrated in Theorem 1.60 is that disjoint cozero-sets are completely separated.

Proposition. *The Boolean algebra of clopen subsets of a totally disconnected compact F-space is DuBois-Reymond separable.*

Proof. Let X be such a space and let $\{A_n\}$ and $\{B_n\}$ be sequences of clopen subsets of X such that

$$A_1 \subset \cdots \subset A_n \subset \cdots \subset B_n \subset \cdots \subset B_1 \ .$$

Then $\bigcup A_n$ and $X \setminus \bigcap B_n$ are disjoint cozero-sets and therefore have disjoint closures since X is an F-space. Hence, there exists a clopen subset H of X containing $\text{cl}(\bigcup A_n)$ and missing $\text{cl}(X \setminus \bigcap B_n)$. Thus, $A_n \subset H \subset B_n$ for each n, and the Boolean algebra of clopen subsets of X is DuBois-Reymond separable. $\quad\square$

2.24. The final result on separability in Boolean algebras is a lemma which will be applied later to the algebra of clopen subsets of $\mathbb{N}^* = \beta\mathbb{N} \setminus \mathbb{N}$. It is a Boolean algebra formulation of a lemma proved in 1956 by W. Rudin and used by him to construct an automorphism of $\mathbb{N}^*$. We will see his original application of the lemma in Chapter 7. The present form of the lemma is taken from the 1963 paper of I. I. Parovičenko who used it to obtain a characterization of $\mathbb{N}^*$. His result will be Theorem 3.31. In both applications, the lemma is employed in the construction of an isomorphism between the algebras of clopen subsets of two compact totally disconnected spaces. The duality between Boolean algebra isomorphisms and the homeomorphisms of the respective Stone spaces is then used to obtain a homeomorphism.

The lemma can be viewed as an extended version of DuBois-Reymond separability. It not only asserts that an element of the Boolean algebra can be interposed between an increasing sequence and a decreasing sequence dominating it, but that the element can also be chosen so that it is not comparable to any member of a given set of elements. The present proof is adapted from the book *The Theory of Ultrafilters* by W. W. Comfort and S. Negrepontis.

Lemma. *Let a Boolean algebra L be dense in itself and be both Cantor and DuBois-Reymond separable. For all l, m, and n let the countable (perhaps finite) subsets $\{A_l\}$, $\{B_m\}$, and $\{C_n\}$ of L satisfy:*

(a) $A_1 < \cdots < A_l < \cdots < B_m < \cdots < B_1$,

(b) $C_n \not\leqslant A_l$,

(c) $B_m \not\leqslant C_n$.

Then there exists D in L such that:

(d) $A_l < D < B_m$,

(e) D *is not comparable to any* C_n.

Proof. Using the Stone Representation Theorem, we will assume that the elements of L are the clopen subsets of its Stone space. The proof will be accomplished by constructing an increasing sequence $\{E_l\}$ and a decreasing sequence $\{F_m\}$ such that $E_l \subset F_m$ for all l and m and such that any set D between them will necessarily be non-comparable to any of the C_n's and will also satisfy $A_l \subset D \subset B_m$.

We begin by using induction to define two sequences (perhaps finite) $\{D(i,j): i \leqslant j\}$ and $\{G(i,j): i \leqslant j\}$ in L with the following properties:

(1) $D(i,j) \supseteq D(i,j+1)$ and $G(i,j) \supseteq G(i,j+1)$, $i \leqslant j$,

(2) $D(i,i) \subset C_i \setminus \bigcup A_l$ and $G(i,i) \subset (\bigcap B_m) \setminus C_i$ for all i,

(3) $D(i,j) \cap G(k,j) = \emptyset$ for all i,j,k with $i \leqslant j$, $k \leqslant j$.

To obtain $D(1,1)$, note that $\{C_1 \setminus \bigcup \{A_l : 1 \leqslant l \leqslant n\} : n \geqslant 1\}$ is a decreasing sequence of non-empty clopen sets. Hence, the intersection of the sequence, $C_1 \setminus \bigcup A_l$, is non-empty by the compactness of the Stone space and therefore has non-empty interior by Cantor separability (Proposition 2.20). Choose $D(1,1)$ to be a non-empty clopen subset of $C_1 \setminus \bigcup A_l$. In a similar fashion, let $G(1,1)$ be a non-empty clopen subset of $(\bigcap B_m) \setminus C_1$.

To continue the construction for $j=2$, $i=1,2$, let H be a non-empty clopen subset of $C_2 \setminus \bigcup A_l$, where the existence of H follows exactly as that of $D(1,1)$ above. Since L is dense in itself, the Stone space contains no isolated points (Proposition 2.19). Hence, we can write H as the union of a disjoint pair H_1 and H_2 of non-empty clopen sets. Thus, either $G(1,1) \setminus H_1 \neq \emptyset$ or $G(1,1) \setminus H_2 \neq \emptyset$ since $H_1 \cap H_2 = \emptyset$. Assume the former and put $D(2,2) = H_1$ and $G(1,2) = G(1,1) \setminus H_1$. Similarly, choose $G(2,2)$ to be a non-empty clopen subset of $(\bigcap B_m) \setminus C_2$ such that $D(1,1) \setminus G(2,2) \neq \emptyset$ and put $D(1,2) = D(1,1) \setminus G(2,2)$. At this stage of the induction it is easy to see that (1)—(3) are satisfied.

Now assume that for some integer $n > 2$ we have chosen $\{D(i,j): 1 \leqslant i \leqslant j \leqslant n\}$ and $\{G(i,j): 1 \leqslant i \leqslant j \leqslant n\}$ satisfying (1)—(3). As for $D(2,2)$, choose K to be a non-empty clopen subset of $C_{n+1} \setminus \bigcup A_l$ and write K as the disjoint union of two non-empty clopen subsets K_1 and

K_2. Then either $G(1,n)\backslash K_1 \neq \emptyset$ or $G(1,n)\backslash K_2 \neq \emptyset$. Assume the former and split K_1 into two disjoint non-empty clopen subsets one of which does not contain $G(2,n)$. Repeat this spitting process for at most n steps until a non-empty clopen set $D(n+1,n+1)$ is obtained such that $D(n+1,n+1) \subset C_{n+1}\backslash \bigcup A_l$ and $G(i,n)\backslash D(n+1,n+1) \neq \emptyset$, for $1 \leqslant i \leqslant n$. Put $G(i,n+1) = G(i,n)\backslash D(n+1,n+1)$ for $1 \leqslant i \leqslant n$. Similarly, choose $G(n+1,n+1)$ to be a non-empty clopen subset of $(\bigcap B_m)\backslash C_{n+1}$ such that $D(i,n)\backslash G(n+1,n+1) \neq \emptyset$ and put $D(i,n+1) = D(i,n)\backslash G(n+1,n+1)$ for $1 \leqslant i \leqslant n$. Then properties (1)—(3) are clearly satisfied by the two sequences for $1 \leqslant i \leqslant j \leqslant n+1$ and the inductive step is complete.

Now by (1), the compactness of the Stone space, and the fact that non-empty G_δ's have non-empty interior we can find non-empty clopen sets D_i and G_i for each $i \geqslant 1$ such that

$$D_i \subset \bigcap \{D(i,j):j \geqslant i\} \quad \text{and} \quad G_i \subset \bigcap \{G(i,j):j \geqslant i\}.$$

By (2), $D_i \subset C_i\backslash \bigcup A_l$ and $G_i \subset (\bigcap B_m)\backslash C_i$ for each i. Further, $D_i \cap G_j = \emptyset$

for all i and j, since we can assume without loss of generality that $i \leqslant j$, in which case we have $D_i \subset D(i,j)$ and $G_j \subset G(i,j)$ and $D(i,j) \cap G(i,j) = \emptyset$ from (3).

Now we will use the newly defined sequences $\{D_i\}$ and $\{G_j\}$ to define two sequences $\{E_l\}$ and $\{F_m\}$ in L such that

$$E_1 \subset \cdots \subset E_l \subset \cdots \subset F_m \subset \cdots \subset F_1$$

and such that any set D occurring between the two sequences will be of the required type, i.e. will satisfy $A_l \subset D \subset B_m$ and will not be comparable to any of the C_n's. For each l and m, put

$$E_l = A_l \cup (\bigcup \{G_i:1 \leqslant i \leqslant l\}) \quad \text{and} \quad F_m = B_m\backslash(\bigcup \{D_j:1 \leqslant j \leqslant m\}).$$

Since $A_l \subset B_m$, $G_i \subset (\bigcap B_m)\backslash C_i$, $D_j \subset C_j\backslash \bigcup A_l$, and $D_j \cap G_i = \emptyset$, for all i, j, l, and m, we have $E_l \subset F_m$. Further, the sequences $\{E_l\}$ and $\{F_m\}$ are easily seen to be increasing and decreasing, respectively.

Now by DuBois-Reymond separability or by the density of the algebra if the sequences $\{E_l\}$ and $\{F_m\}$ are finite, there exists a clopen set D such that

$$A_l \subset E_l \subset D \subset F_m \subset B_m.$$

It remains to show that D is not comparable to any C_n.

For each n, $E_n = A_n \cup (\bigcup \{G_i:1 \leqslant i \leqslant n\}) \subset D$ and G_n contains points of C_n' so that D cannot be contained in C_n. Similarly, points of C_n belong to D_n and $D_n \cap F_n = \emptyset$. Therefore, C_n cannot be contained in D since $D \subset F_n$. Hence, D is the required set. $\square$

Exercises

2A. *Lattice identities*

In any partially ordered set X, the operations of meet and join satisfy the following laws whenever the specified expressions exist:
1. $x \wedge x = x, \ x \vee x = x.$
2. $x \wedge y = y \wedge x, \ x \vee y = y \vee x.$
3. $x \wedge (y \wedge z) = (x \wedge y) \wedge z, \ x \vee (y \vee z) = (x \vee y) \vee z.$
4. $x \wedge (x \vee y) = x \vee (x \wedge y) = x.$

2B. *Lattice inequalities*

1. In any partially ordered set X, the following are equivalent:
   (a) $x \leqslant y,$
   (b) $x \wedge y = x,$
   (c) $x \vee y = y.$

   Thus, the partial order could be surpressed by replacing it with a statement involving meet or join.
2. In any lattice, if $x \leqslant y$, then $x \wedge z \leqslant y \wedge z$ and $x \vee z \leqslant y \vee z$.
3. In any lattice, the distributive inequalities are satisfied:

$$(x \wedge y) \vee (x \wedge z) \leqslant x \wedge (y \vee z),$$

$$(x \vee y) \wedge (x \vee z) \geqslant x \vee (y \wedge z).$$

2C. *Distributive lattices*

1. In any lattice, the following are equivalent for all elements x, y, and z:
   (a) $x \wedge (y \vee z) = (x \wedge y) \vee (x \wedge z),$
   (b) $x \vee (y \wedge z) = (x \vee y) \wedge (x \vee z).$
   [Use 2, 3, and 4 of 2A.]
2. In a distributive lattice, if $z \wedge x = z \wedge y$ and $z \vee x = z \vee y$, then $x = y$.
   [Use 2A.2, 2A.4 and part 1 above.]
3. In a distributive lattice, if an element has a complement, then the complement is unique.

2D. *Demorgan's laws*

Let L be a Boolean algebra and let x, y, and z be arbitrary elements of L.
1. $(x')' = x.$
2. $x \wedge y = 0$ if and only if $x \leqslant y'$. [2B.1, 2B.2.]
3. The bijection $x \mapsto x'$ inverts order, i.e. $x \leqslant y$ implies $y' \leqslant x'$ [2B.2].
4. DeMorgan's Laws are satisfied:

$$(x \vee y)' = x' \wedge y' \quad \text{and} \quad (x \wedge y)' = x' \vee y'.$$

2E. *Suprema of clopen sets*

1. Let $\{U_\alpha\}$ be a family of clopen sets of a zero-dimensional space X. Then the supremum of $\{U_\alpha\}$ exists in $CO(X)$ and is equal to $\mathrm{cl}(\bigcup U_\alpha)$ if and only if $\mathrm{cl}(\bigcup U_\alpha)$ is open. [Proposition 2.5.]
2. State and prove the analoguous result for infima in $CO(X)$.

2F. *Countable neighborhood bases and regularity*

1. A space T is *regular* if every point has a basis of closed neighborhoods. If T is regular and X is dense in T, then a point of X has a countable neighborhood base in X if and only if it has a countable neighborhood base in T. [2.14(e).]
2. If a point of X has a countable neighborhood base in X, it also has a countable neighborhood base in βX, and conversely.

2G. $A \mapsto A^*$

Let X be a space such that $[\mathrm{cl}_X(X \backslash A)]^* = \mathrm{cl}_{X^*}(X^* \backslash A^*)$, for every regular closed subset A of X. Then Proposition 2.16 shows that $A \mapsto A^*$ is a homomorphism of $R(X)$ to $R(X^*)$. The homomorphism is a monomorphism if and only if X admits no non-empty compact regular closed subset.

2H. *More one point compactification*

Let D be an infinite discrete space and let L be the Boolean algebra consisting of the finite subsets of D and their complements.
1. Only one maximal filter in L is not determined by a point.
2. $S(L)$ is αD, the one-point compactification of D.
3. L is not dense in itself, Cantor separable, or DuBois-Reymond separable.

2I. *Boolean rings*

A *Boolean ring* is a ring with identity in which $x^2 = x$ for every x.
1. Every Boolean ring is commutative. $[x + y = (x + y)^2.]$
2. A Boolean algebra L can be made into a Boolean ring be defining addition and multiplication as follows:

$$x + y = (x \wedge y') \vee (x' \wedge y),$$

$$xy = x \wedge y.$$

3. A Boolean ring R can be made into a Boolean algebra by defining meet and join as follows:

$$x \wedge y = xy \quad \text{and} \quad x \vee y = x + y + xy.$$

References: The relationships between Boolean algebras and Boolean rings were first explored by M.H. Stone, 1936, 1937. The Appendix of G. Simmons' book, 1963, contains a discussion of these relationships together with a proof of the Stone Representation Theorem which utilizes Boolean rings.

2J. *Extremally disconnected spaces*

1. A space is extremally disconnected if and only if every pair of disjoint open sets have disjoint closures.
2. A dense subspace of an extremally disconnected space is extremally disconnected.
3. A space X is extremally disconnected if and only if βX is extremally disconnected.
4. If a space X is dense in an extremally disconnected space Y, then every mapping of X into a compact space extends continuously to Y. [Theorem 1.2.]

References: (1)—(3) and other properties of extremally disconnected spaces are outlined in [GJ, ex. 1H, 3N, 6M]. (4) is from the 1958 paper of R.H. McDowell.

2K. *Basically disconnected spaces*

Call a space *basically disconnected* if each cozero-set has open closure.
1. An extremally disconnected space is basically disconnected.
2. A space is basically disconnected if and only if every cozero-set and open set which do not meet have disjoint closures.
3. A Boolean algebra is σ-complete (i.e. every countable subset has a least upper bound) if and only if its Stone space is basically disconnected.

Reference: For additional properties of basically disconnected spaces, see [GJ, ex. 1H, 3N, 6M, 9H].

Chapter 3. On $\beta\mathbb{N}$ and $\mathbb{N}^*$

3.1. The Stone-Čech compactification of the discrete space $\mathbb{N}$ of natural numbers has become one of the most widely studied topological spaces for a number of reasons. First, $\beta\mathbb{N}$ and the growth $\mathbb{N}^* = \beta\mathbb{N}\setminus\mathbb{N}$ admit a seemingly endless flow of interesting subspaces. For example, $\beta\mathbb{N}$ contains a countably compact subspace whose square fails to be pseudocompact. On the other hand, there are 2^c subspaces of $\beta\mathbb{N}$ such that every finite power of each of the subspaces is pseudocompact and no two of the subspaces are homeomorphic. $\mathbb{N}^*$ has a basis of c clopen subspaces each of which is homeomorphic with $\mathbb{N}^*$ itself and yet $\mathbb{N}^*$ also manages to contain 2^c disjoint copies of itself.

Despite the apparent pathologies of the spaces $\beta\mathbb{N}$ and $\mathbb{N}^*$, copies of these two spaces occur frequently in the Stone-Čech compactifications of other spaces and their presence greatly facilitates certain investigations of these compactifications as we shall see in Chapter 4.

Finally, notwithstanding the complexity of $\beta\mathbb{N}$, some of its properties are easily derived from elementary facts about the natural numbers. The effect of this behavior has been to make $\beta\mathbb{N}$ something of a prototype, in that problems involving Stone-Čech compactifications are first examined and solved in this simple case, and then the arguments are adjusted to achieve more general solutions. One particularly evident example where this has been done will be seen in Theorem 5.8 which can be viewed as a modification of the proof (in Theorem 3.22) that $\mathbb{N}^*$ admits a family of c disjoint open sets.

The Cardinality of $\beta\mathbb{N}$

3.2. In his fundamental 1937 paper, E. Čech used the cardinality of $\beta\mathbb{N}$ to estimate the cardinalities of various subspaces of Stone-Čech compactifications. However, at the time he knew only that the cardinality of $\beta\mathbb{N}$ was at least c and at most 2^c and he emphasized the importance of determining the exact cardinality. In the same year, B. Pospíšil answered the question in more general terms by showing that if D is the infinite discrete space of cardinality $\mathfrak{m}$, then the cardinality of βD is $2^{2^{\mathfrak{m}}}$.

Theorem (B. Pospíšil). *If D is the infinite discrete space of cardinality $\mathfrak{m}$, then the cardinality of βD is $2^{2^{\mathfrak{m}}}$.*

Proof. Since the product of no more than $2^{\mathfrak{m}}$ spaces each admitting a dense subset of cardinality $\mathfrak{m}$ or less also has a dense subset of the same cardinality [Hewitt, 1946, or Pondiczery, 1944, contain proofs, and D, p. 175 establishes the special case $\mathfrak{m} = \aleph_0$], D can be mapped onto a dense subspace of the product $I^{2^{\mathfrak{m}}}$. Such a mapping then extends to βD and the image of βD under the extension is compact and contains a dense subspace, and thus is all of $I^{2^{\mathfrak{m}}}$. Hence,

$$|\beta D| \geqslant |I^{2^{\mathfrak{m}}}| = c^{2^{\mathfrak{m}}} = 2^{2^{\mathfrak{m}}}.$$

On the other hand, D admits at most $2^{2^{\mathfrak{m}}}$ ultrafilters, each of which must converge to a point of βD. Hence, the cardinality of βD is at most $2^{2^{\mathfrak{m}}}$. $\square$

Corollary. *The cardinality of $\beta\mathbb{N}$ is $2^{2^{\aleph_0}} = 2^c$.*

The proof of Pospíšil's theorem given here is based on the proof given for $\beta\mathbb{N}$ in 1959 by S. Mrówka.

3.3. The first application which we will make of our knowledge of the cardinality of $\beta\mathbb{N}$ will be to show that every infinite closed subspace of $\beta\mathbb{N}$ contains a copy of $\beta\mathbb{N}$ and therefore has cardinality 2^c. This result is usually attributed to J. Novák in 1953 B, but in that source, Novák remarks that the result was known to Čech as early as 1939. Again the proof is that of S. Mrówka.

Theorem. *The cardinality of each infinite closed subset of $\beta\mathbb{N}$ is 2^c.*

Proof. Let F be an infinite closed subspace of $\beta\mathbb{N}$. Then F contains a countable, discrete subspace E since F is an infinite Hausdorff space. $\mathbb{N} \cup E$ is a countable completely regular space and hence is regular and Lindelöf. But a regular Lindelöf space is paracompact and therefore is normal [D, pp. 163, 174]. Since every point of $\mathbb{N}$ is isolated, E is closed in $\mathbb{N} \cup E$, and thus is C^*-embedded in the normal space $\mathbb{N} \cup E$. Any bounded continuous real-valued function on E can therefore be extended to $\mathbb{N} \cup E$, and from there will extend to $\beta\mathbb{N}$. Consequently, E is dense and C^*-embedded in $\mathrm{cl}_{\beta\mathbb{N}} E$ which is contained in F so that $\mathrm{cl}_{\beta\mathbb{N}} E$ is βE. Since βE is homeomorphic to $\beta\mathbb{N}$, F contains a copy of $\beta\mathbb{N}$ and must have cardinality 2^c. $\square$

3.4. Since every uncountable open subset of $\beta\mathbb{N}$ must contain an infinite closed subset, we have the immediate.

Corollary. *Every uncountable open subset of $\beta\mathbb{N}$ has cardinality 2^c.*

3.5. The proof of Theorem 3.3 contains the proof of the following interesting and important property of $\beta\mathbb{N}$.

Proposition. *Every countable subspace of $\beta\mathbb{N}$ is C^*-embedded.*

Note that the proposition implies that the closure in $\beta\mathbb{N}$ of any discrete, countably infinite subspace of $\beta\mathbb{N}$ is homeomorphic with $\beta\mathbb{N}$.

3.6. The cardinality of $\beta\mathbb{N}$ can also be used to obtain a lower bound for the cardinality of any zero-set contained in the growth of any Stone-Čech compactification.

Theorem. *Every zero-set of βX which misses X contains a copy of $\mathbb{N}^*$ and therefore its cardinal is at least 2^c.*

Proof. Let $\mathbf{Z}(f)$ be a zero-set contained in $\beta X \backslash X$ and put $Y = \beta X \backslash \mathbf{Z}(f)$. Then $h = (f \,|\, Y)^{-1}$ is well-defined and continuous because f does not vanish on Y. Further, h is unbounded. Therefore, the range of h contains an unbounded (and hence closed) copy M of $\mathbb{N}$ and Y therefore contains a copy N of $\mathbb{N}$ which is mapped by h homeomorphically onto M.
(a) N is C-embedded in Y: Let g belong to $C(N)$. Since $\theta = (h \,|\, N)$ is a homeomorphism, $g \circ \theta^{\leftarrow}$ belongs to $C(M)$. Since M is C-embedded in $\mathbb{R}$, there exists a mapping l in $C(\mathbb{R})$ which extends $g \circ \theta^{\leftarrow}$. Then $l \circ h$ belongs to $C(Y)$ and for y in N,

$$(l \circ h)(y) = (g \circ \theta^{\leftarrow} \circ \theta)(y) = g(y)$$

so that $l \circ h$ extends g to all of Y.
It now follows from Proposition 1.48 that $\mathrm{cl}_{\beta Y} N = \beta N$.
(b) $\beta N \backslash N$ is contained in $\mathbf{Z}(f)$: Since f approaches zero along N, $f(p) = 0$ for any p in $\beta N \backslash N$. But such a point p belongs to $\mathbf{Z}(f)$.
Finally, since $\beta N \backslash N$ has the same cardinality as $\mathbb{N}^*$, the cardinality of $\mathbf{Z}(f)$ is at least 2^c. $\square$

3.7. Since every point contained in a G_δ belongs to a zero-set which is also contained in the G_δ, a G_δ point is a zero-set. Hence the following corollary is immediate.

Corollary (Čech). *No point of $\beta X \backslash X$ is a G_δ in βX.*

Čech showed in 1937 that a zero-set contained in a growth must contain a copy of $\mathbb{N}^*$ and from that he deduced Corollary 3.7. However, he did not yet know the cardinality of $\beta\mathbb{N}$. The preceding two results are from [GJ, Chapter 9].

3.8. The same chapter of [GJ] also contains a proof of Theorem 3.2 obtained by showing that an infinite discrete space of cardinality $\mathfrak{m}$ admits $2^{2^{\mathfrak{m}}}$ ultrafilters as well as the following two results which are similar to Theorem 3.6. Note that the zero-set contained in $\beta X \setminus X$ in Theorem 3.6 has been replaced in the following theorem by any non-discrete closed subset of $\beta X \setminus \upsilon X$. The realcompactness of υX is used in the proof to obtain a zero-set of βX which contains a given point of $\beta X \setminus \upsilon X$ and misses υX. The following two results originally appeared in [GJ, 9.11—9.12].

Theorem (Gillman and Jerison). *Every non-discrete closed subspace of $\beta X \setminus \upsilon X$ contains a copy of $\mathbb{N}^*$ and therefore its cardinality is at least 2^c.*

If X is locally compact, then X^* is compact so that any closed discrete subset of X^* is finite. Hence the next result follows immediately.

Corollary. *If X is locally compact and realcompact, then every infinite closed subset of X^* contains a copy of $\mathbb{N}^*$ and therefore its cardinality is at least 2^c.*

The Clopen Sets of $\beta\mathbb{N}$ and $\mathbb{N}^*$

3.9. Proposition 1.42 shows that there is a one-to-one correspondence between the free ultrafilters on $\mathbb{N}$ and the points of $\mathbb{N}^*$. Under this correspondence, a point p of $\mathbb{N}^*$ is identified with the unique free ultrafilter A^p on $\mathbb{N}$ which converges to p in $\beta\mathbb{N}$. A^p consists of precisely those subsets of $\mathbb{N}$ which have p in their closure in $\beta\mathbb{N}$. Thus, every neighborhood of p meets $\mathbb{N}$ in a set belonging to A^p. The points of $\mathbb{N}$ are isolated in $\beta\mathbb{N}$ and are the only isolated points of $\beta\mathbb{N}$ since $\mathbb{N}$ is dense in $\beta\mathbb{N}$.

If X is any space, the closure in βX of any clopen subspace S of X is easily seen to be clopen by observing that $\mathrm{cl}_{\beta X} S$ is the set of points where the extension of the characteristic function of S is equal to 1. By applying this result to $\beta\mathbb{N}$, we will see in the next proposition that the closures in $\beta\mathbb{N}$ of subsets of $\mathbb{N}$ separate points of $\beta\mathbb{N}$ so that $\beta\mathbb{N}$ is totally disconnected.

Proposition. *$\beta\mathbb{N}$ is totally disconnected and therefore is also zero-dimensional.*

Proof. Let p and q be distinct points of $\beta\mathbb{N}$ and choose a subset Z of $\mathbb{N}$ belonging to A^p but not to A^q. Then $\mathrm{cl}_{\beta\mathbb{N}} Z$ is a clopen neighborhood of p which misses q. $\beta\mathbb{N}$ is therefore totally disconnected since the only connected subspaces of $\beta\mathbb{N}$ are the singletons and is zero-dimensional since every compact totally disconnected space is zero-dimensional (Proposition 2.4). □

3.10. Since every subset of $\mathbb{N}$ is closed in $\mathbb{N}$ and also C^*-embedded in $\mathbb{N}$, Proposition 1.48 shows that if A is an infinite subset of $\mathbb{N}$, then $\mathrm{cl}_{\beta\mathbb{N}} A$ is homeomorphic with $\beta\mathbb{N}$ and $A^* = \mathrm{cl}_{\beta\mathbb{N}} A \backslash A$ is homeomorphic with $\mathbb{N}^*$. Further, $\mathrm{cl}_{\beta\mathbb{N}} A$ is clopen in $\beta\mathbb{N}$. The next proposition shows that every infinite clopen subspace of $\beta\mathbb{N}$ is of this form and hence is a copy of $\beta\mathbb{N}$.

Proposition. *Every clopen subspace of $\beta\mathbb{N}$ is of the form $\mathrm{cl}_{\beta\mathbb{N}} A$ for some subset A of $\mathbb{N}$.*

Proof. Let U be a clopen subset of $\beta\mathbb{N}$. Then U is compact, and hence if U is contained in $\mathbb{N}$, U is finite and $U = \mathrm{cl}_{\beta\mathbb{N}} U$. If U meets $\mathbb{N}^*$, then $U \cap \mathbb{N}$ is non-empty since $\mathbb{N}$ is dense in $\beta\mathbb{N}$. Because U is closed, $\mathrm{cl}_{\beta\mathbb{N}}(U \cap \mathbb{N})$ is contained in U. Since U is open, $U \backslash \mathrm{cl}_{\beta\mathbb{N}}(U \cap \mathbb{N})$ is open in $\beta\mathbb{N}$ and misses $\mathbb{N}$. Since $\mathbb{N}$ is dense in $\beta\mathbb{N}$, this difference must be empty and $U = \mathrm{cl}_{\beta\mathbb{N}}(U \cap \mathbb{N})$. □

3.11. Since $\beta\mathbb{N}$ has a base of clopen subsets, we are now able to state the following improvement of Corollary 3.4.

Corollary. *Every open set of $\beta\mathbb{N}$ which meets $\mathbb{N}^*$ has cardinality 2^c.*

3.12. Subspaces of $\beta\mathbb{N}$ have yielded a wealth of interesting examples. In their 1929 memoir on compact spaces, P. Alexandroff and P. Urysohn posed the following question: Does there exist a compact Hausdorff space with no isolated points and such that no point is a limit of a sequence of distinct points? Čech showed in 1937 that the space $\mathbb{N}^*$ is such a space. Recall from 2.11 that a point which is a limit point of a sequence of distinct points is called a *κ-point*.

Proposition. $\mathbb{N}^*$ *is a compact Hausdorff space containing no isolated points and no κ-points.*

Proof. $\mathbb{N}^*$ is clearly a compact Hausdorff space since it is a closed subspace of $\beta\mathbb{N}$. $\mathbb{N}^*$ cannot contain an isolated point since an open subset of $\beta\mathbb{N}$ which meets $\mathbb{N}^*$ contains 2^c points. If a point p of $\mathbb{N}^*$ is a limit of the sequence $\{x_n\}$, then $\{x_n\} \cup \{p\}$ is a countably infinite closed subspace of $\beta\mathbb{N}$ which is impossible by Theorem 3.3. □

3.13. Since $\mathbb{N}$ is normal, any two disjoint subsets of $\mathbb{N}$ have disjoint closures in $\beta\mathbb{N}$ by Corollary 1.15. Thus, if A is an infinite subset of $\mathbb{N}$, we have that

$$[\mathrm{cl}_{\mathbb{N}}(\mathbb{N}\backslash A)]^* = \mathrm{cl}_{\mathbb{N}^*}(\mathbb{N}^*\backslash A^*) \, .$$

By Proposition 2.16, this equation shows that the assignment

$$A \mapsto A^*$$

is a homomorphism of the Boolean algebra of regular closed subsets of $\mathbb{N}$ to the Boolean algebra of regular closed subsets of $\mathbb{N}^*$. We now describe this homomorphism in detail and show that its image is the Boolean algebra of clopen subsets of $\mathbb{N}^*$.

3.14. Proposition. *If A and B are infinite subsets of $\mathbb{N}$, then B^* is contained in A^* if and only if $B \backslash A$ is finite.*

Proof. To show sufficiency, write $B = (B \cap A) \cup (B \backslash A)$ and assume that $B \backslash A$ is finite. Then if $B \cap A$ is finite, B is finite and B^* is empty. Thus, we can assume that $B \cap A$ is infinite. Then:

$$B^* = (\mathrm{cl}_{\beta\mathbb{N}}(B \cap A) \cup \mathrm{cl}_{\beta\mathbb{N}}(B \backslash A)) \backslash B = \mathrm{cl}_{\beta\mathbb{N}}(B \cap A) \backslash B$$

since $B \backslash A$ is its own closure. Hence, B^* is contained in A^*.

To establish necessity, assume that $B \backslash A$ is infinite and exhibit a point in $B^* \backslash A^*$. Since $B \backslash A$ is infinite and misses A, it belongs to some free ultrafilter A^p which does not contain A (Section 3.9). Then p belongs to B^* but not to A^*. $\square$

3.15. Corollary. *If A and B are infinite subsets of $\mathbb{N}$, then $A^* = B^*$ if and only if $(A \backslash B) \cup (B \backslash A)$ is finite. B^* is properly contained in A^* if and only if $A \backslash B$ is infinite and $B \backslash A$ is finite. A^* meets B^* if and only if $A \cap B$ is infinite.*

3.16. We will now see that every clopen set of $\mathbb{N}^*$ is an image under the homomorphism described in Section 3.13 and that sets of the form A^* constitute a base for both the open and the closed sets of $\mathbb{N}^*$.

Proposition. *Every clopen subspace of $\mathbb{N}^*$ is of the form A^* for some infinite subset A of $\mathbb{N}$. The sets A^* form a base for both the open and the closed sets of $\mathbb{N}^*$.*

Proof. Let U be a clopen subspace of $\mathbb{N}^*$. Since $\mathbb{N}^*$ is C^*-embedded in $\beta\mathbb{N}$, the characteristic function of U as a subspace of $\mathbb{N}^*$ is the restriction of some mapping f in $C^*(\beta\mathbb{N})$. Put $A = \{n \in \mathbb{N} : f(n) \geqslant 1/2\}$. Then U is contained in A^* and $\mathbb{N}^* \backslash U$ is contained in $(\mathbb{N} \backslash A)^*$. Since $\mathbb{N}^*$ is the disjoint union of A^* and $(\mathbb{N} \backslash A)^*$, U is equal to A^*.

$\mathbb{N}^*$ is zero-dimensional since $\beta\mathbb{N}$ is zero-dimensional. Thus, every open set of $\mathbb{N}^*$ is a union of sets of the form A^* by the first statement of the proposition. By taking complements, every closed set of $\mathbb{N}^*$ is seen to be the intersection of sets of the same form. $\square$

3.17. Since $\mathbb{N}$ has c infinite subsets and only c subsets in all, the above propositions establish the cardinalities of bases for both $\beta\mathbb{N}$ and $\mathbb{N}^*$.

Corollary. *$\beta\mathbb{N}$ and $\mathbb{N}^*$ each have a base consisting of c clopen sets.*

3.18. The set of automorphisms of $\beta\mathbb{N}$ clearly forms a group under composition. The *orbit* of a point p of $\beta\mathbb{N}$ is the set of points of $\beta\mathbb{N}$ which are images of p under the automorphisms of $\beta\mathbb{N}$. Knowing the form of the clopen sets of $\beta\mathbb{N}$ and $\mathbb{N}^*$ allows a partial description of the orbit of a point of $\mathbb{N}^*$. Since any automorphism must exchange isolated points of the space, it is clear that all automorphisms of $\beta\mathbb{N}$ are extensions of permutations of $\mathbb{N}$. Thus, since there are c permutations of $\mathbb{N}$, there are just c automorphisms of $\beta\mathbb{N}$. A space is called *homogeneous* if for every pair of points of the space, there is an automorphism of the space which exchanges the pair of points. Since $\beta\mathbb{N}$ contains both isolated and non-isolated points, it is immediate that $\beta\mathbb{N}$ *is not homogeneous*.

As we have also seen that $\beta\mathbb{N}$ admits just c automorphisms but contains 2^c points, the non-homogeneity of $\beta\mathbb{N}$ is also clear from cardinality considerations. Observe that this does not imply that $\mathbb{N}^*$ is not homogeneous since there may be automorphisms of $\mathbb{N}^*$ which are not obtained from permutations of $\mathbb{N}$. The question of the non-homogeneity of $\mathbb{N}^*$ is quite complex and will be explored from a cardinality viewpoint later in the chapter.

3.19. Although the orbit of any point in $\mathbb{N}^*$ under automorphisms of $\beta\mathbb{N}$ contains at most c points, we can show that the orbit is dense in $\mathbb{N}^*$.

Proposition. *If A^* and B^* are proper clopen subspaces of $\mathbb{N}^*$, there exists an automorphism of $\beta\mathbb{N}$ and hence of $\mathbb{N}^*$ carrying A^* onto B^*.*

Proof. Since A^* and B^* are both proper subsets of $\mathbb{N}^*$, the complements in $\mathbb{N}$ of the associated infinite sets A and B are both infinite. Let σ be a permutation of $\mathbb{N}$ which carries A onto B and $\mathbb{N}\backslash A$ onto $\mathbb{N}\backslash B$. Then the extension $\beta(\sigma)$ of σ to $\beta\mathbb{N}$ is an automorphism and carries A^* onto B^* by continuity. $\square$

3.20. Since the sets A^* form a base for the open sets of $\mathbb{N}^*$, the following is immediate:

Corollary. *The orbit of any point of $\mathbb{N}^*$ is a dense subspace of $\mathbb{N}^*$.*

The preceding sequence of results is based on [GJ, 6.10(a) and ex. 6 S]. The cited exercise in [GJ] has its origin in the 1956 paper of W. Rudin and the 1953A paper of J. Novák.

3.21. The orbit of a point in $\mathbb{N}^*$ under automorphisms of $\beta\mathbb{N}$ is a dense subset of $\mathbb{N}^*$ of cardinality c. As an application of properties of the basic sets A^* of N*, we will see that the next result concerning subsets of N implies that every dense subset of N* must contain at least c points. The proof is from [GJ, ex. 6Q]. A family of sets is said to be *almost disjoint* if the intersection of any two of the sets is finite.

Proposition. N *admits a family of c almost disjoint infinite subsets.*

Proof. Consider a one-to-one mapping φ of N onto Q. Select an increasing sequence $\{q_n\}$ of rationals converging to each irrational. For each such sequence, define $E = \{\varphi^{\leftarrow}(q_n)\}$ and let $\mathscr{E}$ be the collection of all such sets. Then it is clear that $\mathscr{E}$ has c members and that the intersection of any two members is finite. $\square$

3.22. The *cellularity* of a topological space Y is the smallest cardinal number $\mathfrak{m}$ for which each pairwise disjoint family of open sets of Y has $\mathfrak{m}$ or fewer members. The family of c almost disjoint infinite subsets of N which exist by the preceding proposition will yield a family of c disjoint open subsets of N* and thus provide a lower bound for the cellularity of N*. Actually, this is the largest possible such family.

Theorem. *The cellularity of* N* *is* c.

Proof. Consider the family $\mathscr{E}$ of c almost disjoint infinite subsets of N provided by the preceding proposition. For distinct members E and F of $\mathscr{E}$, $\mathrm{cl}_{\beta N} E \cap \mathrm{cl}_{\beta N} F \cap$ N* $= \emptyset$ since any point in this intersection would belong to the closure of $E \cap F$ which is just $E \cap F$ since $E \cap F$ is finite. Thus, $\{E^* : E \in \mathscr{E}\}$ is a family of c pairwise disjoint non-void open subsets of N*.

On the other hand, it is clear that there can be no such family of larger cardinality since the power set of N has cardinality c. $\square$

Thus, *any dense subset of* N* *must contain at least c points.*

3.23. The previous theorem shows that N* contains a pairwise disjoint family of c clopen subsets. Thus, a copy of N may be obtained in N* by choosing a point from each of countably many members of such a family of clopen sets. By Proposition 3.5, the copy of N just obtained is C^*-embedded in N* and its closure is homeomorphic with βN. Thus, we have shown again a result which was stated earlier in the proof of Theorem 3.3.

Corollary. N* *contains a copy of* βN.

3.24. Since N is both σ-compact and locally compact, Proposition 1.62 implies that N* is an F-space. However, N* actually satisfies a stronger condition in that disjoint cozero-sets are separated by a partition of N*, i.e. they are contained in disjoint clopen sets whose union is all of N*.

Theorem 1.60 requires only that disjoint cozero-sets be completely separated. We will shortly see an example which will show that the same statement fails for arbitrary open sets by showing that disjoint open sets of $\mathbb{N}^*$ need not have disjoint closures.

Proposition. *Two disjoint cozero sets of $\mathbb{N}^*$ are separated by a partition and hence $\mathbb{N}^*$ is an F-space.*

Proof. Let $\mathbf{Cz}(g)$ and $\mathbf{Cz}(h)$ be disjoint cozero-sets in $\mathbb{N}^*$. We may assume that both g and h are non-negative so that $\mathbf{Cz}(g) = \mathrm{pos}(g-h)$ and $\mathbf{Cz}(h) = \mathrm{neg}(g-h)$. Since $\mathbb{N}^*$ is C^*-embedded in $\beta\mathbb{N}$, $g-h$ has a continuous extension f to all of $\beta\mathbb{N}$. By the continuity of f, $\mathrm{cl}_{\beta\mathbf{N}}(\mathrm{pos}(f))$ contains $\mathbf{Cz}(g)$ and misses $\mathbf{Cz}(h)$. Hence, $\mathrm{cl}_{\beta\mathbf{N}}(\mathrm{pos}(f)) \cap \mathbb{N}^*$ and $\mathbb{N}^* \backslash \mathrm{cl}_{\beta\mathbf{N}}(\mathrm{pos}(f))$ yield the required partition of $\mathbb{N}^*$.

Since sets which are separated by a partition are completely separated, disjoint cozero-sets of $\mathbb{N}^*$ are completely separated and $\mathbb{N}^*$ is therefore an F-space. $\square$

3.25. Since $\mathbb{N}^*$ is also totally disconnected and compact, the following corollary is immediate from Proposition 2.23.

Corollary. *The Boolean algebra of clopen subsets of $\mathbb{N}^*$ is DuBois-Reymond separable.*

3.26. The next result will ultimately lead to a proof that the Boolean algebra of clopen sets of $\mathbb{N}^*$ also is Cantor separable.

Proposition. *If a sequence of clopen subsets of $\mathbb{N}^*$ has the finite intersection property, then the intersection of the sequence of sets contains a non-empty clopen set.*

Proof. Let $\{A_n^*\}$ be a sequence of clopen sets of $\mathbb{N}^*$ having the finite intersection property. Then we may assume that the sequence is nested, i.e., that

$$A_1^* \supset A_2^* \supset \cdots \supset A_n^* \supset \cdots$$

and therefore that the cardinality of $A_{n+1} \backslash A_n$ is finite for all n. Choose a sequence of distinct points $A = \{x_n\}$ such that x_n belongs to $\bigcap\{A_i : i \leqslant n\}$. Then $|A \backslash A_n| \leqslant n-1$ for each n so that A^* is contained in A_n^* and A^* is therefore contained in $\bigcap A_n^*$. $\square$

3.27. Corollary. *Every non-empty G_δ in $\mathbb{N}^*$ has non-empty interior.*

3.28. Corollary. *The zero-sets of $\mathbb{N}^*$ are regular closed sets.*

Proof. Since a zero-set is a G_δ, the preceding corollary shows that a non-empty zero-set of $\mathbb{N}^*$ has non-empty interior. Now assume that a zero-set Z_1 of $\mathbb{N}^*$ is not regular closed, i.e. that there is a point p in $Z_1\backslash\mathrm{cl}_{\mathbb{N}^*}(\mathrm{int}_{\mathbb{N}^*}Z_1)$. Then there exists a zero-set neighborhood Z_2 of p such that Z_2 misses $\mathrm{int}_{\mathbb{N}^*}Z_1$. Thus, $Z_1\cap Z_2$ is a non-empty zero-set of $\mathbb{N}^*$ which is contained in the boundary of Z_1 and hence has empty interior. But this is a contradiction. $\square$

3.29. All of the hypotheses of Proposition 2.22 have now been verified, and we can make the following observation:

Corollary. *The Boolean algebra of clopen subsets of $\mathbb{N}^*$ is Cantor separable.*

3.30. We have seen in Proposition 3.5 that every countable subspace of $\mathbb{N}^*$ is C^*-embedded in $\mathbb{N}^*$. Further, $\mathbb{N}^*$ contains many copies of $\mathbb{N}^*$ and $\beta\mathbb{N}$, each of which is C^*-embedded because each is compact. (Proposition 1.47.) However, we now see that no dense subspace of $\mathbb{N}^*$ is C^*-embedded and therefore, $\mathbb{N}^*$ is not the Stone-Čech compactification of any of its dense subspaces. The following proposition appears in L. Gillman's 1967 paper in which he discusses results about the spaces $\mathbb{N}^*$ and $\mathbb{R}^*$ which have been obtained with the aid of the Continuum Hypothesis. It is a good example of the way in which the Continuum Hypothesis arises naturally in considerations involving $\mathbb{N}^*$. (Results which use the Continuum Hypothesis will be indicated by [CH].)

Proposition [CH]. *Dense subsets of $\mathbb{N}^*$ are not C^*-embedded.*

Proof. By Proposition 1.49, it is sufficient to show that $\mathbb{N}^*\backslash\{p\}$ is not C^*-embedded in $\mathbb{N}^*$. We will show that $\mathbb{N}^*\backslash\{p\}$ is the union of disjoint open sets A and B each of which contains p in its closure. Thus, there exists a two-valued mapping on $\mathbb{N}^*\backslash\{p\}$ which will not extend continuously to $\mathbb{N}^*$.

Assuming the Continuum Hypothesis, the basis of c zero-set neighborhoods of p can be indexed by ω_1 and written $\{Z_\alpha:\alpha<\omega_1\}$. Proceeding by transfinite induction, assume for a given $\alpha<\omega_1$ that cozero sets A_σ and B_σ have been defined for all $\sigma<\alpha$ such that

$$p\notin A_\gamma\cup B_\tau \quad \text{and} \quad A_\gamma\cap B_\tau=\emptyset$$

for all $\gamma,\tau<\alpha$. Since a countable union of cozero sets is a cozero set, Proposition 3.24 shows that there exist complementary clopen sets A'_α and B'_α such that

$$\bigcup_{\sigma<\alpha} A_\sigma\subset A'_\alpha \quad \text{and} \quad \bigcup_{\sigma<\alpha} B_\sigma\subset B'_\alpha.$$

The set $Z_\alpha\cap(\bigcap\{\mathbb{N}^*\backslash(A_\sigma\cup B_\sigma):\sigma<\alpha\})$ contains p and thus is a non-void zero-set of $\mathbb{N}^*$ and has a non-void interior by Corollary 3.28. Since $\mathbb{N}^*$ contains no isolated points, the set

$$Z_\alpha \cap (\bigcap\{\mathbb{N}^* \setminus (A_\sigma \cup B_\sigma) : \sigma < \alpha\}) \setminus \{p\}$$

contains disjoint non-void cozero-sets A_α'' and B_α''. Now define

$$A_\alpha = (A_\alpha' \setminus Z_\alpha) \cup A_\alpha'' \quad \text{and} \quad B_\alpha = (B_\alpha' \setminus Z_\alpha) \cup B_\alpha''.$$

Then A_α and B_α are disjoint cozero sets both of which fail to contain p and the induction hypothesis is satisfied for all $\sigma, \tau \leq \alpha$.

Now define

$$A = \bigcup_{\alpha < \omega_1} A_\alpha \quad \text{and} \quad B = \bigcup_{\alpha < \omega_1} B_\alpha.$$

A and B are disjoint open sets, and neither contains p. If q is a point of $\mathbb{N}^*$ other than p, then some neighborhood Z_α of p misses q so that q is in $A_\alpha \cup B_\alpha$ by construction. Hence, $A \cup B = \mathbb{N}^* \setminus \{p\}$. Since each basic neighborhood Z_α of p contains $A_\alpha'' \cup B_\alpha''$, every neighborhood of p meets both A and B so that p is in the closures of both A and B. $\square$

In Proposition 3.24 we saw that disjoint cozero-sets of $\mathbb{N}^*$ are separated by a partition. The preceding proof shows that it was necessary to assume in Proposition 3.24 that the sets were cozero-sets since A and B above are disjoint open sets of $\mathbb{N}^*$ which not only are not separated by a partition but also fail to have disjoint closures.

A Characterization of $\mathbb{N}^*$

3.31. Among the properties satisfied by $\mathbb{N}^*$ and its Boolean algebra of clopen sets are six which characterize $\mathbb{N}^*$ as a topological space. $\mathbb{N}^*$ is clearly compact and in Proposition 3.9 we saw that it is totally disconnected. Corollaries 3.25 and 3.29 establish that the Boolean algebra of clopen sets of $\mathbb{N}^*$ is both DuBois-Reymond and Cantor separable. The *weight* of a space is the least cardinal number of a basis for the space, and thus Corollary 3.17 shows that $\mathbb{N}^*$ has weight c. Corollary 3.12 established that $\mathbb{N}^*$ has no isolated points and this result together with Proposition 2.19 shows that the Boolean algebra of clopen sets is dense in itself. In 1963, I. I. Parovičenko showed that in the presence of the Continuum Hypothesis, these six properties characterize the space $\mathbb{N}^*$. The Continuum Hypothesis is used in the proof to index the c clopen sets of $\mathbb{N}^*$ by the countable ordinals. The proof is a modification of an argument used in 1956 by W. Rudin and we will see the argument in its original context later in Theorem 7.11.

In addition, without using the Continuum Hypothesis, Parovičenko showed that $\mathbb{N}^*$ maps continuously onto every compact space having weight at most $\aleph_1$. We will see in Chapter 6 that this result will aid in the description of the growths of other compactifications of $\mathbb{N}$.

Theorem [CH] (Parovičenko). *A totally disconnected compact space Y having weight c and without isolated points and such that $CO(Y)$ is both Cantor and DuBois-Reymond separable is homeomorphic to $\mathbb{N}^*$. Further, $\mathbb{N}^*$ maps onto any compact space having weight at most $\aleph_1$.*

Proof. The first statement will be proven first. We have just seen that $\mathbb{N}^*$ satisfies all of the conditions stated in the hypothesis. The proof will be accomplished by constructing a Boolean algebra isomorphism σ of $CO(Y)$ onto $CO(\mathbb{N}^*)$. Then our remarks following the Stone Representation Theorem, 2.10, show that σ induces a homeomorphism of the respective Stone spaces. But because both $\mathbb{N}^*$ and Y are compact and totally disconnected, they are easily seen to be homeomorphic to the Stone spaces of $CO(\mathbb{N}^*)$ and $CO(Y)$, respectively, so that $\mathbb{N}^*$ and Y are therefore homeomorphic.

It remains to construct the isomorphism σ. We will accomplish the construction by transfinite induction. The induction will be carried out by showing that if σ has been defined for a countable field of clopen subsets of Y which does not include a clopen set U, then because σ preserves the Boolean operations, σ can be extended to a countable field containing U. We need only consider countable fields because of the assumption of the Continuum Hypothesis.

Using the Continuum Hypothesis, we can index the c members of the respective Boolean algebras by the countable ordinals, i.e. we can write $CO(Y) = \{U_\alpha : \alpha < \omega_1\}$ and $CO(\mathbb{N}^*) = \{V_\alpha : \alpha < \omega_1\}$. We can also accomplish the indexing so that $U_0 = Y$ and $V_0 = \mathbb{N}^*$. Now define $\sigma(U_0) = V_0$ and $\sigma(\emptyset) = \emptyset$ so that the family for which σ is initially defined forms a field. Now assume that σ has been defined for a countable field $\mathscr{C}$ contained in $CO(Y)$ and that α is the least ordinal for which U_α has not been defined. Write $\mathscr{C}$ as the union of the families $\{F_i\}$, $\{G_i\}$, and $\{C_i\}$ such that $F_i \subset U_\alpha$, $U_\alpha \subset G_i$, and no inclusion relation holds between C_i and U_α for all values of i. Put $A_n = F_1 \cup \cdots \cup F_n$ and $B_n = G_1 \cap \cdots \cap G_n$ so that

$$A_1 \subset \cdots \subset A_n \subset \cdots \subset U_\alpha \subset \cdots \subset B_n \subset \cdots \subset B_1 ,$$

and no C_i is contained in any A_n and no C_i contains any B_n. Because σ preserves containment, we have

$$\sigma(A_1) \subset \cdots \subset \sigma(A_n) \subset \cdots \subset \sigma(B_n) \subset \cdots \subset \sigma(B_1)$$

and no $\sigma(C_i)$ is contained in any $\sigma(A_n)$ and no $\sigma(C_i)$ contains any $\sigma(B_n)$. Now by Lemma 2.24, there exists V_β in $CO(\mathbb{N}^*)$ such that no inclusion relation holds between any $\sigma(C_i)$ and V_β and

$$\sigma(A_1) \subset \cdots \subset \sigma(A_n) \subset \cdots \subset V_\beta \subset \cdots \subset \sigma(B_n) \subset \cdots \subset \sigma(B_1).$$

Define $\sigma(U_\alpha) = V_\beta$ and $\sigma(Y \setminus U_\alpha) = \mathbb{N}^* \setminus V_\beta$, and let $\mathscr{C}'$ be the field generated by $\mathscr{C}$ and U_α. The requirement that σ preserve the Boolean algebra operations dictates a unique extension of σ to $\mathscr{C}'$. This accomplished, $\sigma^\leftarrow$ satisfies the same induction hypothesis and we can define $\sigma^\leftarrow(V_\delta)$ where δ is the least ordinal such that V_δ is not in the range of σ. The process is continued to yield the required Boolean algebra isomorphism.

We now establish the second statement of the theorem. Observe that in the construction of the isomorphism σ above, we required all of the hypotheses on both $\mathbb{N}^*$ and Y in order to apply Lemma 2.24. The hypotheses on $\mathbb{N}^*$ were required to construct σ and on Y to construct $\sigma^\leftarrow$. If we did not wish to construct an inverse, then $CO(Y)$ could be replaced with any Boolean algebra L having cardinality at most $\aleph_1$ and the construction would yield a one-to-one homomorphism of L into $CO(\mathbb{N}^*)$. Thus, if X is a compact space having weight at most $\aleph_1$, since X has a base of regular closed sets there is a one-to-one homomorphism σ of $R(X)$ into $CO(\mathbb{N}^*)$. We will use this homomorphism to construct a mapping of $\mathbb{N}^*$ onto X.

For each point p of $\mathbb{N}^*$, the members of $CO(\mathbb{N}^*)$ containing p form a maximal filter. Let $\mathscr{F} = \{\sigma^\leftarrow(U) : p \in U\}$. Since σ is a homomorphism, $\sigma^\leftarrow(U) \wedge \sigma^\leftarrow(V) \neq 0$ for any two members of $\mathscr{F}$. Since $\mathscr{F}$ is clearly closed under supersets in $R(X)$, $\mathscr{F}$ is a filter. Further, if F in $R(X)$ is such that $F \wedge \sigma^\leftarrow(U) \neq 0$ for all members $\sigma^\leftarrow(U)$ of $\mathscr{F}$, then $\sigma(F) \wedge U \neq 0$ for all U containing p since σ is a monomorphism. Hence, $\sigma(F)$ contains p and F belongs to $\mathscr{F}$. Thus, $\mathscr{F}$ is a maximal filter. Since X has a base of regular closed sets and is compact, $\bigcap \mathscr{F}$ is a singleton. Hence, we can define a function f from $\mathbb{N}^*$ to X by putting $f(p) = \bigcap \mathscr{F}$. The function f is onto since the image of every maximal filter in $R(X)$ has non-empty intersection in $\mathbb{N}^*$.

We now show that f is continuous. Let $f(p) = x$ and let U be a neighborhood of x in X. Then there exists an open neighborhood V of x such that $x \in V \subset \text{cl}\, V \subset U$. Since $\text{cl}\, V$ is regular closed, $\sigma(\text{cl}\, V) = W$ with W in $CO(\mathbb{N}^*)$. Because $f(p) = x$, p is in W and for any q in W, $f(q)$ belongs to $\sigma^\leftarrow(W) = \text{cl}\, V$. Hence, $f[W]$ is contained in U, and f is continuous. $\quad \square$

Note that in the proof of the second part of the theorem, we showed that $CO(\mathbb{N}^*)$ contains an isomorphic image of any Boolean algebra having cardinality at most $\aleph_1$. By considering the dual of this statement, we saw that any compact space of weight at most $\aleph_1$ is a continuous image of $\mathbb{N}^*$. In his 1969 D paper, S. Negrepontis shows that if the Continuum Hypothesis is assumed, then $\mathbb{N}^*$ is the unique compact space of weight $\aleph_1$ which has this property together with the following one: If K is any compact space of weight less than $\aleph_1$ and f and g are mappings of $\mathbb{N}^*$ onto K, then there exists an automorphism h of $\mathbb{N}^*$ such that $g \circ h = f$.

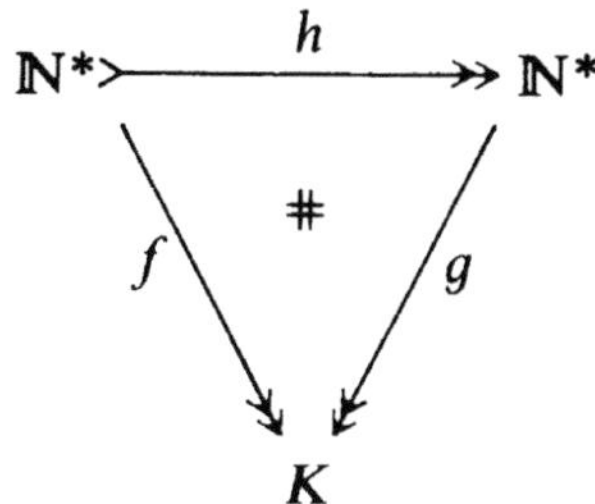

This characterization of $\mathbb{N}^*$ is a special case of a more general theorem of Negrepontis. The key step in his proof is to obtain a generalization of Lemma 2.24 which we used in the proof of the previous theorem.

3.32. By Proposition 2.22, the condition of Cantor separability in the theorem can be replaced by requiring that the zero-sets be regular closed sets. Proposition 2.23 shows that the requirement that the space be an F-space can replace DuBois-Reymond separability. Thus, we can restate Parovičenko's characterization without involving the separability conditions. This restatement appears in R. G. Woods 1971 B paper.

Theorem [CH]. *A totally disconnected compact F-space without isolated points and having weight c and such that every zero-set is a regular closed set is homeomorphic with $\mathbb{N}^*$.*

3.33. Examples. Each of the four conditions in addition to total disconnectivity and compactness is necessary in the above characterization. There exists a totally disconnected compact space which satisfies any three of the conditions but not the fourth.

(a) Consider the disjoint union $\mathbb{N}^* \cup \{x\}$ of $\mathbb{N}^*$ and an isolated point. $\mathbb{N}^* \cup \{x\}$ is easily seen to satisfy all the requirements except the non-existence of isolated points.

(b) Let D be the discrete space of cardinality 2^c and consider $D^* = \beta D \setminus D$. D^* is clearly compact and is easily seen to be totally disconnected and without isolated points. Since D^* is C^*-embedded in the F-space βD, it is itself an F-space. By Lemma 4.21 the zero-sets of D^* are regular closed sets. However, an argument similar to that of Theorem

3.22 will show in Chapter 5 that the weight of D^* is strictly greater than $\mathfrak{c}$.

(c) Consider a strictly increasing sequence $\{A_n^*\}$ of clopen subsets of $\mathbb{N}^*$ and put $X = \mathrm{cl}_{\mathbb{N}^*}(\bigcup A_n^*)$. X is clearly totally disconnected, compact, and without isolated points. Since X is C^*-embedded in the F-space $\mathbb{N}^*$, X is an F-space and the weight of X is easily seen to be $\mathfrak{c}$. Since X is the supremum of $\{A_n^*\}$ in the algebra of clopen sets of X, the clopen sets of X cannot have Cantor separability and hence the zero-sets of X are not all regular closed sets.

(d) Finally, $\mathbb{N}^* \times \mathbb{N}^*$ fails to satisfy only the F-space condition. $\mathbb{N}^* \times \mathbb{N}^*$ is clearly totally disconnected, compact, and without isolated points and it is easy to see that $\mathbb{N}^* \times \mathbb{N}^*$ has weight $\mathfrak{c}$. By Proposition 1.66, if the product of two F-spaces is an F-space, one of the spaces must be a P-space. Since a compact P-space must be finite, $\mathbb{N}^* \times \mathbb{N}^*$ fails to be an F-space. It remains to show either that the zero-sets of $\mathbb{N}^* \times \mathbb{N}^*$ are regular closed or, from the point of view of Theorem 3.31, that the clopen sets of $\mathbb{N}^* \times \mathbb{N}^*$ have Cantor separability. We will show the latter. Let $\{A_n\}$ be a strictly increasing sequence of clopen subsets of $\mathbb{N}^*$ which is dominated by a clopen subset B, i.e.

$$A_1 \subset \cdots \subset A_n \subset \cdots \subset B.$$

A basic open set of $\mathbb{N}^* \times \mathbb{N}^*$ is a rectangle $R = U \times V$ formed by the product of two basic clopen sets of $\mathbb{N}^*$. Since B is compact, B is the union of finitely many basic clopen rectangles,

$$B = \bigcup\{R_i : 1 \leqslant i \leqslant m\}$$

and we can assume that the R_i are mutually disjoint. Then each member of the sequence $\{A_n\}$ can be written

$$A_n = \bigcup\{(A_n \cap R_i) : 1 \leqslant i \leqslant m\}.$$

Since a projection map is always open and the projection parallel to a compact factor is also closed [D, p. 227], for $j = 1,2$ and $i = 1,\ldots,m$, we have that

$$\pi_j[A_1 \cap R_i] \subseteq \cdots \subseteq \pi_j[A_n \cap R_i] \subseteq \cdots \subseteq \pi_j[R_i],$$

is a sequence of clopen sets of $\mathbb{N}^*$. The properties of the clopen sets of $\mathbb{N}^*$ can now be applied to each of these sequences. There are three cases. First, if infinitely many of the containments of the sequence are proper, then the Cantor separability of the clopen sets of $\mathbb{N}^*$ yields a clopen subset $C_{j,i}$ of $\mathbb{N}^*$ such that

$$\pi_j[A_n \cap R_i] \subset C_{j,i} \subset \pi_j[R_i]$$

for every $n = 1, 2, \ldots$. Second, if the sequence has only finitely many distinct elements but $\pi_j[A_n \cap R_i]$ is never equal to $\pi_j[R_i]$ for any n, then the density of the algebra of clopen sets provides a clopen set $C_{j,i}$ such that

$$\pi_j[A_n \cap R_i] \subset C_{j,i} \subset \pi_j[R_i]$$

for every $n = 1, 2, \ldots$. Finally, if $\pi_j[A_n \cap R_i] = \pi_j[R_i]$ for some n, then put $C_{j,i} = \pi_j[R_i]$.

The hypothesis on the original sequence in $\mathbb{N}^* \times \mathbb{N}^*$ make it impossible that the final case can occur for every choice of j and i. Thus, we have that

$$A_1 \subset \cdots \subset A_n \subset \cdots \subset \bigcup_{1 \leq i \leq m} (C_{1,i} \times C_{2,i}) \subset \bigcup_{1 \leq i \leq m} R_i = B$$

and the clopen sets of $\mathbb{N}^* \times \mathbb{N}^*$ have Cantor separability.

(e) Note that (a), (c), and (d) show that no two of the conditions dense in itself, Cantor separable, and DuBois-Reymond separable imply the third. Also note that (d) shows that the converse to Proposition 2.21 is false. No sequence in $\mathbb{N}^* \times \mathbb{N}^*$ can converge, although $\mathbb{N}^* \times \mathbb{N}^*$ fails to be DuBois-Reymond separable.

3.34. The following proposition characterizes the class of spaces which have zero-dimensional Stone-Čech compactifications, and, together with Parovičenko's characterization of $\mathbb{N}^*$, it will allow us to recognize certain homeomorphs of $\mathbb{N}^*$. A space is called *strongly zero-dimensional* if for every pair of disjoint zero-sets of the space, there is a clopen set containing one zero-set and missing the other. Any discrete space is clearly strongly zero-dimensional.

Proposition. *A space X is strongly zero-dimensional if and only if βX is zero-dimensional (or equivalently, βX is totally disconnected).*

Proof. The equivalence of the two conditions on βX is Proposition 2.4. Let X be strongly zero-dimensional. If p and q are distinct points of βX, then we can choose disjoint zero-set neighborhoods Z_1 and Z_2 of p and q, respectively, so that $Z_1 \cap X$ and $Z_2 \cap X$ are disjoint zero-sets of X. Then there exists a clopen set U containing $Z_1 \cap X$ and missing $Z_2 \cap X$. Thus, $\mathrm{cl}_{\beta X} U$ is a clopen set containing p but not q so that βX is totally disconnected.

Conversely, assume that Z_1 and Z_2 are disjoint zero-sets of X and that βX is zero-dimensional. Then $\mathrm{cl}_{\beta X} Z_1$ and $\mathrm{cl}_{\beta X} Z_2$ are disjoint closed sets of βX and Proposition 2.13 shows that X is strongly zero-dimensional since disjoint closed sets in a zero-dimensional Lindelöf space are separated by a partition. ☐

3.35. Since any point has a base of cozero-set neighborhoods and every neighborhood contains a zero-set containing the point, it is easy to see that a strongly zero-dimensional space is zero-dimensional. Proposition 2.13 shows that the converse is true if the space is also Lindelöf. Thus, we have proven the

Proposition. *Zero dimensionality and strong zero-dimensionality are equivalent in Lindelöf spaces.*

3.36. In metric spaces, Lindelöf and second countability are equivalent [D, p. 187]. Hence, every subspace of the line $\mathbb{R}$ is Lindelöf since second countability is an hereditary property. It is easy to see that a subspace of $\mathbb{R}$ has a base of clopen sets if it contains no intervals. Since no interval can be zero-dimensional, we have characterized the strongly zero-dimensional subspaces of $\mathbb{R}$:

Proposition. *A subspace of $\mathbb{R}$ is strongly zero-dimensional if and only if it contains no intervals.*

3.37. Parovičenko's characterization of $\mathbb{N}^*$ now enables us to show that a zero-set in the growth of a strongly zero-dimensional space which admits only continuously many real-valued mappings is a copy of $\mathbb{N}^*$. The next result appears in the 1968 paper of W. W. Comfort and S. Negrepontis.

Proposition [CH]. *If X is a strongly zero-dimensional space for which the cardinality of $C^*(X)$ is $\mathfrak{c}$, then any zero-set of βX which misses X is homeomorphic with $\mathbb{N}^*$.*

Proof. Any such zero-set Z is clearly compact and it is immediate from Proposition 3.34 that it is also totally disconnected. Z has weight $\mathfrak{c}$ since the cozero-sets of the C^*-embedded space Z all arise from members of $C^*(X)$ and form a base for Z. Corollary 1.63 shows that Z is an F-space. If p is an isolated point of Z, then there exists a zero-set neighborhood Z_1 of p in βX such that $Z_1 \cap Z = \{p\}$. But then $\{p\}$ is a zero-set and Theorem 3.6 shows that a zero-set contained in X^* must contain $2^{\mathfrak{c}}$ points, which is a contradiction. Finally, consider the space $\beta X \backslash Z$. It is clear from Proposition 1.59 and Theorem 1.53 that $\beta X \backslash Z$ is locally compact and realcompact. In Lemma 4.21, to which the reader may refer directly, we will see that a zero-set in the growth of a locally compact and realcompact space is a regular closed set.

Hence, we have shown that Z satisfies all the conditions of Theorem 3.32, and Z is necessarily homeomorphic with $\mathbb{N}^*$. ☐

3.38. Examples. Since $\mathbb{Q}$ contains no intervals, $\mathbb{Q}$ *is strongly zero-dimensional.* Note that we have also verified and used this fact in Example 2.15 and that we have shown there that *the space Y of Example 2.14 is strongly zero-dimensional.* Therefore, any zero-set of $\beta\mathbb{Q}$ or βY which is contained in $\mathbb{Q}^*$ or Y^*, respectively, is a copy of $\mathbb{N}^*$.

3.39. Proposition 3.34 indicates that the conjecture that βX has a base of clopen sets whenever X does is false. The following example was introduced in 1955 by C.H. Dowker and shows that a zero-dimensional space need not be strongly zero-dimensional. The discussion of the example is adapted from the text of R. Engelking.

Example (Dowker). Consider the ordered space $\omega_1 + 1$ of all ordinals not greater than the first uncountable ordinal. Since for any $\alpha < \omega_1$ the set $\{\gamma : \gamma < \alpha\}$ is clopen in the order topology, $\omega_1 + 1$ is zero-dimensional. Dowker's example is *a zero-dimensional subspace M of $(\omega_1 + 1) \times I$ such that the growth M^* of M contains a copy of I, therefore showing that βM is not zero-dimensional.*

Begin the construction of M by defining two points of I to be equivalent if their difference is rational. Each equivalence class is dense and countable and so there are c equivalence classes. Choose a collection $\{P_\gamma : \gamma < \omega_1\}$ of distinct equivalence classes other than the rationals and put $S_\alpha = I \setminus \bigcup \{P_\gamma : \gamma \geqslant \alpha\}$. Then each S_α is zero-dimensional since each fails to contain an interval. Define the following subspaces of $(\omega_1 + 1) \times I$:

$$M_\alpha = \bigcup_{\gamma \leqslant \alpha} (\{\gamma\} \times S_\gamma), \qquad M = \bigcup_{\alpha < \omega_1} M_\alpha, \quad \text{and} \quad M^+ = M \cup (\{\omega_1\} \times I).$$

(a) M is zero-dimensional: Each M_α is a clopen subspace of M and is zero-dimensional since it is a subspace of the zero-dimensional space $\omega_1 \times S_\alpha$. Thus, M is zero-dimensional since it is the union of clopen zero-dimensional subspaces.

(b) M^+ is normal: First note that since each M_α is regular and second countable, the Urysohn Metrization Theorem [D, p. 195] shows that M_α is metrizable and hence, normal. Now let F_1 and F_2 be disjoint closed subsets of M^+. Then $F_1 \cap (\{\omega_1\} \times I)$ and $F_2 \cap (\{\omega_1\} \times I)$ are disjoint compact sets and therefore are contained in disjoint open sets U_1 and U_2 of M^+, respectively. Since F_1 and F_2 are closed and neighborhoods of ω_1 in $\omega_1 + 1$ are "tails", there is some $\alpha < \omega_1$ such that $F_i \cap (M^+ \setminus M_\alpha) \subset U_i$ for each i. It now follows from the fact that M_α is normal and open in M^+ that the sets F_1 and F_2 are contained in disjoint open sets of M^+.

(c) Disjoint closed sets of M have disjoint closures in M^+: Let F_1 and F_2 be disjoint closed sets in M. If a point (ω_1, r) belongs to the closures in

M^+ of both sets, then there exist two sequences $\{(\alpha_i, x_i)\}$ and $\{(\beta_i, y_i)\}$ contained in F_1 and F_2, respectively, such that $\{x_i\}$ and $\{y_i\}$ each converge to r, $\sigma < \alpha_i < \beta_i < \alpha_{i+1}$ for each i, and r is in S_σ. But then $\sup\{\alpha_i\} = \sup\{\beta_i\} = \gamma < \omega_1$ and (γ, r) belongs to $F_1 \cap F_2$, which is impossible.

As an immediate consequence of (b) and (c), we have that M is normal. Further,

(d) M is C^*-embedded in M^+: By Theorem 1.2, we need only show that completely separated sets of M are also completely separated in M^+. But completely separated sets in M have disjoint closures in M and also in M^+ by (c). But then their closures in M^+ are completely separated since M^+ is normal.

Finally, since M is dense and C^*-embedded in M^+, we see that $\beta M = \beta(M^+)$. Then $\{\omega_1\} \times I$ is a copy of I contained in βM so that βM cannot be zero-dimensional.

Types of Ultrafilters and the Non-Homogeneity of $\mathbb{N}^*$

3.40. If a space X is homogeneous, does it necessarily follow that the growth $X^* = \beta X \setminus X$ is also homogeneous? This question was raised during a seminar at the University of Wisconsin in 1955 and in 1956 W. Rudin showed that the space $\mathbb{N}^*$ provides a negative answer in the presence of the Continuum Hypothesis. Rudin's solution sparked a sequence of results in the area of non-homogeneity of growths which will be discussed in the next chapter.

In the present chapter, we consider Z. Frolík's proof from 1967 B that $\mathbb{N}^*$ is not homogeneous. Frolík's method does not require the Continuum Hypothesis and can be modified to show that X^* fails to be homogeneous whenever X is not pseudocompact. This modification will be treated in the next chapter.

3.41. Every permutation σ of $\mathbb{N}$ extends to a homeomorphism

$$\beta(\sigma): \beta\mathbb{N} \to \beta\mathbb{N}$$

and the restriction of $\beta(\sigma)$ to $\mathbb{N}^*$, denoted by σ^*, is an automorphism of $\mathbb{N}^*$. For a pair of points p and q of $\mathbb{N}^*$, define $p \sim q$ if $\sigma^*(p) = q$ for some permutation σ. Clearly, $\sim$ is an equivalence relation. Let T be the set of equivalence classes and let

$$\tau: \mathbb{N}^* \to T$$

be the function which assigns to each free ultrafilter its equivalence class.

The elements of T are called *types of ultrafilters*. If $t = \tau(p)$, t is called the type of p and p is said to be of type t. The following result was attributed to H. Kenyon by W. Rudin in his 1956 paper.

Theorem. *There are 2^c types of ultrafilters in $\mathbb{N}^*$ and there is a dense set of c ultrafilters of each type.*

Proof. Corollary 3.20 established that the images of any point in $\mathbb{N}^*$ under automorphisms of the form σ^* comprise a dense subspace of $\mathbb{N}^*$. Since $\mathbb{N}^*$ contains c disjoint open sets by Theorem 3.22, there must be at least c ultrafilters of each type. On the other hand, there are only c permutations of $\mathbb{N}$, so there must be exactly c. Since there are 2^c ultrafilters in $\mathbb{N}^*$, it is clear that $\mathbb{N}^*$ must contain 2^c types of ultrafilters. $\square$

3.42. We saw earlier that any countably infinite discrete subspace of $\beta\mathbb{N}$ is C^*-embedded in $\beta\mathbb{N}$. Therefore, if X is such a subspace, then $\mathrm{cl}_{\beta\mathbb{N}} X \approx \beta\mathbb{N}$ and if we put $X^* = \mathrm{cl}_{\beta\mathbb{N}} X \backslash X$, then $X^* \approx \mathbb{N}^*$. A point p in X^* must then have a type as a point of $\mathrm{cl}\, X$ as well as a type as a point of $\beta\mathbb{N}$. Let h be a homeomorphism of $\mathrm{cl}\, X$ onto $\beta\mathbb{N}$ and define the *type of p relative to X* to be $\tau(h(p))$ and denote this relative type by $\tau(p, X)$. Observe that this definition is independent of the homeomorphism chosen since for any other homeomorphism g of $\mathrm{cl}\, X$ onto $\beta\mathbb{N}$, $g \circ h^{\leftarrow}$ sends $h(p)$ to $g(p)$ and is an automorphism of $\beta\mathbb{N}$ so that $h(p)$ and $g(p)$ are of the same type. The following lemma describes the most important properties of relative types and is based on the treatment of relative types given by A. K. and E. F. Steiner in their 1971 paper.

Lemma. *Let X and Y be countable discrete subspaces of $\beta\mathbb{N}$. Then:*
  (a) *If Y is contained in X and p belongs to $X^* \cap Y^*$, then $\tau(p, X) = \tau(p, Y)$.*
  (b) *If p and q belong to X^* and Y^*, respectively, then $\tau(p, X) = \tau(q, Y)$ if and only if there is a homeomorphism h of $\mathrm{cl}\, X$ onto $\mathrm{cl}\, Y$ such that $h(p) = q$.*
  (c) *If h is an automorphism of $\beta\mathbb{N}$ or $\mathbb{N}^*$ and p belongs to X^*, then $\tau(p, X) = \tau(h(p), h[X])$.*

Proof. (a) Let g be a homeomorphism of $\mathrm{cl}\, Y$ onto $\mathrm{cl}\, X$ such that $g(p) = p$ and let h be a homeomorphism of $\mathrm{cl}\, X$ onto $\beta\mathbb{N}$. Then

$$\tau(p, Y) = \tau(h \circ g(p)) = \tau(h(p)) = \tau(p, X).$$

(b) Let $\tau(p, X) = \tau(q, Y)$. Then there exist homeomorphisms f and g of $\mathrm{cl}\, X$ and $\mathrm{cl}\, Y$, respectively, onto $\beta\mathbb{N}$ such that $\tau(f(p)) = \tau(g(q))$. Hence, there exists an automorphism k of $\beta\mathbb{N}$ which sends $f(p)$ to $g(q)$, and $h = g^{\leftarrow} \circ k \circ f$ is the required homeomorphism. The converse is clear from the definition of relative type and (c) follows from (b). $\square$

Note that if p is a point of $\mathbb{N}^*$ and Z belongs to the ultrafilter A^p, then p is in Z^* and it follows from (a) of the lemma that $\tau(p) = \tau(p, Z)$. In the remainder of the chapter, it will be helpful to identify a point p of $\mathbb{N}^*$ with the unique ultrafilter A^p on $\mathbb{N}$ which converges to p. Thus, *we will think of p both as a point of $\mathbb{N}^*$ and as a collection of subsets of $\mathbb{N}$,* i. e. as the ultrafilter A^p.

Part (b) of the lemma can be restated in terms of bijections of sets instead of homeomorphisms. Observe that in light of the properties of the Stone-Čech extension of a mapping, the existence of a homeomorphism of cl X onto cl Y which sends p to q is equivalent to the existence of a bijection of X onto Y which sends the neighborhood traces of p on X onto neighborhood traces of q on Y. This is occasionally a more convenient way in which to approach the property of relative types described in (b).

3.43. In order to become more familiar with the notion of the type of an ultrafilter and with the meaning of relative types, consider the following example of a mapping of $\beta\mathbb{N}$ into $\mathbb{N}^*$. The example will depend on the observation that if t is any type and $X = \{x_n\}$ is a discrete sequence contained in $\beta\mathbb{N}$, then there exists p in X^* such that $\tau(p, X) = t$ since cl X is a copy of $\beta\mathbb{N}$.

Example. *There exists a mapping f of $\beta\mathbb{N}$ into $\mathbb{N}^*$ such that $f[\mathbb{N}]$ is discrete and $f^2[\mathbb{N}]$ is a singleton which does not belong to $f[\mathbb{N}]$.* Choose a type t and a decomposition $\{N_n\}$ of $\mathbb{N}$ with each N_n an infinite set. Choose an x_n in N_n^* such that each x_n is of type t and choose a point p in $\mathbb{N}^*$ whose type relative to $X = \{x_n\}$ is t. Then there exists a bijection of each N_n onto X which sends the neighborhood traces of x_n onto the neighborhood traces of p on X. If f is the Stone-Čech extension of the mapping of $\mathbb{N}$ defined in this way, then f sends each x_n to p, and sends $\mathbb{N}$ to X. Hence, $f^2[\mathbb{N}] = \{p\}$.

3.44. One proof that $\beta\mathbb{N}$ is not homogeneous (Section 3.18) was based on cardinality considerations and consisted of showing that a point of $\mathbb{N}^*$ cannot be mapped to every other point of $\mathbb{N}^*$ under automorphisms of $\beta\mathbb{N}$ simply because there are too few such automorphisms. However, this does not eliminate the possibility that there might be enough automorphisms of $\mathbb{N}^*$ to map every point to every other point. To show that this is not the case, we will use Lemma 3.42(c) which shows that the set of relative types of a point in $\mathbb{N}^*$ is an invariant which is preserved under automorphisms of $\mathbb{N}^*$. In order to further describe this invariant, we introduce the following notation and terminology. For S a subset of $\beta\mathbb{N}$ and p a point of $\mathbb{N}^*$, define $\tau[p, S] = \{\tau(p, X) : X \subset S, X \approx \mathbb{N}\}$. Thus, $\tau[p, S]$ is the set of relative types of p which occur relative to countable discrete subspaces of S. Our invariant then becomes $\tau[p,\mathbb{N}^*]$, i.e. the set of relative types in $\mathbb{N}^*$. Also, using this notation, it follows from Lemma 3.42(a) that $\tau[p,\mathbb{N}]=\{\tau(p)\}$.

If p is a point of $\mathbb{N}^*$, we will say that a type t *produces* $\tau(p)$ or that $\tau(p)$ *is produced* by t if $\tau(p,X)=t$ for any countable discrete subspace X of $\mathbb{N}^*$. This producing relation on the set of types is summarized in the following

Proposition. *The set $\tau[p,\mathbb{N}^*]$ of relative types of a point p of $\mathbb{N}^*$ is the set of types which produce the type of the ultrafilter p. A type s produces a type t if there exists a countable discrete subspace X of $\mathbb{N}^*$ and a point p of type t in X^* such that $\tau(p,X)=s$.*

3.45. The central result in Frolík's theory of types is the calculation of the cardinalities of the sets of types which produce or are produced by a given type. The proof is accomplished through an examination of the possible number of relative types of a point p in $\mathbb{N}^*$.

Theorem (Frolík). *Any type is produced by at most $\mathfrak{c}$ types and any type produces $2^{\mathfrak{c}}$ types.*

Proof. The proof of the second statement is accomplished by showing that $\mathbb{N}^*$ contains $2^{\mathfrak{c}}$ disjoint copies of $\beta\mathbb{N}$.

(a) Let $\{M_n\}$ be any countable partition of $\mathbb{N}$ into infinite sets and let x_n and y_n be in M_n^* for each n. If $x_n \neq y_n$ for each n, then $\mathrm{cl}_{\beta\mathbb{N}}\{x_n\} \cap \mathrm{cl}_{\beta\mathbb{N}}\{y_n\}=\emptyset$: To show this, write $M_n=R_n \cup S_n$, a disjoint union with R_n in x_n and S_n in y_n for each n. For any p in $\mathbb{N}^*$ and Z in p, write $Z=(Z\cap(\bigcup R_n))\cup(Z\cap(\bigcup S_n))$. Exactly one of these sets is in p. Suppose that p belongs to $\mathrm{cl}_{\beta\mathbb{N}}\{x_n\}$. Then we have

$$p \in \mathrm{cl}_{\beta\mathbb{N}}\{x_n\} \subset \mathrm{cl}_{\beta\mathbb{N}}(\bigcup R_n).$$

Hence, $\bigcup R_n$ belongs to p and therefore $\bigcup S_n$ is not in p so that p cannot belong to $\mathrm{cl}_{\beta\mathbb{N}}\{y_n\}$.

(b) Since the cardinality of M_n^* is $2^{\mathfrak{c}}$ for each n, (a) shows that $\mathbb{N}^*$ contains $2^{\mathfrak{c}}$ disjoint sets of the form $\mathrm{cl}_{\beta\mathbb{N}}\{x_n\}$. Each is homeomorphic to $\beta\mathbb{N}$ and so for any type t, each contains a point p such that $\tau(p,\{x_n\})=t$. Then the type $\tau(p)$ is produced by t from the preceding proposition. Since only $\mathfrak{c}$ of these $2^{\mathfrak{c}}$ points can be of any given type, t must produce $2^{\mathfrak{c}}$ types.

The first statement of the theorem will follow after completing two more partial steps.

(c) For any point p in $\mathbb{N}^*$, there is a set $\mathscr{X}$ of at most $\mathfrak{c}$ discrete, countable subsets of $\mathbb{N}^*$ such that

  (i) Each member of $\mathscr{X}$ contains p in its closure, and

(ii) If Y is any discrete countable set with p in Y^*, then Y contains
    some member of $\mathscr{X}$:

For each partition $\{M_n\}$ of $\mathbb{N}$, choose y_n in M_n^* to form Y with p in Y^*,
if possible. The same partition may yield many such Y's. Now for each
Y that can be formed in this way when all partitions are considered, take
all sets X contained in Y such that p belongs to $\mathrm{cl}_{\beta\mathbb{N}} X \setminus Y$. By (a), any two
choices of Y from the same partition meet in an infinite set containing
p in its closure, so we need consider only one choice of Y from each
partition. Since each Y contains only $\mathfrak{c}$ infinite subsets and since there
are only $\mathfrak{c}$ partitions, we have $|\mathscr{X}| \leqslant \mathfrak{c} \cdot \mathfrak{c} = \mathfrak{c}$.

By Proposition 3.44, a type t is produced by a type r exactly when
there is an ultrafilter p of type t and a discrete sequence X in $\mathbb{N}^*$ with
p in X^* and the type of p relative to X is r. We now see that any point can
have at most $\mathfrak{c}$ distinct relative types.

(d) If X is contained in a discrete sequence Y and p belongs to $\mathrm{cl}_{\beta\mathbb{N}} X \setminus Y$,
then $\tau(p,X) = \tau(p,Y)$. This follows immediately from Lemma 3.42(a).

Thus, (d) shows that in computing the relative types of p, we need ·
only consider the members of $\mathscr{X}$ and $\mathscr{X}$ contains at most $\mathfrak{c}$ sets. Since
there are only $\mathfrak{c}$ ultrafilters of each type (Theorem 3.41), the set of ultrafilters of a given type can have a total of at most $\mathfrak{c}$ relative types. Hence,
each type can be produced by at most $\mathfrak{c}$ types. $\square$

The first estimate in Frolík's Theorem cannot be improved. In 1971,
the Steiners introduced a method for constructing types and showed the
existence of a point in $\mathbb{N}^*$ having $\mathfrak{c}$ relative types. Their description of
such a point will be given later in Example 7.9.

3.46. We can now show that $\mathbb{N}^*$ is not homogeneous as an application
of the producing relation.

Corollary (Frolík). $\mathbb{N}^*$ *is not homogeneous.*

Proof. Let h be an automorphism of $\mathbb{N}^*$ and p and q be points of $\mathbb{N}^*$
such that $h(p) = q$. If X is any copy of $\mathbb{N}$ in $\mathbb{N}^*$ having p as a limit point,
then we have $\tau[p,X] = \tau[q,h[X]]$ from Lemma 3.42(c). Thus, the two
sets $\tau[p,\mathbb{N}^*]$ and $\tau[q,\mathbb{N}^*]$ of relative types are identical. The family
$\{\tau[p,\mathbb{N}^*] : p \in \mathbb{N}^*\}$ of all such sets of relative types covers the set T of
types. However, the cardinality of T is $2^{\mathfrak{c}}$ and that of each element of this
cover is at most $\mathfrak{c}$ since each type is produced by at most $\mathfrak{c}$ types. Thus,
there must exist points r and s of $\mathbb{N}^*$ such that $\tau[r,\mathbb{N}^*] \neq \tau[s,\mathbb{N}^*]$.
But then no automorphism of $\mathbb{N}^*$ can map r to s and $\mathbb{N}^*$ cannot be
homogeneous. $\square$

3.47. The orbits of two points p and q of $\mathbb{N}^*$ under automorphisms of
$\mathbb{N}^*$ are disjoint exactly when no automorphism carries p to q. Thus,

the set of all such orbits decomposes $\mathbb{N}^*$ into a union of disjoint sets. Since any two points belonging to the same orbit have the same set of at most $\mathfrak{c}$ relative types, there must be $2^{\mathfrak{c}}$ distinct orbits. We have verified the

Corollary. *For any point p of $\mathbb{N}^*$, there are $2^{\mathfrak{c}}$ points of $\mathbb{N}^*$ which cannot be mapped to p by automorphisms of $\mathbb{N}^*$.*

Exercises

3 A. *Cardinality of Stone-Čech compactifications*

1. If X is separable, $|\beta X| \leqslant 2^{\mathfrak{c}}$.
2. If X contains a closed C^*-embedded copy of $\mathbb{N}$, then $|\beta X| \geqslant 2^{\mathfrak{c}}$.
3. The cardinalities of $\beta\mathbb{R}$, $\beta\mathbb{Q}$, and $\beta\mathbb{P}$ are all $2^{\mathfrak{c}}$.

3 B. *βX and sequential compactness*

A space is said to be *sequentially compact* if every sequence in the space has a convergent subsequence.
1. $\beta\mathbb{N}$ and $\mathbb{N}^*$ are not sequentially compact. [Theorem 3.3.]
2. A closed subspace of a sequentially compact space is sequentially compact.
3. If $\{x_n\}$ is a sequence in a space X, then either every subsequence of $\{x_n\}$ has a cluster point or X contains a closed copy of $\mathbb{N}$.
4. If X is normal and βX is sequentially compact, then X is sequentially compact. [Proposition 1.48 and (3).]
References: One place in which (4) has been used is in the thesis of S. Purisch in which he characterizes the class of suborderable pseudo-compact spaces as the class of spaces having orderable Stone-Čech compactifications. In the context of the orderability of βX, see also the paper of Venkataraman, Rajagopalan, and Soundararajan.

3 C. *Discrete subspaces of $\beta\mathbb{N}$*

If X is a countable discrete subspace of $\beta\mathbb{N}$ and Y is a countable discrete subspace of X^*, then Y is discrete as a subspace of $\beta\mathbb{N}$.

3 D. *$\mathbb{N}^*$ is not basically disconnected*

1. If $\{A_n\}$ is a strictly increasing sequence in $CO(\mathbb{N}^*)$, then $\mathrm{cl}(\bigcup A_n)$ is not open. [Exercise 2 E or Corollary 3.29.]
2. $\mathbb{N}^*$ is not basically disconnected, hence an F-space need not be basically disconnected. [Exercise 2 K.]
Reference: This is a variant of [GJ, Ex. 6 W].

3E. *More on M*

1. A closed pseudocompact subspace of a realcompact space is compact. [Corollary 1.58.]
2. *M* of Example 3.39 contains closed copies of ω_1 and therefore is not realcompact.
3. Each of the subspaces M_α of *M* is Lindelöf. However, $M = \bigcup M_\alpha$ is not Lindelof. [(2) or Proposition 3.35.]

References: In addition to the discussion in Engelking's book, *M* is discussed in [GJ, Ex. 16MN] and in the thesis of P. Nyikos.

3F. *Sums of ultrafilters*

Let X be a countable discrete subspace of $\mathbb{N}^*$ and let $\mathcal{U}$ be an ultrafilter on X. Define the *sum of X with respect to $\mathcal{U}$*, denoted by $\Sigma_\mathcal{U} X$, to be the family of subsets of $\mathbb{N}$ having the following form: $\bigcup\{M_x : x \in P_\alpha\}$ where P_α is in $\mathcal{U}$ and M_x is in x for each x in P_α.

1. $\Sigma_\mathcal{U} X$ is a filter on $\mathbb{N}$.
2. If Z is a subset of $\mathbb{N}$ which meets every member of $\Sigma_\mathcal{U} X$, then every element of $\mathcal{U}$ contains an element x such that Z is in x.
3. If Z meets every member of $\Sigma_\mathcal{U} X$ and $S = \{x \in X : Z \in x\}$, then S belongs to $\mathcal{U}$. [Proposition 2.6.]
4. $\Sigma_\mathcal{U} X$ is an ultrafilter on $\mathbb{N}$. [Proposition 1.18.]
5. If p is in X^* and $\mathcal{U}$ is the trace of the neighborhoods of p on X, then $p = \Sigma_\mathcal{U} X$ and $\tau(p, X)$ produces $\tau(p)$.
6. If $\mathcal{U}$ is any free ultrafilter on X and $p = \Sigma_\mathcal{U} X$, then p belongs to X^* and $\mathcal{U}$ is the trace of the neighborhoods of p on X.

Reference: Frolík, 1967B.

3G. *Sums of types*

Let $X = \{x_n\}$ and $Y = \{y_n\}$ be countable discrete subspaces of $\mathbb{N}^*$.

1. If $\{t_n\}$ is any sequence of types of ultrafilters, then there exists X such that $\tau(x_n) = t_n$. [Theorems 3.22 and 3.41.]
2. There exist pairwise disjoint families $\{R_n\}$ and $\{S_n\}$ of subsets of $\mathbb{N}$ such that R_n belongs to x_n and S_n belongs to y_n for each n.
3. If $\tau(x_n) = \tau(y_n)$ for each n, then there exists a permutation σ of $\mathbb{N}$ such that $\sigma^*(x_n) = \sigma^*(y_n)$ for all n. [(2) and Lemma 3.42.]

Let $h : \beta\mathbb{N} \to \mathrm{cl}_{\beta\mathbb{N}} X$ be the homeomorphism obtained by extending the assignment $n \mapsto x_n$ to all of $\beta\mathbb{N}$. For p in $\mathbb{N}^*$, define $\Sigma_p X$ to be $\Sigma_\mathcal{U} X$ where $\mathcal{U}$ is the ultrafilter of neighborhood traces of $h(p)$ on X. For a sequence $\{t_n\}$ of types, define $\Sigma_p\{t_n\}$ to be $\tau(\Sigma_p X)$ where X is such that $\tau(x_n) = t_n$ for each n.

4. Then $\Sigma_p\{t_n\}$ is well-defined, i.e. if $\tau(x_n) = \tau(y_n)$ for each n and p is in $\mathbb{N}^*$, then $\tau(\Sigma_p X) = \tau(\Sigma_p Y)$. [(3)]

Reference: Frolík, 1967B.

Chapter 4. Non-Homogeneity of Growths

4.1. This chapter will be the first of three to be devoted to the relationships between a space X and its growth $X^* = \beta X \backslash X$. In order to be certain that in considering growths of Stone-Čech compactifications we will not be discussing a restricted class of spaces, we will first follow the outline of [GJ, ex. 9K] to show that *any* completely regular space can be expressed in the form $\beta S \backslash S$ for a suitable choice of S.

In order to derive this result, we will need to consider spaces of ordinal numbers. The basic material on ordinal numbers can be found in [D]. Recall that every ordinal number is the set of all its predecessors. If α is any ordinal number, the symbol α will also be used to denote the topological space obtained by imposing the interval topology on α. If α is a non-limit ordinal, i.e. if α has an immediate predecessor, the space α is compact [GJ, 5.11]. With this notation, the celebrated counterexample known as the Deleted Tychonoff Plank is written

$$\mathbf{T} = (\omega_1 + 1) \times (\omega_0 + 1) \backslash \{(\omega_1, \omega_0)\}$$

where ω_0 is the first infinite ordinal and ω_1 is the first uncountable ordinal. The space $\mathbf{T}$ was introduced by Tychonoff in 1930 as an example of a completely regular space which is not normal. In his 1949 paper, H. Tong established that the Stone-Čech compactification of T is obtained by adding the point (ω_1, ω_0) so that

$$\beta\mathbf{T} = (\omega_1 + 1) \times (\omega_0 + 1).$$

This result is also discussed in [GJ, 8.20] and will be obtained below as Proposition 4.4.

The critical properties of ω_1 used in the proof of this fact are that no countable subset of ω_1 is cofinal and that every real-valued continuous function on ω_1 is constant on a tail of ω_1. By similar properties of larger ordinals, we will show that if Y is a compact space, an ordinal ω_α can be chosen so that

$$\beta(\omega_\alpha \times Y) = (\omega_\alpha + 1) \times Y.$$

This result can then be used to show that any space can be expressed as a growth of some Stone-Čech compactification.

4.2. If Y is a compact space, let $\aleph_\alpha$ be a cardinal number greater than the cardinality of Y and such that α is a non-zero non-limit ordinal. Using this notation, we have the following

Proposition.

$$\beta(\omega_\alpha \times Y) = (\omega_\alpha + 1) \times Y.$$

Proof. It is clear that $(\omega_\alpha + 1) \times Y$ is compact and contains $\omega_\alpha \times Y$ densely. It remains to show that $\omega_\alpha \times Y$ is C^*-embedded in $(\omega_\alpha + 1) \times Y$. Let f be a bounded, real-valued mapping on $\omega_\alpha \times Y$. Then for each point y in Y, the restriction $f|\omega_\alpha \times \{y\}$ is constantly equal to some value r_y on a tail $[\sigma_y, \omega_\alpha) \times \{y\}$ of the "row" $\omega_\alpha \times \{y\}$. Because the cardinality of Y is less than $\aleph_\alpha$, $\sigma = \sup\{\sigma_y : y \in Y\}$ is less than ω_α since α is a non-limit ordinal. Since $\sigma \geqslant \sigma_y$, each restriction $f|[\sigma, \omega_\alpha) \times \{y\}$ is constant. Define $\beta(f)$ to be f on $\omega_\alpha \times Y$ and to be r_y at (ω_α, y). Then $\beta(f)$ is continuous and extends f. $\square$

4.3. We can now show that every space is the growth of some Stone-Čech compactification.

Proposition. *For any completely regular space X there exists a space S such that X is homeomorphic to $\beta S \backslash S$.*

Proof. Let Y be a compactification of X and choose ω_α as above. Now define S to be the following subspace of $(\omega_\alpha + 1) \times Y$:

$$S = (\omega_\alpha + 1) \times (Y \backslash \{(\omega_\alpha, x) : x \in X\}).$$

Since S contains $\omega_\alpha \times Y$, the previous result and Proposition 1.49 imply that $\beta S = (\omega_\alpha + 1) \times Y$. The result now follows since $S^* = \{(\omega_\alpha, x) : x \in X\}$ is a copy of X. $\square$

4.4. By a technique similar to that just used we will now show that $(\omega_1 + 1) \times (\omega_0 + 1)$ is the Stone-Čech compactification of the Deleted Tychonoff Plank.

Proposition. $\beta((\omega_1 + 1) \times (\omega_0 + 1) \backslash \{(\omega_1, \omega_0)\}) = (\omega_1 + 1) \times (\omega_0 + 1)$.

Proof. Let $Y = \omega_0 + 1$. Then we have seen that

$$\beta(\omega_1 \times (\omega_0 + 1)) = (\omega_1 + 1) \times (\omega_0 + 1).$$

Since $\mathbf{T}=(\omega_1+1)\times(\omega_0+1)\setminus\{(\omega_1,\omega_0)\}$ contains $\omega_1\times(\omega_0+1)$, the result follows. $\square$

4.5. In Corollary 3.46 we saw that the growth $\mathbb{N}^*$ is not homogeneous. In this chapter we will build upon our previous results for $\mathbb{N}^*$ to show in two distinct ways that X^* will fail to be homogeneous whenever X fails to be pseudocompact, i.e. whenever X admits an unbounded, real-valued mapping. The class of pseudocompact spaces was first investigated by E. Hewitt in 1948 and the next two results appear in [GJ, 1.20, 1.21].

Lemma. *Any subspace of a space X on which a real-valued mapping of X is unbounded contains a C-embedded copy of $\mathbb{N}$ which is closed in X.*

Proof. Observe that the main portion of the proof is given in part (a) of the proof of Theorem 3.6. It remains only to show that the C-embedded copy of $\mathbb{N}$ is closed. For any cluster point of a C-embedded copy N of $\mathbb{N}$, there is a real-valued mapping on N which is unbounded on every neighborhood of the cluster point. Since such a mapping can have no real-valued extension to the cluster point, N must be closed. $\square$

4.6. Now the corollary characterizing the class of pseudocompact spaces is immediate.

Corollary (Hewitt). *The pseudocompact spaces are those spaces which contain no C-embedded copy of $\mathbb{N}$.*

Taking the closure in βX of a C-embedded copy of $\mathbb{N}$ contained in a space X produces a copy of $\mathbb{N}^*$ contained in X^*. The existence of such a copy of $\mathbb{N}^*$ is the key to showing that the growth of a non-pseudocompact space fails to be homogeneous. We will frequently identify a C-embedded copy of $\mathbb{N}$ with $\mathbb{N}$ and refer to the points of the copy by n.

We will first consider Z. Frolík's adaptation of his argument for $\mathbb{N}^*$ to show that the growth of any non-pseudocompact space fails to be homogeneous. While Frolík's method will demonstrate the non-homogeneity of such growths, no specific kinds of points will be exhibited through this approach. In the remainder of the chapter, we will consider the classification of several kinds of points which can be identified in growths of Stone-Čech compactifications. In particular, we will investigate the existence of P-points and non-P-points in growths.

Types of Points in X^*

4.7. We first establish a notion of types for an arbitrary space Y. For a point y in Y, let $\mathcal{M}_y$ denote the set of all C^*-embedded copies of $\mathbb{N}$ in

Y with y in $\mathrm{cl}_Y \mathbb{N} \backslash \mathbb{N}$. For such a copy of $\mathbb{N}$, $\mathrm{cl}_{\beta Y} \mathbb{N}$ is homeomorphic to $\beta \mathbb{N}$ so that $\mathrm{cl}_Y \mathbb{N}$ is a subspace of $\beta \mathbb{N}$. Thus, if y is in $\mathrm{cl}_Y \mathbb{N} \backslash \mathbb{N}$, the trace of the neighborhoods of y on $\mathbb{N}$ forms an ultrafilter on $\mathbb{N}$. Hence, we can speak of the type of y with respect to $\mathbb{N}$ as discussed in Chapter 3. If S is a subspace of Y, types of a point y in Y with respect to sets of $\mathscr{M}_y$ whose closures are contained in S are called *types in S*. In the special case that Y is βX and S is X^*, types in S are called *ideal types*.

4.8. The heart of Frolík's argument lies in showing that if X is a non-pseudocompact space, then certain points of X^* have "too few" ideal types to allow X^* to be homogeneous. The following preliminary results are needed to relate countable sets of X^* to C-embedded copies of $\mathbb{N}$ in X.

Lemma. *Let $\mathbb{N}$ be C-embedded in X and let Y be a countable subspace of βX. Then if neither subspace meets the βX-closure of the other, the two subspaces have disjoint closures in βX.*

Proof. As in the proof of Proposition 1.64, there exist disjoint cozero-sets U and V of βX which contain $\mathbb{N}$ and Y, respectively. Thus, $X \backslash U$ is a zero-set of X missing $\mathbb{N}$. Since $\mathbb{N}$ is C-embedded in X, Theorem 1.3 implies that $\mathbb{N}$ and $X \backslash U$ are completely separated. Then

$$\mathrm{cl}_{\beta X}(X \backslash U) \cap \mathrm{cl}_{\beta X} \mathbb{N} = \emptyset \,.$$

Since $\mathrm{cl}_{\beta X} Y$ is contained in $\mathrm{cl}_{\beta X}(X \backslash U)$, $\mathrm{cl}_{\beta X} Y$ must miss $\mathrm{cl}_{\beta X} \mathbb{N}$. $\square$

4.9. Corollary. *Let $\mathbb{N}$ be C-embedded in X and let Y be a countable subspace of βX whose βX-closure is contained in X^*. Then*

$$\mathrm{cl}_{\beta X} \mathbb{N} \cap \mathrm{cl}_{\beta X} Y \subset \mathrm{cl}_{\beta X}(Y \cap \mathrm{cl}_{\beta X} \mathbb{N}) \,.$$

Proof. Suppose that p is in $\mathrm{cl}_{\beta X} \mathbb{N} \cap \mathrm{cl}_{\beta X} Y$ but not in $\mathrm{cl}_{\beta X}(Y \cap \mathrm{cl}_{\beta X} \mathbb{N})$. Then there is a neighborhood U of p such that $Y \cap U \cap \mathrm{cl}_{\beta X} \mathbb{N} = \emptyset$. Since $\mathrm{cl}_{\beta X}(Y \cap U) \cap \mathbb{N} = \emptyset$, the lemma gives $\mathrm{cl}_{\beta X}(Y \cap U) \cap \mathrm{cl}_{\beta X} \mathbb{N} = \emptyset$. But this is a contradiction since p is in both these closures. Hence, p belongs to $\mathrm{cl}_{\beta X}(Y \cap \mathrm{cl}_{\beta X} \mathbb{N})$. $\square$

4.10. The following special case of the preceding corollary shows that if $\mathbb{N}$ is C-embedded in X, then every ideal type of a point in $\mathbb{N}^*$ is also a type in $\mathbb{N}^*$.

Corollary. *Let $\mathbb{N}$ be C-embedded in X and let Y be a countable discrete subspace of βX whose βX-closure is contained in X^*. Then a point in $Y^* \cap \mathbb{N}^*$ is in Z^* for some Z contained in $Y \cap \mathrm{cl}_{\beta X} \mathbb{N}$.*

4.11. An upper bound for the cardinality of the ideal types of a point of X^* belonging to a copy of $\mathbb{N}^*$ can now be obtained by applying Theorem 3.45.

Proposition. *If* $\mathbb{N}$ *is C-embedded in* X, *the cardinal of the ideal types of a point in* $\mathbb{N}^*$ *is at most* $\mathfrak{c}$.

Proof. Let t be the ideal type of a point p of $\mathbb{N}^*$ with respect to some Y contained in X^*. By the preceding corollary, there exists $Z \subset Y \cap \mathrm{cl}_{\beta X}\mathbb{N}$ such that p is in $\mathrm{cl}_{\beta X}Z$. From Lemma 3.42 involving $\beta\mathbb{N}$, it is clear that the types of p with respect to Y and Z coincide. Thus, since Z is contained in $\mathbb{N}^*$, every ideal type of p is an ideal type of p considered as a point of $\mathbb{N}^* = \beta\mathbb{N}\backslash\mathbb{N}$. But now the result follows from the fact that any type of ultrafilter on $\mathbb{N}$ is produced by at most $\mathfrak{c}$ types. $\square$

4.12. The upper bound on the cardinality of ideal types can now be used to show that a growth can only be homogeneous when the space is pseudocompact. The proof is very similar to the proof that $\mathbb{N}^*$ is not homogeneous in Corollary 3.46.

Theorem (Frolík). *The growth of a non-pseudocompact space fails to be homogeneous.*

Proof. Let X be a non-pseudocompact space. For each point p in X^*, let T_p be the set of all ideal types of p. It is clear that if h is any automorphism of X^* such that $h(p) = q$, then $T_p = T_q$. Now let $\mathbb{N}$ be C-embedded in X and choose p in $\mathbb{N}^*$. It is now sufficient to show that there exists q in $\mathbb{N}^*$ such that $T_p \neq T_q$. Since there are $2^{\mathfrak{c}}$ types of points in $\mathbb{N}^*$, by the previous proposition it is possible to choose a type t not in T_p. Then any q in $\mathbb{N}^*$ having type t with respect to a countable discrete subspace of $\mathbb{N}^*$ will do. $\square$

This approach to non-homogeneity appears in Frolík's 1967C paper.

4.13. The preceding theorem shows that a sufficient condition for X^* to be non-homogeneous is that X is not pseudocompact. However, this condition is not necessary.

Example. Let N contained in $\mathbb{N}^*$ be a copy of $\mathbb{N}$ and choose a point p in N^*. Let $X = \beta\mathbb{N}\backslash(N \cup \{p\})$. Then X^* contains both isolated points and a cluster point and therefore is not homogeneous. X is countably compact, and therefore pseudocompact, since any countable set of X has $2^{\mathfrak{c}}$ cluster points in $\beta\mathbb{N}$ and only countably many points have been removed.

4.14. We now consider another approach to the question of the non-homogeneity of growths. This approach differs from the one just considered in that the Continuum Hypothesis is required. Despite this shortcoming, the technique has the added advantage of singling out two interesting kinds of points which occur in growths, the P-points and the remote points, and showing their existence along parallel lines. P-points were defined in Section 1.65 and will be studied beginning in Section 4.29. Remote points will be defined and discussed beginning in Section 4.37.

The existence of P-points and remote points will first be demonstrated through an algebraic approach developed by D. Plank in 1969. Unfortunately, Plank's method requires restrictive hypotheses. To show the existence of P-points in X^* for X belonging to a less restricted class of spaces, we will consider the more powerful methods of W. Rudin and Isiwata. To show the existence of remote points under less restrictive conditions, we will utilize a construction developed by Fine and Gillman.

C-Points and C^*-Points in X^*

4.15. In Chapter 1 we saw that βX is homeomorphic to the spaces of maximal ideals $\mathcal{M}(X)$ of $C(X)$ and $\mathcal{M}^*(X)$ of $C^*(X)$, each taken with the hull-kernel topology. The respective maximal ideals are of the form

$$M^p = \{f \in C(X) : p \in \mathrm{cl}_{\beta X} \mathbf{Z}(f)\}$$

and

$$M^{*p} = \{f \in C^*(X) : \beta(f)(p) = 0\}$$

for p in βX. The homeomorphisms $p \mapsto M^p$ of βX with $\mathcal{M}(X)$ and $p \mapsto M^{*p}$ of βX with $\mathcal{M}^*(X)$ are denoted by τ and τ^*, respectively. The above descriptions of the maximal ideals of the two rings enable us to express $\mathrm{cl}_{\beta X} \mathbf{Z}(f)$ and $\mathbf{Z}(\beta(f))$ in terms of τ and τ^*, respectively. If f is in $C(X)$,

$$\mathrm{cl}_{\beta X} \mathbf{Z}(f) = \tau^{\leftarrow}(\{M^p \in \mathcal{M}(X) : f \in M^p\})$$

and if f is in $C^*(X)$,

$$\mathbf{Z}(\beta(f)) = (\tau^*)^{\leftarrow}(\{M^{*p} \in \mathcal{M}^*(X) : f \in M^{*p}\}).$$

The investigation of the different kinds of points of $\mathbb{N}^*$ will be carried out by studying subsets of the above form. We will frequently need to consider closures, interiors, and boundaries in the three spaces X, βX, and X^*, and one must be careful to note the subscripts on the operators denoted by cl, int, and ∂.

4.16. A point p of X^* is called a *C-point* if

$$p \notin \partial_{X^*}(\mathrm{cl}_{\beta X} \mathbf{Z}(f) \cap X^*)$$

for all f in $C(X)$. Similarly, a point p of X^* is called a *C*-point* if

$$p \notin \partial_{X^*}(\mathbf{Z}(\beta(f)) \cap X^*)$$

for all f in $C^*(X)$.

The significance of the C^*-points is easily seen when the definition of C^*-point is compared with that of P-point. Recall from Section 1.65 that a *P-point* is one which is in the interior of every zero-set which contains it. Hence, a P-point cannot belong to the boundary of any zero-set. By definition, a C^*-point is a point which cannot lie in the X^*-boundary of the trace on X^* of any zero-set of βX. Thus, it is clear that a P-point of X^* is a C^*-point. The converse is also true.

Proposition. *A point in X^* is a C^*-point if and only if it is a P-point of X^*.*

Proof. We have just seen that any P-point of X^* must be a C^*-point.

We will prove the converse by establishing the contrapositive. If a point q of X^* is not a P-point, then there is a zero-set Z_1 in X^* such that q lies in $\partial_{X^*} Z_1$. Because Z_1 is a G_δ in the subspace X^*, there is a $G_\delta S$ of βX such that $Z_1 = X^* \cap S$. But now there also exists a zero-set Z_2 of βX such that $q \in Z_2 \subset S$. Since q is in the X^*-boundary of Z_1 and $Z_2 \cap X^*$ is contained in Z_1, we must have that q belongs to $\partial_{X^*}(Z_2 \cap X^*)$. Thus, q cannot be a C^*-point. $\square$

4.17. The characterization of C-points is more difficult and cannot be accomplished without making assumptions about X. We will first prove an existence theorem for C^*-points and C-points and then undertake separate examinations of the two kinds of points. The relationship between C-points and remote points will be described in Theorem 4.40.

4.18. The definitions just given indicate that the existence of such points will depend upon the intersection of families of dense open sets being non-void. For example, it is clear that the C^*-points are precisely the points belonging to

$$\bigcap \{X^* \setminus \partial_{X^*}(\mathbf{Z}(\beta(f)) \cap X^*) : f \in C^*(X)\}\,.$$

The following analogue of the Baire Category Theorem is thus important in showing the existence of such points. The proof given is based on a method used by W. Rudin in 1956.

Proposition. *In a locally compact space in which every non-empty G_δ has non-empty interior, the intersection of any family of no more than $\aleph_1$ dense open sets is dense. If the space also has no isolated points, then the intersection contains at least $2^{\aleph_1}$ points.*

Proof. Write $\mathscr{D} = \{U_\alpha : \alpha < \omega_1\}$ for the collection of dense open sets and let G be any non-empty open set in such a space Y. We must show that G meets $\bigcap \mathscr{D}$. For $\alpha < \omega_1$, suppose that we have defined a collection $\{V_\beta : \beta < \alpha\}$ of open subsets of G such that for any $\beta < \alpha$,

    (a) $\mathrm{cl}_Y V_\beta$ is compact,

    (b) $\mathrm{cl}_Y V_\beta \subset U_\beta \cap V_\tau$ for all $\tau < \beta$.

If α is a non-limit ordinal, i.e. if $\alpha = \gamma + 1$, then $\bigcap_{\beta < \alpha} V_\beta = V_\gamma \neq \emptyset$. If α is a limit ordinal, then $\bigcap_{\beta < \alpha} V_\beta = \bigcap_{\beta < \alpha} \mathrm{cl}_Y V_\beta$. But the latter set is the intersection of a decreasing family of non-empty compact sets and hence is non-empty. In either case, $\bigcap_{\beta < \alpha} V_\beta$ is a non-empty G_δ.

Thus, $\bigcap_{\beta < \alpha} V_\beta$ has a non-void interior which must meet U_α since U_α is dense. Since U_α is also open, there is an open set V_α with compact closure such that

$$\mathrm{cl}_Y V_\alpha \subset U_\alpha \cap \left(\bigcap_{\beta < \alpha} V_\beta \right) \subset U_\alpha \cap G .$$

Thus, $\{V_\alpha : \alpha < \omega_1\}$ is defined inductively so that $\{\mathrm{cl}_Y V_\alpha : \alpha < \omega_1\}$ is a family of compact sets with the finite intersection property. Therefore

$$\emptyset \neq \bigcap \mathrm{cl}_Y V_\alpha \subset \left(\bigcap \mathscr{D} \right) \cap G ,$$

and $\bigcap \mathscr{D}$ is dense.

For the second statement, note that if Y has no isolated points, then at each step of the induction, there are two disjoint choices for V_α, so that $|\mathscr{D}| \geq 2^{\aleph_1}$. $\square$

4.19. The next sequence of results will conclude with the proof that a non-empty G_δ in the growth of a locally compact realcompact space has non-empty interior. This will be the major step in applying the preceding proposition. The first lemma appears in the 1962 paper of Fine and Gillman and will be employed to show that a C-embedded copy of $\mathbb{N}$ is completely separated from a disjoint closed subset.

Lemma. *Let f be a non-negative and unbounded member of $C(X)$ and let $\{J_n\}$ be a sequence of disjoint closed intervals in $\mathbb{R}$ such that each interval contains a value of f in its interior. Then the family of sets*

$$E_n = f^{\leftarrow}(J_n), \qquad n \geq 1,$$

satisfies:

(a) *Each E_n is a zero-set with non-void interior.*

(b) *If Z_n is any sequence of zero-sets with Z_n contained in E_n, then $\bigcup Z_n$ is a zero set.*

Proof. (a) is clear since each J_n is a zero-set with non-void interior and these properties will be shared by $E_n = f^{\leftarrow}(J_n)$.

To prove (b), we exhibit a map g in $C^*(X)$ such that $\mathbf{Z}(g) = \bigcup Z_n$. For each $n \geqslant 1$, there exists s_n in $C^*(\mathbb{R})$ such that $s_n^{\leftarrow}(1) = J_n$ and $s_n^{\leftarrow}(0) = \bigcup_{m \neq n} J_m$. Now for Z_n given as in (b), write $Z_n = Z(h_n)$, and assume that $0 \leqslant h_n \leqslant 2^{-n}$. Consider the sequence of functions defined by

$$g_n(x) = h_1(x) \cdot (s_1 \circ f)(x) + \cdots + h_n(x) \cdot (s_n \circ f)(x) .$$

The sequence $\{g_n\}$ converges uniformly to a mapping g in $C^*(X)$. It is clear that g is never zero outside of the union $\bigcup E_n$, and that for each integer n, g agrees with h_n on E_n, so that $\mathbf{Z}(g) = \bigcup Z_n$. $\square$

4.20. The next result is [GJ, ex. 9 M1].

Corollary. *A copy of $\mathbb{N}$ which is C-embedded in a space X is completely separated from any closed subset disjoint from it.*

Proof. Let $\mathbb{N}$ be C-embedded in X and let the closed set F miss $\mathbb{N}$. For each n in $\mathbb{N}$, there is a zero-set W_n containing F which does not contain n. Then $Z = \bigcap W_n$ is a zero-set containing F and missing $\mathbb{N}$. There exists a member of $C(X)$ which is a homeomorphism on $\mathbb{N}$ so that by the preceding lemma we can choose zero-sets E_n with n contained in E_n for each n. Further, for each n in $\mathbb{N}$, choose a zero-set Z_n that is contained in E_n, misses Z, and contains n. Then F and $\mathbb{N}$ are completely separated since they are contained in the disjoint zero-sets Z and $\bigcup Z_n$, respectively. $\square$

4.21. Note that the next proposition shows that every non-empty G_δ in the growth of a locally compact realcompact space has non-empty interior since every non-empty G_δ contains a non-empty zero-set. The result is taken from the 1960 paper of Fine and Gillman.

Proposition. *The zero-sets in the growth of a locally compact and realcompact space are regular closed sets.*

Proof. Let X be a locally compact and realcompact space that is not compact. First observe that the proof of Corollary 3.28 shows that it is sufficient to show that a non-empty zero-set has non-empty interior.

Since X is locally compact, X^* is closed (Theorem 1.59) and therefore is C^*-embedded in βX. Thus, a zero-set Z of X^* is of the form

$Z = \mathbf{Z}(f) \cap X^*$ for some f in $C(\beta X)$. Let p belong to Z. Since X is real-compact and p lies outside of X, by Theorem 1.53 there is a mapping g in $C(\beta X)$ such that $g(p) = 0$ and g does not vanish on X. Let $h = |f| + |g|$. Then $\mathbf{Z}(h)$ is a subset of Z containing p, i.e. $p \in \mathbf{Z}(h) \subset Z$. Let $\{x_n\}$ be a discrete sequence of distinct points of X on which h approaches zero. Since the reciprocal of $h|X$ is continuous and unbounded on $\{x_n\}$, by Lemma 4.5 we can assume that the sequence is C-embedded in X. Choose a compact neighborhood V_n of x_n for each n such that the family $\{V_n\}$ is pairwise disjoint and

$$|h(x) - h(x_n)| < 1/n$$

for each x in V_n. Then the preceding corollary shows that there exists a mapping l in $C^*(X)$ such that $l(x_n) = 1$ for each n and l vanishes outside of $\bigcup V_n$. If $\beta(l)$ is the extension of l to βX and q is any point of X^* such that $\beta(l)(q) \neq 0$, then every neighborhood of q meets infinitely many of the compact sets V_n and hence $h(q) = 0$. Thus, we have shown that

$$X^* \setminus \mathbf{Z}(\beta(l)) \subset \mathbf{Z}(h) \subset Z,$$

and Z therefore has non-empty interior in X^*. $\quad\Box$

4.22. The local compactness is critical in the preceding proposition as the following example shows.

Example. *The space of rationals* $\mathbf{Q}$ *is realcompact but not locally compact.* $\mathbf{Q}^*$ *contains non-empty zero-sets which have empty interior.* For example, let h embed $\mathbf{Q}$ as a dense subspace of the closed interval I such that $h[\mathbf{Q}]$ does not contain 0. Then the zero-set $\mathbf{Z}(\beta(h))$ is non-empty and is contained in $\mathbf{Q}^*$. However, we will show that for any point p in $\mathbf{Q}^*$, $\beta(h)$ is not constant on any $\mathbf{Q}^*$-neighborhood of p so that in particular, $\mathbf{Z}(\beta(h))$ has empty interior in $\mathbf{Q}^*$. Let V be any closed $\beta\mathbf{Q}$-neighborhood of p. Then $V \cap \mathbf{Q}$ contains disjoint closed intervals J and L. Because neither interval is compact, $\mathrm{cl}_{\beta\mathbf{Q}} J \cap \mathbf{Q}^*$ and $\mathrm{cl}_{\beta\mathbf{Q}} L \cap \mathbf{Q}^*$ are disjoint non-empty subsets of $V \cap \mathbf{Q}^*$. But because h is a homeomorphism, the values which $\beta(h)$ takes on $\mathrm{cl}_{\beta\mathbf{Q}} L \cap \mathbf{Q}^*$ are distinct from those taken on $\mathrm{cl}_{\beta\mathbf{Q}} J \cap \mathbf{Q}^*$. Hence, $\beta(h)$ is not constant on $V \cap \mathbf{Q}^*$. This example was suggested in [GJ, ex. 6O].

4.23. Proposition 4.21 shows that the growth of a locally compact and realcompact space will satisfy the hypotheses of Proposition 4.18. Now by assuming the Continuum Hypothesis and restricting the cardinality of the ring of continuous functions, we obtain the central existence theorem for C-points and C^*-points.

Theorem [CH] (Plank). *The growth of a locally compact and real-compact but non-compact space will have a dense set of 2^c C-points (resp. C^*-points) if the cardinality of $C(X)$ (resp. $C^*(X)$) is c.*

Proof. If X is a non-compact, locally compact and realcompact space, then the space X^* is a non-empty compact space and every non-empty G_δ of X^* has non-empty interior by Proposition 4.21. Further, since X is realcompact, X^* can have no isolated points. If a point p in X^* were isolated, then there would exist a zero-set neighborhood Z_1 of p in βX such that $Z_1 \cap X^* = \{p\}$. Since X is realcompact, there is a zero-set Z_2 containing p but missing X. Thus, $Z_1 \cap Z_2 = \{p\}$. But Theorem 3.6 shows that a zero-set of βX contained in X^* must have cardinality at least 2^c, giving a contradiction.

Using the Continuum Hypothesis, $\mathscr{D} = \{X^* \setminus \partial_{X^*} \mathbf{Z}(\beta(f)) : f \in C^*(X)\}$ is a family of $\aleph_1$ dense open subsets of X^*. Since X^* is a compact space in which the non-empty G_δ's have non-empty interiors, $\bigcap \mathscr{D}$ is a dense subset of at least 2^c C^*-points. Since $|\beta X| = |\mathscr{M}^*(X)| \leqslant 2^c$, $\bigcap \mathscr{D}$ must contain exactly 2^c points. The proof for C-points is similar. $\quad\square$

4.24. The following corollary is immediate from the observation that $C^*(X)$ is contained in $C(X)$.

Corollary [CH]. *The growth of a locally compact and realcompact space X which fails to be compact has a dense subset of 2^c points which are both C-points and C^*-points if the cardinality of $C(X)$ is c.*

4.25. Note that the hypothesis that the cardinality of $C^*(X)$ or $C(X)$ be no more than c is redundant if X is separable. The following example shows that Theorem 4.23 applies to some non-separable spaces.

Example (Plank). Let X be a non-closed cozero-set in $\mathbb{N}^*$. Such a co-zero-set must exist in $\mathbb{N}^*$ because $\mathbb{N}^*$ is not a P-space (Proposition 1.65). X is not separable since X contains an open copy of $\mathbb{N}^*$ which in turn contains a family of c disjoint open sets by Theorem 3.22. This family of sets must then be open in X since the copy of $\mathbb{N}^*$ was open in X. The space X is realcompact because a cozero-set in a realcompact space is realcompact [GJ, 8.14]. Since X is open in the compact space $\mathbb{N}^*$, X is locally compact. Proposition 3.24 shows that $\mathbb{N}^*$ is an F-space. Therefore the cozero-set X is C^*-embedded in $\mathbb{N}^*$ and hence is also C^*-embedded in $\beta\mathbb{N}$. Because $\beta\mathbb{N}$ is separable, the cardinality of $C(X)$ is just c.

4.26. A sufficient condition for a point p of X^* to be a C-point or a C^*-point can be stated in terms of the corresponding ideals M^p and M^{*p}.

Gelfand and Kolmogoroff characterized the maximal ideal M^p of $C(X)$ by showing (Theorem 1.30) that

$$M^p = \{ f \in C(X) : p \in \mathrm{cl}_{\beta X} \mathbf{Z}(f) \} \, .$$

A subfamily of M^p is the set O^p of those mappings f in $C(X)$ such that p belongs to the interior of $\mathrm{cl}_{\beta X} \mathbf{Z}(f)$, i.e.

$$O^p = \{ f \in C(X) : p \in \mathrm{int}_{\beta X} \mathrm{cl}_{\beta X} \mathbf{Z}(f) \} \, .$$

O^p is easily seen to be an ideal contained in M^p. The corresponding ideals of $C^*(X)$ are given (Theorem 1.26) by

$$M^{*p} = \{ f \in C^*(X) : p \in \mathbf{Z}(\beta(f)) \}$$

and

$$O^{*p} = \{ f \in C^*(X) : p \in \mathrm{int}_{\beta X} \mathbf{Z}(\beta(f)) \} \, .$$

4.27. Now consider the case where O^p is equal to M^p and p is in X^*. For every mapping f in $C(X)$, if p is in $\mathrm{cl}_{\beta X} \mathbf{Z}(f)$, then p must be in $\mathrm{int}_{\beta X} \mathrm{cl}_{\beta X} \mathbf{Z}(f)$. Thus, p is a C-point. The corresponding argument for O^{*p} is similar and we have verified the following

Proposition. *If p is a point of X^* and $M^p = O^p$ (resp. $M^{*p} = O^{*p}$), then p is a C-point (resp. C^*-point).*

4.28. Example. The converse of the proposition is false. We have from Theorem 4.23 that $\mathbf{N}^*$ contains a dense set of C^*-points. The function j in $C^*(\mathbf{N})$ defined by $j(n) = 1/n$ for each n in $\mathbf{N}$ belongs to M^{*p} for all p in $\mathbf{N}^*$, but $\mathrm{int}_{\beta X} \mathbf{Z}(\beta(j)) = \emptyset$ so that j cannot be in O^{*p} for any p in $\mathbf{N}^*$.

P-Points in X^*

4.29. In Proposition 4.16 we saw that the P-points of X^* are just the C^*-points. Thus, by combining that result with Theorem 4.23, we obtain the following two corollaries.

Corollary [CH]. *The growth of a locally compact and realcompact space X which is not compact has a dense set of $2^{\mathfrak{c}}$ P-points if the cardinality of $C^*(X)$ is $\mathfrak{c}$.*

4.30. The second corollary is the original result in the investigation of the non-homogeneity of growths and was obtained by W. Rudin in 1956.

Corollary [CH] (W. Rudin). $\mathbb{N}^*$ *has a dense set of* 2^c *P-points.*

4.31. Any homeomorphism must carry *P*-points to *P*-points. Thus, any space containing both *P*-points and non-*P*-points must fail to be homogeneous. Because a pseudocompact *P*-space must be finite, $\mathbb{N}^*$ must also contain non-*P*-points.

Proposition [CH] (W. Rudin). $\mathbb{N}^*$ *contains a dense set of* 2^c *non-P-points and fails to be homogeneous if the Continuum Hypothesis is assumed.*

Proof. $\mathbb{N}^*$ clearly contains non-*P*-points as we have just seen. If the Continuum Hypothesis is assumed, it also contains *P*-points and therefore fails to be homogeneous.

Because every G_δ containing a *P*-point is a neighborhood of the *P*-point, no *P*-point can be a cluster point of a sequence. Every discrete sequence in $\mathbb{N}^*$ has 2^c cluster points, so $\mathbb{N}^*$ therefore contains 2^c non-*P*-points. Any homeomorphism of $\beta\mathbb{N}$ must send a non-*P*-point of $\mathbb{N}^*$ to another non-*P*-point. Because the orbit of a point of $\mathbb{N}^*$ is dense in $\mathbb{N}^*$, the non-*P*-points form a dense subset. $\square$

The investigation of *P*-points and their role in the non-homogeneity of growths of Stone-Čech compactifications now becomes an exercise in eliminating the restrictive hypotheses of Corollary 4.29.

4.32. We have seen that if X fails to be pseudocompact, then X contains a *C*-embedded copy of $\mathbb{N}$ (Lemma 4.5), and that X^* therefore contains a copy of $\mathbb{N}^*$. The next result shows that if X is also locally compact, then the *P*-points of $\mathbb{N}^*$ are also *P*-points of X^*.

Lemma. *If* $\mathbb{N}$ *is C-embedded in a locally compact space* X, *then every P-point of* $\mathbb{N}^*$ *is a P-point of* X^*.

Proof. Let p be a *P*-point of $\mathbb{N}^*$ and suppose that f in $C(X^*)$ vanishes at p. We must show that f vanishes on an X^* neighborhood of p. Since X^* is closed in βX, f can be extended to a mapping g in $C(\beta X)$. Since g vanishes at p, there is an infinite subset E of $\mathbb{N}$ such that p belongs to $\mathrm{cl}_{\beta X} E$ and g approaches zero along E. Now choose a neighborhood V_n of each n in E such that the closures in X of the neighborhoods form a pairwise disjoint family of compact sets and

$$|g(x) - g(n)| < 1/n$$

for every x in V_n. Corollary 4.20 shows that there exists a non-negative mapping l in $C^*(X)$ such that $l(n) = 1$ for each n in E and l vanishes outside of $\bigcup V_n$. Then it is clear that p belongs to the cozero-set $\mathbf{Cz}(\beta(l))$. Since l vanishes outside of $\bigcup V_n$, no point of $X^* \cap \mathbf{Cz}(\beta(l))$ is a limit point of $X\backslash\bigcup V_n$. Further, since each V_n has compact closure in X, every neighborhood of a point in $X^* \cap \mathbf{Cz}(\beta(l))$ meets infinitely many of the sets $\{V_n\}$. Thus by the choice of the neighborhoods V_n, it is clear that for any point q in $X^* \cap \mathbf{Cz}(\beta(l))$, $f(q)=g(q)=0$. Thus, we have shown that

$$X^* \cap \mathbf{Cz}(\beta(l)) \subset \mathbf{Z}(f)$$

so that f vanishes on a X^*-neighborhood of p. □

4.33. The preceding lemma shows that every C-embedded copy of $\mathbb{N}$ in a locally compact space will lead to the existence of P-points in the growth of the space. The next lemma shows that for a realcompact space there will be sufficiently many C-embedded copies of $\mathbb{N}$ to produce a dense set of P-points in the growth. The following result is contained in [GJ, ex. 9 D].

Lemma. *If V is a neighborhood of a point in $\beta X \backslash \upsilon X$, then $V \cap \upsilon X$ contains a C-embedded copy of $\mathbb{N}$, and hence V contains a copy of $\beta \mathbb{N}$.*

Proof. Since βX is completely regular, we can assume that a neighborhood of a point p of X^* is a zero-set neighborhood, Z_p. Since p is in $\beta X \backslash \upsilon X$, there is a zero-set $\mathbf{Z}(f)$ such that $p \in \mathbf{Z}(f) \subset \beta X \backslash \upsilon X$. Since $Z_p \cap \mathbf{Z}(f) \subset Z_p$, there exists g in $C(X)$ which is unbounded on Z_p and hence, $Z_p \cap \upsilon X$ contains a C-embedded copy of $\mathbb{N}$ by Lemma 4.5. Thus, $Z_p \supset \mathrm{cl}_{\beta X} \mathbb{N}$. □

4.34. The first statement in the following theorem was established by W. Rudin in 1956 with the additional hypothesis that X be normal. The proofs of the next two theorems follow the outline of [GJ, ex. 9 M] as did that of Lemma 4.32.

Theorem [CH]. *The growth of a non-pseudocompact locally compact space contains both P-points and non-P-points. If the space is also realcompact, then both sets of points are dense.*

Proof. Since X is not pseudocompact, X contains a C-embedded copy of $\mathbb{N}$. If the Continuum Hypothesis is assumed, then $\mathbb{N}^*$ contains P-points which are also P-points of X^* by Lemma 4.32. The non-P-points of $\mathbb{N}^*$ will clearly not be P-points of X^*.

If in addition X is realcompact, then every X^*-neighborhood of a point q in X^* contains a copy of $\mathbb{N}^*$ by Lemma 4.33 and thus contains both P-points and non-P-points of X^*. □

Theorem 4.34 in an improvement over Corollary 4.29 since the restriction on the cardinality of $C^*(X)$ has been dropped. For example, Corollary 4.29 will not apply to an uncountable discrete space.

4.35. Note that the first statement of Theorem 4.34 implies that the growth of a non-pseudocompact, locally compact space fails to be homogeneous. If we drop the local compactness hypothesis, we will still be able to show the non-homogeneity of X^* but will not be able to show the existence of *P*-points in X^*. The technique will utilize one member of the family of spaces $v_f X$ introduced in Section 1.53. Recall that for f in $C(X)$,

$$v_f X = \beta X \setminus \{ p \in \beta X : f^\alpha(p) = \infty \}$$

where f^α is the extension of f to βX and takes its values in $\alpha \mathbb{R}$, the one point compactification of $\mathbb{R}$. Since $v_f X$ is an open subspace of βX, it is clearly locally compact. The family of spaces $\{ v_f X \}$ was introduced by M. Henriksen and is studied in [GJ, ex. 8 B]. The following theorem was proven by T. Isiwata in 1957C and utilizes the subspace $v_f X$ for a well-chosen f.

Theorem [CH] (Isiwata). *The growth of a non-pseudocompact space fails to be homogeneous.*

Proof. If a space X is not pseudocompact, then X contains a *C*-embedded copy of $\mathbb{N}$. In order to take advantage of the copy of $\mathbb{N}^*$ which $\mathbb{N}$ yields in X^*, we will define a convenient locally compact space Y with $X \subset Y \subset \beta X$, to which Lemma 4.32 can be applied. We shall then show that if φ is an automorphism of X^* such that $\varphi(p)$ is in $\mathbb{N}^*$ for some *P*-point p of $\mathbb{N}^*$, then $\varphi(p)$ is also a *P*-point of $\mathbb{N}^*$. This will complete the proof since it will show that a *P*-point of $\mathbb{N}^*$ cannot be mapped to all points of $\mathbb{N}^*$ by homeomorphisms of X^*.

Let f be a mapping in $C(X)$ such that $f(n) = n$ for all n in $\mathbb{N}$. Then consider

$$Y = v_f X \setminus \varphi^\leftarrow(\mathbb{N}^*).$$

Since Y is obtained by deleting a compact subspace of βX, Y is locally compact. Since the restriction $f^\alpha | Y$ is unbounded on $\mathbb{N}$, Lemma 4.5 shows that we can assume that $\mathbb{N}$ is *C*-embedded in Y. It is clear that $\mathbb{N}^*$ is contained in Y^*. By Lemma 4.32, every *P*-point of $\mathbb{N}^*$ is a *P*-point of Y^*. Further, $\varphi | Y^*$ is a homeomorphism onto its range, so that $\varphi(p)$ is a *P*-point of $\varphi[Y^*]$. Since $\varphi(p) \in \mathbb{N}^* \subset \varphi[Y^*]$, $\varphi(p)$ is a *P*-point of $\mathbb{N}^*$. □

4.36. Note that the theorem states only that X^* is not homogeneous. While the proof utilizes *P*-points, it does not necessarily imply that there exist *P*-points in X^*. The *P*-points need only be *P*-points of a subspace of X^*.

Example. Since $\mathbb{Q}$ is clearly not pseudocompact, the theorem shows that $\mathbb{Q}^*$ *is not homogeneous. However* $\mathbb{Q}^*$ *contains no P-points.* We first investigate another characterization of P-points. If f is in $C(X)$ for any X, then $f^{\leftarrow}(r)$ is a zero-set because every point of $\mathbb{R}$ is a zero-set. Thus, if p is a P-point and belongs to $f^{\leftarrow}(r)$, f is constant on a neighborhood of p since every zero-set containing p is a neighborhood of p. Returning to $\mathbb{Q}^*$, in Example 4.22 we exhibited a mapping $\beta(h)$ on $\mathbb{Q}^*$ which is not constant on any open set of $\mathbb{Q}^*$. Hence, $\mathbb{Q}^*$ can contain no P-point.

Remote Points in X^*

4.37. A *remote point* of βX is a point which does not belong to the closure of any discrete subspace of X. It is clear that any remote point must lie in X^*. Remote points are related to C-points in somewhat the same manner that P-points are related to C^*-points. However, the existence theorems for remote points require much more stringent hypotheses. The existence of remote points seems to be a phenomenon confined to metric spaces, but by no means common to all metric spaces. For example, if D is an infinite discrete space, it is clear from the definition of remote point that βD can contain no such point. To eliminate this behavior, we shall find it necessary to restrict the existence of isolated points in order to obtain remote points. One property of metric spaces which we shall find useful is that every closed set is a zero-set. The other useful properties will be outlined in the next two lemmas.

4.38. The first lemma uses the following property of realcompact spaces. If X is a realcompact space, the intersection of the free maximal ideals of $C(X)$ is the subring of $C(X)$ consisting of those mappings having compact support, i.e. those maps f in $C(X)$ for which $\operatorname{cl}_X \mathbf{Cz}(f)$ is compact [GJ, 8.19]. Denote this subring by $C_K(X)$. Since the free maximal ideals in $C(X)$ correspond to the points of X^*, we can write

$$\bigcap \{M^p : p \in X^*\} = C_K(X)$$

if X is realcompact.

Recall that a point p of X^* is a *C-point* if $p \notin \operatorname{int}_{X^*}(\operatorname{cl}_{\beta X} \mathbf{Z}(f) \cap X^*)$ for all f in $C(X)$.

Lemma. *If* X *is realcompact, then*

$$\operatorname{int}_{X^*}(\operatorname{cl}_{\beta X}(\mathbf{Z}(f)) \cap X^*) = \operatorname{int}_{\beta X}(\operatorname{cl}_{\beta X} \mathbf{Z}(f)) \cap X^*$$

for every mapping f *in* $C(X)$.

Proof. The containment

$$\operatorname{int}_{\beta X}(\operatorname{cl}_{\beta X}\mathbf{Z}(f))\cap X^* \subset \operatorname{int}_{X^*}(\operatorname{cl}_{\beta X}(\mathbf{Z}(f))\cap X^*)$$

is clear. Now let p be a point in $\operatorname{int}_{X^*}(\operatorname{cl}_{\beta X}(\mathbf{Z}(f))\cap X^*)$. Since the closures in βX of the zero-sets of X form a base for the closed sets of βX, there exists g in $C(X)$ such that

$$p\in X^*\backslash(\operatorname{cl}_{\beta X}(\mathbf{Z}(g))\cap X^*)\subset\operatorname{cl}_{\beta X}(\mathbf{Z}(f))\cap X^*.$$

Then g fails to belong to M^p but the product fg belongs to $\bigcap\{M^q:q\in X^*\}$ since every point of X^* belongs to either $\operatorname{cl}_{\beta X}\mathbf{Z}(f)$ or $\operatorname{cl}_{\beta X}\mathbf{Z}(g)$. Thus, since X is realcompact, fg is in $C_K(X)$ and $K=\operatorname{cl}_X\mathbf{Cz}(fg)$ is compact. Since K is contained in X, p cannot belong to K and since p also does not belong to $\operatorname{cl}_{\beta X}\mathbf{Z}(g)$, we have that

$$p\in\beta X\backslash(K\cup\operatorname{cl}_{\beta X}\mathbf{Z}(g))\subset\operatorname{cl}_{\beta X}\mathbf{Z}(f),$$

so that p belongs to $\operatorname{int}_{\beta X}\operatorname{cl}_{\beta X}\mathbf{Z}(f)$. $\quad\square$

We saw in Example 4.28 that the converse of Proposition 4.27 was false for C^*-points. The importance of the preceding lemma will be to show that the converse is true for C-points when X is a metric space. The lemma can actually be proven in greater generality. In 1966, S. M. Robinson showed that if X is completely uniformizable, then $\bigcap\{M^p:p\in X^*\}=C_K(X)$. Since if a measurable cardinal exists, there are completely uniformizable spaces which are not realcompact, Robinson's result is more general than the one cited for realcompact spaces. However, the realcompact hypothesis includes all metric spaces of non-measurable cardinal and in particular, all separable metric spaces. For a discussion of the role of the measurable cardinal and to see how unlikely it is that any metric space of interest fails to be realcompact, see [GJ, Chapter 12 and Theorems 15.20, 15.24].

4.39. The second lemma is important in the study of remote points since it provides many well-placed discrete subspaces in a metric space. The lemma is taken from F. Hausdorff's book, *Set Theory*.

Lemma. *In a metric space, the boundary of an open set can be represented as the set of cluster points of a discrete subspace contained in the open set.*

Proof. Let X be a metric space with metric d and let F be the boundary of an open subset G of X. We can assume that F is non-empty, since the discrete subspace can be chosen to be the empty set if F is empty. For each x in G, let $\delta(x)$ be the distance of x from F. Then $\delta(x)>0$. Now consider all subsets S of G which satisfy

$$d(x,y)\geqslant(1/2)\delta(x)+(1/2)\delta(y)$$

for x and y in S. Let A be a maximal set with respect to this property.

The cluster points of A are contained in F: A cluster point x of A is the limit of a sequence $\{x_n\}$ of distinct points of A. Since δ is continuous, $\delta(x_n)\to\delta(x)$, and in the inequality

$$d(x_n,x_{n+1})\geqslant(1/2)\delta(x_n)+(1/2)\delta(x_{n+1}),$$

the left-hand side converges to 0 and the right-hand side to $\delta(x)$. Hence, $\delta(x)=0$ and x is in F.

Every point of F is a cluster point of A: For a point z in F and $\varepsilon>0$, choose y in G such that $d(z,y)<\varepsilon/3$. If y is in A, we are done. If not, the maximality of A implies that there is a point x in A for which

$$d(x,y)<(1/2)\delta(x)+(1/2)\delta(y).$$

Then we have

$$(*)\qquad d(x,y)<(1/2)\delta(x)+(1/2)\delta(y)\leqslant(1/2)d(z,x)+(1/2)d(z,y)$$

from the definition of δ. Now we must show that x is close to z. Using the triangle inequality and $(*)$, we have

$$d(z,x)\leqslant d(z,y)+d(x,y)<(1/2)d(z,x)+(3/2)d(z,y)$$

which yields

$$d(z,x)\leqslant 3d(z,y)<\varepsilon.$$

Finally, the discreteness of A follows from the fact that all cluster points of A are contained in F. $\quad\square$

4.40. We now relate remote points to C-points. Recall that in Section 1.32 we introduced the following notation for the z-ultrafilters on a space X: If p is in βX,

$$A^p=\{\mathbf{Z}(f)\in\mathbf{Z}[X]:f\in M^p\}.$$

Theorem. *Let p be in X^* where X is a metric space of non-measurable cardinal, and consider the following four conditions:*
    (a) *p is a C-point of X^*.*

(b) *A^p has no member which is nowhere dense.*
(c) *$M^p = O^p$.*
(d) *p is a remote point in βX.*
Conditions (a), (b), and (c) are equivalent and are implied by (d). All four conditions are equivalent if the set of isolated points of X has compact closure in X.

Proof. (a)$\Rightarrow$(b): Let p be a C-point and let Z be in A^p. Then p is in $\text{int}_{X^*}(\text{cl}_{\beta X} Z \cap X^*)$. Thus by Lemma 4.38, p is in $V = \text{int}_{\beta X} \text{cl}_{\beta X} Z$. But then $V \cap X$ is a non-empty open set contained in Z and Z is not nowhere dense.

(b)$\Rightarrow$(c): Let f be in M^p. Since X is a metric space, we can find a g in $C(X)$ such that $\mathbf{Z}(g) = \text{cl}_X(X \setminus \mathbf{Z}(f))$; thus $X = \mathbf{Z}(f) \cup \mathbf{Z}(g)$. If p is in $\text{cl}_{\beta X}(\mathbf{Z}(g))$, g must be in M^p and we must then have p in

$$\text{cl}_{\beta X}(\mathbf{Z}(f) \cap \mathbf{Z}(g)) = \text{cl}_{\beta X} \partial_X(\mathbf{Z}(f)) .$$

But this contradicts (b) since $\partial_X(\mathbf{Z}(f))$ is nowhere dense. Thus, p is in $\beta X \setminus \text{cl}_{\beta X} \mathbf{Z}(g) \subset \text{cl}_{\beta X} \mathbf{Z}(f)$ which shows that f is in O^p.

(c)$\Rightarrow$(a): This follows from Proposition 4.27.

(d)$\Rightarrow$(b): Suppose that A^p has a nowhere dense member Z. Then Z is the boundary of its complement and by Lemma 4.39, there exists a discrete space $D \subset X \setminus Z$ such that $D \cup Z = \text{cl}_X D$. Thus, p is in $\text{cl}_{\beta X} Z \subset \text{cl}_{\beta X} D$, so p is not a remote point.

Now assuming in addition that the set L of isolated points of X has compact closure in X, we can show (b)$\Rightarrow$(d):

It is sufficient to show that if p is not a remote point, then some member of A^p is nowhere dense. Let D be a discrete subspace of X such that p is in $\text{cl}_{\beta X} D$. Then $\text{cl}_X D$ is in A^p and if $\text{cl}_X D$ is not nowhere dense, then there exists a point in $D \cap \text{int}_X \text{cl}_X D$. But such a point must be isolated because D is discrete. Since $\text{cl}_{\beta X}(\text{cl}_X L \cap \text{cl}_X D)$ is contained in X because $\text{cl}_X L$ is compact, p belongs to $\text{cl}_{\beta X}(D \setminus L)$, and thus $\text{cl}_X(D \setminus L)$ is a nowhere dense member of A^p. $\square$

The equivalence of (b) and (d) was shown in 1962 by Fine and Gillman and the proof that (b) implies (c) was contributed by M. Mandelker. The theorem appears in D. Plank's 1969 paper.

4.41. By combining the hypotheses of the preceding theorem with those of Theorem 4.23, we obtain the following result concerning the existence of remote points. The separability hypothesis shows that the cardinality of $C(X)$ is c and also that the cardinality of X is non-measurable so that X is realcompact. (Note that Exercise 1D also shows that X is realcompact.)

Proposition [CH]. *If X is a separable, locally compact, non-compact metric space in which the set of isolated points has compact closure, then βX contains a set of 2^c remote points which is a dense subspace of X^*.*

4.42. The proposition establishes the existence of 2^c remote points of $\beta\mathbb{R}$ which form a dense subspace of $\mathbb{R}^*$, but the local compactness hypothesis prevents an application to such basic examples as $\mathbb{Q}$. In order to eliminate the need for local compactness, it is necessary to use methods developed by Fine and Gillman. The following proposition is a portion of a result which appeared in their 1962 paper.

Proposition (Fine-Gillman). *If X is a non-pseudocompact space which contains no more than $\aleph_1$ dense open sets, then there is a free z-ultrafilter on X no member of which is nowhere dense.*

Proof. Because X is not pseudocompact, $C(X)$ contains an unbounded, non-negative function and therefore there exists a family $\{E_n : n \geqslant 1\}$ of zero-sets of X which satisfy (a) and (b) of Lemma 4.19. We will use the properties of this family to construct a family $\mathscr{F}$ of non-empty zero-sets of X such that $\mathscr{F}$ has the finite intersection property and every dense open subset of X contains a member of $\mathscr{F}$. Thus $\mathscr{F}$ will be contained in a z-ultrafilter but no z-ultrafilter containing $\mathscr{F}$ can contain a nowhere dense member.

Let $\{U_\alpha : \omega_0 \leqslant \alpha < \omega_1\}$ be the family of dense open subsets of X. We will construct the family $\mathscr{F} = \{F_\alpha : \alpha < \omega_1\}$ by transfinite induction in such a way that $F_\alpha \subset U_\alpha$ for $\omega_0 \leqslant \alpha < \omega_1$. The induction is accomplished for $n < \omega_0$ by putting $F_n = \bigcup \{E_i : i \geqslant n\}$. Lemma 4.19 shows that each F_n is a zero-set, $\mathrm{int}\, F_n \neq \emptyset$, and $\mathrm{int}\, F_n = \bigcup \{\mathrm{int}\, E_i : i \geqslant n\}$. Note that since $\bigcap F_n = \emptyset$, any z-ultrafilter containing $\mathscr{F}$ must be free.

To complete the induction, consider any ordinal α with $\omega_0 \leqslant \alpha < \omega_1$ and assume that for each $\beta < \alpha$ we have defined a zero-set F_β such that the family $\{\mathrm{int}\, F_\beta : \beta < \alpha\}$ has the finite intersection property and such that for $\beta \geqslant \omega_0$, $F_\beta \subset U_\beta$. To define F_α, first rearrange the countable family $\{\mathrm{int}\, F_\beta : \beta < \alpha\}$ into a sequence $\{T_n : n < \omega_0\}$. Next define an increasing sequence of integers $\{n_i\}$ and a sequence of zero-sets $\{Z_j\}$ as follows: Let $k < \omega_0$ and assume that the n_i have been defined for all $i < k$ and that the Z_j have been defined for all $j \leqslant n_{k-1}$. By the induction hypothesis, $T_1 \cap \cdots \cap T_k$ meets $\mathrm{int}\, F_{n_{k-1}+1}$ and hence meets $\mathrm{int}\, E_{n_k}$ for some $n_k > n_{k-1}$. Since U_α is open and dense, we can choose a zero-set Z_{n_k} with non-void interior such that

$$Z_{n_k} \subset T_0 \cap \cdots \cap T_k \cap E_{n_k} \cap U_\alpha.$$

Finally, put $Z_j = \emptyset$ for $n_{k-1} < j < n_k$ to complete the definition of the sequence $\{Z_j\}$. Now define $F_\alpha = \bigcup Z_j$. Since Z_j is contained in E_j for each j, Lemma 4.19 shows that F_α is a zero-set having non-empty interior. To complete the induction step, note that any finite intersection of the sets $\{\operatorname{int} F_\beta\}_{\beta < \alpha}$ contains a set of the form $T_0 \cap \cdots \cap T_k$. Therefore it contains Z_{n_k} and hence meets $\operatorname{int} F_\alpha$. Thus, the collection $\{\operatorname{int} F_\beta : \beta \leqslant \alpha\}$ has the finite intersection property and is therefore contained in a z-ultrafilter which must be free and cannot contain a nowhere dense member. $\square$

4.43. When combined with Theorem 4.40, the preceding result is sufficient to show the existence of a single remote point. The following modification of the preceding proof was developed by Plank in his thesis to demonstrate the existence of infinitely many remote points and thereby obtain the main.

Theorem [CH] (Plank). *If X is a non-compact separable metric space in which the set of isolated points has compact closure, then βX contains 2^c remote points which form a dense subspace of X^*.*

Proof. Since X is a separable metric space, it has a countable base of open sets [D, p. 187] and therefore under the assumption of the Continuum Hypothesis there can be no more than $\aleph_1$ dense open subsets of X. Let $\{U_\alpha : \omega_0 \leqslant \alpha < \omega_1\}$ be the family of dense open sets. Since a separable metric space has non-measurable cardinal, X is realcompact. Because the set of isolated points of X has compact closure, a point p of X^* is a remote point if and only if no member of A^p is nowhere dense (Theorem 4.40).

We will modify the proof of the previous proposition in order to obtain 2^c z-ultrafilters in such a way that the corresponding remote points in X^* form a dense subspace. Let V be a closed βX-neighborhood of a point q in X^*. In order to exhibit a remote point in V, it is sufficient to show the existence of a free z-ultrafilter A^p on X such that no member of A^p is nowhere dense and p belongs to V. We will show that infinitely many such z-ultrafilters exist by constructing the family $\mathscr{F}$ of zero-sets as in the proof of the preceding proposition so that they are contained in $V \cap X$ and so that $\mathscr{F}$ is contained in infinitely many z-ultrafilters.

Because X is realcompact, there exists a non-negative h in $C(X)$ such that its extension h^α to βX is infinite at q (Section 1.53). Choose a non-negative g in $C(\beta X)$ such that $g(q)=1$ and g vanishes on $\beta X \backslash V$. Define a function f by $f = h \cdot (g|X)$. Then f is unbounded on $V \cap X$. Choose a family $\{E_n : n \geqslant 1\}$ of zero-sets of X which satisfy (a) and (b) of Lemma 4.19 and are contained in $V \cap X$. Divide the sequence $\{E_n\}$ into infinitely many subsequences, $\{E_n^k : n \geqslant 1\}$, $k = 1, 2, \ldots,$ by putting

$E_n^k = E_{2^{k-1}(2n-1)}$. Each sequence $\{E_n^k\}$ also satisfies the conditions of Lemma 4.19. Thus, for each k we may apply the method used in the proof of the previous proposition to obtain a family $\{Z_\alpha^k : \alpha < \omega_1\}$ of non-empty zero-sets which has the finite intersection property and is such that $Z_\alpha^k \subset U_\alpha$ for each α, $\omega_0 \leqslant \alpha < \omega_1$. Now define $Z_\alpha = \bigcup \{Z_\alpha^k : k \geqslant 1\}$ for each $\alpha < \omega_1$. Thus, $\mathscr{F} = \{Z_\alpha : \alpha < \omega_1\}$ has the finite intersection property and satisfies $Z_\alpha \subset U_\alpha$ for $\omega_0 \leqslant \alpha < \omega_1$. Therefore, $\mathscr{F}$ is contained in a free z-ultrafilter A^p and A^p can contain no nowhere dense set. Further, since the family $\{Z_0^k : k \geqslant 1\}$ is a family of pairwise disjoint sets and the family $\{Z_0^k\} \cup \mathscr{F}$ also has the finite intersection property for each k, $\mathscr{F}$ is contained in infinitely many free z-ultrafilters.

Each such z-ultrafilter corresponds to a remote point which must belong to V. This set S of points can be expressed as the intersection

$$S = \{p \in \beta X : \mathscr{F} \subset A^p\} = \bigcap \{\mathrm{cl}_{\beta X} Z : Z \in \mathscr{F}\} \, .$$

Thus, S is a non-void, infinite, compact subset of $V \cap X^*$ and therefore contains at least 2^c points by Theorem 3.8. On the other hand, since X is separable, βX is a continuous image of $\beta \mathbb{N}$ and therefore can contain no more than 2^c points since the cardinality of $\beta \mathbb{N}$ is 2^c.

Finally, since q was selected as an arbitrary point of X^* and V as an arbitrary closed neighborhood of q, the remote points of βX form a dense subset of X^*. $\square$

4.44. Example. Theorem 4.43 shows that $\mathbb{Q}^*$ *contains a dense set of* 2^c *remote points of* $\beta \mathbb{Q}$. Note that Lemma 4.33 shows that $\mathbb{Q}^*$ *also contains a dense set of* 2^c *points which fail to be remote points of* $\beta \mathbb{Q}$. *If* p *is a remote point of* $\beta \mathbb{Q}$, *then the subspace* $\mathbb{Q} \cup \{p\}$ *shows that there exists a countable space without isolated points such that one of the points is not the limit of any discrete subspace.*

A similar example can be obtained by regarding $\mathbb{Q}$ as a subspace of $\beta \mathbb{R}$ and considering $\mathbb{Q} \cup \{q\}$ where q is a remote point of $\beta \mathbb{R}$.

The Example of $\beta \mathbb{R}$

4.45. The remote points of $\beta \mathbb{R}$ and the P-points of $\mathbb{R}^*$ make the example of $\beta \mathbb{R}$ particularly interesting. Since $\mathbb{R}$ is both locally compact and σ-compact, $\mathbb{R}^*$ is an F-space by Proposition 1.62. The set of remote points of $\mathbb{R}^*$ will be seen to contain an infinite compact F-space and the following result will be used to show the existence of remote points of $\beta \mathbb{R}$ which fail to be P-points of $\mathbb{R}^*$.

Proposition. *An infinite compact F-space contains at least 2^c non-P-points.*

Proof. Proposition 1.64 guarantees that an infinite compact F-space contains a copy of $\mathbb{N}^*$. We have seen (Proposition 4.31) that $\mathbb{N}^*$ contains 2^c non-P-points. The result follows since the non-P-points of the embedded copy of $\mathbb{N}^*$ must also be non-P-points of the compact F-space. □

Thus we see immediately that $\mathbb{R}^*$ has 2^c non-P-points.

4.46. Now consider the space $\mathbb{R}$. Let R and P denote the sets of remote points of $\beta\mathbb{R}$ and of P-points of $\mathbb{R}^*$, respectively. Then R' and P' will denote the non-remote points and the non-P-points. The following result appears in D. Plank's 1969 paper and yields four disjoint dense subsets of $\mathbb{R}^*$ each having cardinality 2^c.

Theorem [CH] (Plank). *The sets $P \cap R$, $P \cap R'$, $P' \cap R$, and $P' \cap R'$ are each dense subsets of $\mathbb{R}^*$ and each has cardinal 2^c.*

Proof. The result follows for $P \cap R$ from Corollary 4.24.

For $P \cap R'$ and $P' \cap R'$, let V be a closed $\beta\mathbb{R}$-neighborhood of any point p in $\mathbb{R}^*$. Then $V \cap \mathbb{R}$ is non-pseudocompact and C-embedded in $\mathbb{R}$. But then $V \cap \mathbb{R}$ contains a copy N of $\mathbb{N}$ which is C-embedded in $\mathbb{R}$ by Proposition 4.6. Then N^* is homeomorphic with $\mathbb{N}^*$ and is contained in $V \cap \mathbb{R}^*$. Since N is discrete, no point of N^* can be a remote point. Earlier we have seen that a point of N^* is a P-point of N^* if and only if it is a P-point of $\mathbb{R}^*$. Since N^* is homeomorphic to $\mathbb{N}^*$, it contains 2^c P-points if the Continuum Hypothesis is assumed and by the preceding Proposition, N^* contains 2^c non-P-points.

For $P' \cap R$, again let V be a closed $\beta\mathbb{R}$-neighborhood of a point p in $\mathbb{R}^*$. As in the proof of Theorem 4.43, construct an infinite compact set S of remote points in $V \cap \mathbb{R}^*$. Since $\mathbb{R}^*$ is a compact F-space, the C^*-embedded subspace S is also an F-space. Therefore, S has 2^c non-P-points and each of these is a non-P-point of $\mathbb{R}^*$. Since S is made up of remote points, $V \cap \mathbb{R}^*$ has 2^c points which are non-P-points of $\mathbb{R}^*$ but are remote points in $\beta\mathbb{R}$. □

Exercises

4A. *P-points of $\mathbb{N}^*$*

Let p be in $\mathbb{N}^*$. Then p is a P-point of $\mathbb{N}^*$ if and only if for every countable family $\{Z_n\}$ contained in A^p there exists Z in A^p such that $|Z \setminus Z_n|$ is finite for every n.
Reference: N. Hindman, Thesis, 1969 A.

4B. *Compact F-spaces*

If X is a compact F-space and S is a subspace of X such that $|X\setminus S| < 2^c$, then S is pseudocompact. [Proposition 1.64.]
Reference: Fine and Gillman, 1960.

4C. *Pseudocompactness and $|X^*|$*

1. If $|X^*| < 2^c$, then X is pseudocompact. [Lemma 4.5.]
2. If X is not pseudocompact, $|\beta X\setminus \upsilon X| \geqslant 2^c$. [Lemma 4.5 and Corollary 1.58.]
Reference: Based on [GJ, ex. 9D].

4D. *Round subsets*

A subset S of βX is said to be *round* if for any zero-set Z of X such that $\mathrm{cl}_{\beta X} Z$ contains S, $\mathrm{cl}_{\beta X} Z$ is a neighborhood of S.
1. S is round if and only if

$$\bigcap \{M^p : p \in S\} = \bigcap \{O^p : p \in S\}. \quad \text{[Proposition 4.27.]}$$

2. If $\mathrm{cl}_{\beta X} S$ is a round subset of βX, then S is a round subset of βX.
3. Every open subset of βX is round.
4. Any union of round subsets of βX is round.
5. If p belongs to X^* and $\{p\}$ is round, then p is a C-point. [Proposition 4.26.]
6. A subset S of βX is round if and only if, for any zero-set Z of X,

$$\mathrm{int}_S(\mathrm{cl}_{\beta X}(Z) \cap S) = \mathrm{int}_{\beta X}(\mathrm{cl}_{\beta X} Z) \cap S.$$

[For $p \in \mathrm{int}_S(\mathrm{cl}_{\beta X}(Z) \cap S)$, $p \in S\setminus \mathrm{cl}_{\beta X} W \subset \mathrm{cl}_{\beta X} Z$ for some zero-set W of X.]
References: (1)—(4) are from Mandelker, 1969, and (6) from Mandelker, 1971.

4E. *μ-compact spaces*

In Lemma 4.38, we used the fact that if X is realcompact, then $C_K(X) = \bigcap \{M^p : p \in X^*\}$. Call any space for which this equality holds a *μ-compact space*.
1. A zero-set Z is non-compact if and only if Z belongs to some free z-ultrafilter. [Apply Theorem 1.50 to Z, then use Proposition 1.7.]
2. The following are equivalent for any f in $C(X)$:
    (a) f belongs to every free maximal ideal of $C(X)$.
    (b) $\mathbf{Z}(f)$ meets every non-compact zero-set of X.

(c) $\mathrm{cl}_{\beta X} Z$ contains X^*.
   [(1), Proposition 1.17, and Theorem 1.30.]
3. The following are equivalent:
   (a) X is μ-compact.
   (b) X^* is a round subset of βX.
   (c) For any zero-set Z of X,

$$\mathrm{int}_{X^*}(\mathrm{cl}_{\beta X}(Z) \cap X^*) = \mathrm{int}_{\beta X}(\mathrm{cl}_{\beta X} Z) \cap X^*.$$

   (d) Every cozero-set with non-compact closure contains a non-compact zero-set.
4. The point ω_1 in $\omega_1 + 1$ is a round subset of $\beta \omega_1$. [4.1]
5. Not every μ-compact space is realcompact. [ω_1, Exercise 4C.]
6. The intersection of the μ-compact subspaces of βX which contain X is μ-compact.
7. For any X, there is a smallest μ-compact subspace of βX which contains X.

Reference: (1) and (2) are based on [GJ, 4.10, 4E]. (3)—(5), Mandelker, 1971. (6)—(7) Johnson and Mandelker, 1973.

4F. *Round subsets and $\beta \mathbb{R}$*

1. $\mathbb{R}^*$ is a round subset of $\beta \mathbb{R}$.
2. A point p of $\mathbb{R}^*$ is a remote point of $\beta \mathbb{R}$ if and only if $\{p\}$ is round. [Theorem 4.40]
3. $\mathbb{R}^*$ is the union of four disjoint round subsets of $\beta \mathbb{R}$, each of which is dense in $\mathbb{R}^*$. [Exercise 4D, Theorem 4.46.]

Reference: (2) is from Mandelker, 1969.

4G. *ψ-compact spaces*

Let $C_\psi(X)$ denote the set of all members of $C(X)$ which have pseudocompact support. Call any space for which $C_\psi(X) = C_K(X)$ a *ψ-compact space*.

1. Every realcompact space is ψ-compact. [Exercise 1C and Corollary 1.58.]
2. Every P-space is ψ-compact. [Proposition 1.65.]
3. Not every μ-compact space is ψ-compact. [ω_1, Exercise 4E.5.]
4. For any X, there is a smallest ψ-compact subspace of βX which contains X.

Reference: Johnson and Mandelker, 1973.

Chapter 5. Cellularity of Growths

5.1. In this chapter we consider relationships between the existence of families of open subsets of a space X and the cellularity of $X^* = \beta X \setminus X$. Recall that the *cellularity* of a space Y is the smallest cardinal number $\mathfrak{m}$ for which each pairwise disjoint family of non-empty open sets of Y has $\mathfrak{m}$ or fewer members. The *density* of a space Y is the smallest cardinal number which can be the cardinal number of some dense subspace of Y. It is clear that the density of a space is at least as great as the cellularity. Most of the results will take the form of providing a lower bound for the cellularity of X^* by demonstrating the existence of families of pairwise disjoint open sets in X^*. The methods used will be reminiscent of that used in Chapter 3 to show that the cellularity of $\mathbb{N}^*$ is $\mathfrak{c}$. In the last sections of the chapter, we will show that there exists a point in $\mathbb{N}^*$ which belongs to the closure of each member of a family of $\mathfrak{c}$ pairwise disjoint open sets of $\mathbb{N}^*$.

Lower Bounds for the Cellularity of X^*

5.2. The central results on the cellularity of X^* were obtained by W. W. Comfort and H. Gordon in 1964. The following theorem appears in a more general form in the 1928 paper of A. Tarski and will play a role similar to that played by Proposition 3.21 in the determination of the cellularity of $\mathbb{N}^*$. Recall that a family of sets is said to be *almost disjoint* if any two members of the family meet in a finite set.

Theorem (Tarski). *If $\mathfrak{m}$ and $\mathfrak{n}$ are cardinal numbers with $\mathfrak{n}$ infinite, then $\mathfrak{m} \leqslant \mathfrak{n}^{\aleph_0}$ if and only if a set having cardinality $\mathfrak{n}$ admits an almost disjoint collection of cardinality $\mathfrak{m}$ consisting of infinite subsets.*

Proof. Let D be a set with cardinality $\mathfrak{n}$ and assume that $\mathscr{E}$ is a collection of subsets satisfying the required conditions. With each E in $\mathscr{E}$, associate a countably infinite subset S_E of E. Since the members of $\mathscr{E}$ are almost disjoint, the assignment $E \mapsto S_E$ is one-to-one from $\mathscr{E}$ into the set of countably infinite subsets of D, and $\mathfrak{m} \leqslant \mathfrak{n}^{\aleph_0}$.

To establish the converse, it is sufficient to show that there exists an almost disjoint family of cardinality $\mathfrak{n}^{\aleph_0}$ of infinite subsets of D. Let $\mathscr{F}$ be the set of all finite sequences of distinct elements of D and let $\mathscr{G}$ be the set of all countably infinite sequences of distinct elements of D. Then the cardinality of $\mathscr{F}$ is $\mathfrak{n}$ and of $\mathscr{G}$ is $\mathfrak{n}^{\aleph_0}$. The proof will be completed by finding a collection of almost disjoint infinite subsets of $\mathscr{F}$ indexed by $\mathscr{G}$. For each G in $\mathscr{G}$, let I_G be the set of all F in $\mathscr{F}$ such that F is an initial segment of G. Then each set I_G is infinite and any distinct pair of sets of the form I_G have only finitely many elements of $\mathscr{F}$ in common. Thus, $\mathscr{E} = \{I_G : G \in \mathscr{G}\}$ is an almost disjoint family of infinite subsets of $\mathscr{F}$ and the cardinality of $\mathscr{E}$ is $\mathfrak{n}^{\aleph_0}$. $\square$

5.3. A family of subsets of a space X is called *locally finite* if every point of X has a neighborhood which meets only finitely many members of the family. A subset is called *relatively compact* if its closure is compact. We now apply Tarski's Theorem to obtain a lower bound for the cellularity of X^* related to the existence of a locally finite family of open subsets of X.

Theorem (Comfort and Gordon). *If X admits an infinite locally finite collection of $\mathfrak{n}$ non-empty relatively compact open sets, then the cellularity of X^* is at least $\mathfrak{n}^{\aleph_0}$.*

Proof. Let $\mathscr{U}$ be the locally finite collection of relatively compact open sets. We first show that the members of $\mathscr{U}$ can be shrunk to obtain a pairwise disjoint family of $\mathfrak{n}$ compact sets with non-void interiors. Let g be a function from $\mathscr{U}$ to X such that $g(T)$ is in T for T in $\mathscr{U}$, and let $A = g[\mathscr{U}]$. For each a in A, choose T_a in $\mathscr{U}$ such that $g(T_a) = a$. The function $a \mapsto T_a$ is one-to-one and onto a subset of $\mathscr{U}$ and the local finiteness of $\mathscr{U}$ guarantees that the cardinality of A is $\mathfrak{n}$. Using the local finiteness of $\mathscr{U}$ again and the regularity of X, we can find compact sets U_a and Y_a for each a in A such that

$$a \in \operatorname{int} U_a \subset U_a \subset \operatorname{int} Y_a \subset Y_a \subset T_a,$$

and such that $Y_a \cap A = \{a\}$. Now define

$$V_a = U_a \setminus \bigcup \{\operatorname{int} Y_b : b \in A, b \neq a\}.$$

Then V_a is a compact neighborhood of a. Further, if a and b are distinct points of A and if x is in V_a, then x is not in $\operatorname{int} Y_b$ so that V_a and V_b are disjoint since V_b is contained in $\operatorname{int} Y_b$. Thus,

$$\mathscr{V} = \{V_a : a \in A\}$$

is a pairwise disjoint and locally finite collection of $\mathfrak{n}$ compact sets each having non-void interior.

Now apply Tarski's Theorem to find a collection $\mathscr{E}$ of $\mathfrak{n}^{\aleph_0}$ infinite subsets of A such that any two members of $\mathscr{E}$ have finite intersection. For each E in $\mathscr{E}$, define

$$W_E = \beta X \backslash \operatorname{cl}_{\beta X}\left(X \backslash \bigcup_{a \in E} V_a\right).$$

Since each set W_E is open, the proof will be completed by showing that

$$\{W_E \cap X^* : E \in \mathscr{E}\}$$

is a pairwise disjoint family of non-void sets.

We begin by showing that the family is pairwise disjoint. Choose distinct elements E_1 and E_2 of $\mathscr{E}$. Then X can be written

$$(*) \qquad X = \left(X \backslash \bigcup_{a \in E_1} V_a\right) \cup \left(X \backslash \bigcup_{a \in E_2} V_a\right) \cup \left(\bigcup_{a \in E_1} V_a \cap \bigcup_{a \in E_2} V_a\right).$$

Call the last of these sets K. Then K can be written $K = \bigcup\{V_a : a \in E_1 \cap E_2\}$ and is therefore a compact subset of X. Taking the closure in βX of $(*)$ yields

$$\beta X = \operatorname{cl}_{\beta X}\left(X \backslash \bigcup_{a \in E_1} V_a\right) \cup \operatorname{cl}_{\beta X}\left(X \backslash \bigcup_{a \in E_2} V_a\right) \cup K.$$

Taking complements in βX now gives

$$W_{E_1} \cap W_{E_2} \cap (\beta X \backslash K) = \emptyset$$

so that $W_{E_1} \cap W_{E_2} \subset K \subset X$. Thus, $(W_{E_1} \cap X^*) \cap (W_{E_2} \cap X^*) = \emptyset$.

To show that $W_E \cap X^*$ is non-empty, begin by associating to each a in A a continuous function f_a of X to $[0,1]$ such that $f_a(a) = 1$ and $f_a(x) = 0$ for all x in $X \backslash V_a$. For E in $\mathscr{E}$, define $f_E = \Sigma\{f_a : a \in E\}$. The local finiteness of $\mathscr{V}$ guarantees that f_E is continuous. Let

$$F_E = \{x \in X : f_E(x) = 1\}$$

and note that $E \subset F_E \subset X$. Since F_E is the union of a locally finite family of closed subsets of X, F_E is closed in X. But E contains the infinite discrete subset E which is not closed in βX. Hence, there is a cluster point p_E of F_E in X^* and $\beta(f_E)(p_E)=1$. Since f_E vanishes on $X\setminus\bigcup\{V_a:a\in E\}$, $\beta(f_E)$ vanishes on the βX-closure of this set. Hence, p_E is in $W_E\cap X^*$. $\quad\square$

5.4. Examples. The theorem shows that the locally compact spaces ω_1 and the Deleted Tychonoff Plank $\mathbf{T}=(\omega_1+1)\times(\omega_0+1)\setminus\{(\omega_1,\omega_0)\}$ admit no infinite locally finite collections of open subsets. More generally, this will be the case for any space X for which X^* contains at most one point.

On the other hand, an infinite discrete space D of cardinality $\mathfrak{n}$ admits a locally finite family of $\mathfrak{n}$ finite open sets, and hence D^* has cellularity at least $\mathfrak{n}^{\aleph_0}$.

5.5. The property of not admitting infinite locally finite collections of open sets is characteristic of the class of pseudocompact spaces and will allow us to obtain a lower bound for the cellularity of the growth of a non-pseudocompact locally compact space.

Proposition. *A space is pseudocompact if and only if it admits no infinite locally finite collection of non-empty open subsets.*

Proof. Suppose that a space X admits a countably infinite locally finite collection of non-empty open sets $\{U_n\}$. Choose a sequence of distinct points $\{x_n\}$ such that x_n belongs to U_n. Choose a real-valued continuous function f_n such that $f_n(x_n)=n$ and $f_n(y)=0$ for y in $X\setminus U_n$. Then $f=\Sigma f_n$ is continuous because $\{U_n\}$ is locally finite and X is not pseudocompact since f is unbounded.

On the other hand, if X is not pseudocompact, then there exists an unbounded real-valued continuous function f on X such that $f[X]$ contains a sequence $\{r_n\}$ which has no accumulation point in $\mathbb{R}$. Choose a locally finite family of open intervals $\{V_n\}$ in $\mathbb{R}$ such that r_n is in V_n. Then $\{f^{\leftarrow}(V_n)\}$ is an infinite locally finite sequence of non-empty open sets of X. $\quad\square$

5.6. In a locally compact space, any locally finite family of open sets can be assumed to be composed of relatively compact sets. From this observation, the following corollary is immediate since $\mathfrak{c}=\aleph_0^{\aleph_0}$.

Corollary. *The cellularity of the growth of a locally compact but non-pseudocompact space is at least $\mathfrak{c}$.*

Thus, $\mathbb{R}^*$ has cellularity of at least $\mathfrak{c}$. Theorem 5.8 below will show that $\mathfrak{c}$ is also an upper bound for the cellularity of $\mathbb{R}^*$.

5.7. Theorem 5.3 shows that the existence of a locally finite family of n open subsets of a space X implies the existence of a family of $n^{\aleph_0}$ pairwise disjoint open subsets of X^*. The following theorem shows that the existence of such a locally finite family is not a necessary condition by showing that for any cardinal number m, there exists a pseudocompact space whose growth has cellularity m. However, by Proposition 5.5, the pseudocompact space can admit only a finite locally finite collection of open subsets.

Theorem (Comfort and Gordon). *If m is any cardinal number, there exists a locally compact space S such that:*
(1) *The cellularity of S^* is m.*
(2) *Any space having the same growth as S is pseudocompact.*

Proof. If m is finite, choose X to be the discrete space consisting of m points. Otherwise, choose X to be the one point compactification of the discrete space of cardinality m. In either case, the cellularity of X is m. In Proposition 4.3 we saw that there exists a space S such that S^* is homeomorphic to X. It remains to show that any space with growth X is pseudocompact. Theorem 4.34 shows that the growth of a non-pseudocompact locally compact space contains infinitely many non-P-points. Since X can contain at most one non-P-point, any space with X as its growth must be pseudocompact. □

5.8. In Theorem 3.22, we used an almost disjoint family of c infinite subsets of $\mathbb{N}$ in showing that the cellularity of $\mathbb{N}^*$ is c. The following theorem describes an analogous situation. The requirement that the members of the family of sets be infinite is replaced by requiring each set of the family to contain a non-compact zero-set and the almost disjoint requirement is replaced by asking that any two members of the family meet in a relatively compact set, i.e. in a set having compact closure.

Theorem (Comfort and Gordan). *If m is a cardinal number, then the cellularity of X^* is at least m if and only if X admits a collection of m cozero-sets $\{U_\alpha\}$ such that:*
(1) *Each U_α contains a non-compact zero-set, and*
(2) *$U_\alpha \cap U_\beta$ is relatively compact for $\alpha \neq \beta$.*

Proof. Assume that there is a collection $\{V_\alpha\}$ of m pairwise disjoint non-empty open subsets of X^*. For each α, choose p_α in V_α and an open subset W_α of βX such that $W_\alpha \cap X^* = V_\alpha$. Next, for each α choose a mapping f_α of βX into I with $f_\alpha(p_\alpha)=1$ and $f_\alpha(x)=0$ for all x not in W_α. Define

$$U_\alpha = \{x \in X : f_\alpha(x) > 1/2\}$$

and

$$Z_\alpha = \{x \in X : f_\alpha(x) \geqslant 2/3\} \,.$$

To show that Z_α is the non-compact zero-set required in (1), it suffices to show that p_α is in Z_α^*. But this is clear since p_α is not in $(X \backslash Z_\alpha)^*$ because $f_\alpha(p_\alpha)=1$ and $f_\alpha \leqslant 2/3$ on $(X \backslash Z_\alpha)^*$. Now suppose that $\{U_\alpha\}$ fails to satisfy (2). Then for some pair α and β, $\alpha \neq \beta$, there exists a point p of X^* for which

$$p \in \mathrm{cl}_{\beta X}\, \mathrm{cl}_X(U_\alpha \cap U_\beta) = \mathrm{cl}_{\beta X}(U_\alpha \cap U_\beta) \,.$$

But then $f_\alpha(p) \geqslant 1/2$ and $f_\beta(p) \geqslant 1/2$ and it follows that p is in $W_\alpha \cap W_\beta \cap X^* = V_\alpha \cap V_\beta = \emptyset$, which is a contradiction.

To show the converse, assume that X admits a collection $\{U_\alpha\}$ of cozero-sets satisfying (1) and (2). For each α, let Z_α be a non-compact zero-set contained in U_α. Since βX is normal and disjoint zero-sets of X have disjoint closures in βX, we can choose a mapping f_α of βX into I for each α such that $f_\alpha(p)=0$ for all p in $\mathrm{cl}_{\beta X}(X \backslash U_\alpha)$ and $f_\alpha(q)=1$ for all q in $\mathrm{cl}_{\beta X} Z_\alpha$. Let

$$V_\alpha = \{p \in \beta X : f_\alpha(p) > 0\} \,.$$

To complete the proof, we need to show that $V_\alpha \cap X^* \neq \emptyset$ for each α and that $V_\alpha \cap V_\beta \cap X^* = \emptyset$ for $\alpha \neq \beta$. First, note that $\emptyset \neq Z_\alpha^* \subset V_\alpha \cap X^*$ so that $V_\alpha \cap X^*$ is non-empty. Now suppose that p is in $V_\alpha \cap V_\beta$ for distinct α and β. Then $(f_\alpha f_\beta)(p) > 0$ so that

$$p \notin \mathrm{cl}_{\beta X}(X \backslash U_\alpha) \cup \mathrm{cl}_{\beta X}(X \backslash U_\beta) \,.$$

Hence, p is in $\mathrm{cl}_{\beta X}(U_\alpha \cap U_\beta)$ so that p belongs to X since $U_\alpha \cap U_\beta$ is relatively compact. Thus, $V_\alpha \cap V_\beta \cap X^*$ is empty. $\quad \square$

5.9. Examples. The only relatively compact open subset of $\mathbb{Q}$ is the empty set, hence the only families of open sets of $\mathbb{Q}$ satisfying the hypothesis of Theorem 5.8 are pairwise disjoint families. Because $\mathbb{Q}$ will admit only countable families of pairwise disjoint open sets, the theorem shows that the *cellularity of $\mathbb{Q}^*$ is $\aleph_0$.*

Because $\mathbb{R}$ is locally compact and not pseudocompact, the cellularity of $\mathbb{R}^*$ is at least $\mathfrak{c}$ by Corollary 5.6. Since $\mathbb{R}$ has a countable base for the open sets, Theorem 5.8 shows that *the cellularity of $\mathbb{R}^*$ is exactly $\mathfrak{c}$.*

5.10. Since every subset of a discrete space is both a zero-set and a co-zero-set and the only compact subsets are finite, the preceding theorem and Tarski's Theorem yield the following corollary for discrete spaces.

Corollary. *For the infinite discrete space D of cardinality $\mathfrak{n}$, the cellularity of D^* is $\mathfrak{n}^{\aleph_0}$.*

Thus, we see again that the cellularity of $\mathbb{N}^*$ is $\mathfrak{c}$. Since $\mathfrak{c}^{\aleph_0} = \mathfrak{c}$, we have the interesting result that the cellularity of D^* is no larger when the cardinality of D is $\mathfrak{c}$.

5.11. Corollary. *If D is the discrete space of cardinality $\mathfrak{c}$, the cellularity of D^* is $\mathfrak{c}$.*

5.12. It is particularly simple to construct a dense subspace of the growth of a discrete space and hence we can determine the density of such a growth. The following result showing that the density and cellularity are equal was obtained in 1965 by W. W. Comfort.

Theorem (Comfort). *The density of the growth of the discrete space of infinite cardinality $\mathfrak{n}$ is $\mathfrak{n}^{\aleph_0}$.*

Proof. Let D be the discrete space of cardinality $\mathfrak{n}$. In Corollary 5.10, we saw that the cellularity of D^* is $\mathfrak{n}^{\aleph_0}$ so that the density of D^* is at least $\mathfrak{n}^{\aleph_0}$.

To obtain the reverse inequality, first observe that for any open set of D^* there is a countable set M of D such that the open set contains M^*. Let $\mathscr{C}$ be the collection of all countably infinite subsets of D and choose a point p_M in M^* for each M belonging to $\mathscr{C}$. Then $\{p_M : M \in \mathscr{C}\}$ is a dense subset of D^* having cardinality $\mathfrak{n}^{\aleph_0}$. $\square$

$\mathfrak{n}$-Points and Uniform Ultrafilters

5.13. If $\mathfrak{n}$ is a cardinal number, a point is called an $\mathfrak{n}$-*point* if the point lies simultaneously in the closures of a pairwise disjoint family of $\mathfrak{n}$ open sets which do not contain the point. In 1967, R. S. Pierce asked if the existence of a 3-point in $\mathbb{N}^*$ could be demonstrated without using the Continuum Hypothesis. Pierce's question was motivated by his use of sheaf theory to relate questions involving regular rings and modules to topological problems. In 1969 B, N. Hindman showed without the Continuum Hypothesis that not only did a 3-point exist in $\mathbb{N}^*$, but that in fact there is a $\mathfrak{c}$-point in $\mathbb{N}^*$. This section will be devoted to Hindman's work and the following section to Pierce's investigation of 2- and 3-points in totally disconnected compact spaces.

An ultrafilter on a discrete space is called a *uniform ultrafilter* if every member of the ultrafilter has the same cardinality as the space. If D is an infinite discrete space, let uD denote the subspace of βD consisting of the points p of βD such that A^p is a uniform ultrafilter. It is clear that uD is contained in D^* and that $u\mathbb{N}=\mathbb{N}^*$. The following theorem shows that the cardinality of uD is the same as that of βD.

Theorem. *If D is the infinite discrete space of cardinality $\mathfrak{n}$, uD has cardinality $2^{2^{\mathfrak{n}}}$.*

Proof. We first show that there exists a uniform ultrafilter on D. Let $\mathscr{F}$ be the collection of all subsets of D whose complements have cardinality strictly less than $\mathfrak{n}$. Then $\mathscr{F}$ is a filter and any ultrafilter containing $\mathscr{F}$ must be uniform.

In order to show that $|uD|=2^{2^{\mathfrak{n}}}$, we will show that uD contains a copy of βD. It is first necessary to show that uD is closed. If p does not belong to uD, then p is in $\mathrm{cl}_{\beta D}Z$ for some Z of cardinality less than $\mathfrak{n}$. But then $\mathrm{cl}_{\beta D}Z$ is a neighborhood of p which misses uD, and uD is closed in βD. Now express D as the union of $\mathfrak{n}$ disjoint subsets F_α where each F_α has cardinality $\mathfrak{n}$. Since $\mathrm{cl}_{\beta D}F_\alpha=\beta F_\alpha$ and is homeomorphic to βD, there exists a point in uF_α and such a point must belong to uD. Now we may obtain a copy of βD in uD by choosing x_α in $uD\cap F_\alpha^*$ and setting $B=\{x_\alpha\}$. A mapping f in $C^*(B)$ can be extended to βD by defining a function g on D by letting g be constantly equal to $f(x_\alpha)$ on F_α. Then the extension of g to βD extends f. Thus, $\mathrm{cl}_{\beta D}B$ is a copy of βD and is contained in uD since uD is closed. $\quad\square$

5.14. In the proof of the preceding theorem, we have verified the following useful result.

Proposition. *Every subset of the infinite discrete space D having the same cardinality as D belongs to some uniform ultrafilter on D.*

5.15. If A_1 and A_2 are distinct members of an almost disjoint family of infinite subsets of $\mathbb{N}$, then A_1^* and A_2^* are disjoint open subsets of $\mathbb{N}^*$. This property of almost disjoint families of subsets of $\mathbb{N}$ has an analogue in the case of uD when D is the discrete space of infinite cardinality $\mathfrak{n}$. If $\mathscr{A}$ is a family of subsets of D each having cardinality $\mathfrak{n}$ and such that any two members of $\mathscr{A}$ meet in a set of cardinality smaller than $\mathfrak{n}$, then $A_1^*\cap A_2^*\cap uD$ is empty for A_1 and A_2 in $\mathscr{A}$. Such a family thus yields disjoint open subsets of uD. The following lemma is the first of a sequence of three results which will ultimately demonstrate the existence of a $2^{\mathfrak{n}}$-point in uD for certain cardinal numbers $\mathfrak{n}$. The lemma shows that for any family of $\mathfrak{m}$ disjoint open subsets of uD obtained from a family $\mathscr{A}$ as above, there is a point p in uD such that every neighborhood of p meets $\mathfrak{m}$ of the disjoint open sets.

Lemma. *If an infinite collection $\mathscr{A}$ of subsets of the infinite discrete space D of cardinality $\mathfrak{n}$ satisfies*
   (1) $|A| = \mathfrak{n}$ for all A in $\mathscr{A}$, and
   (2) $|A_1 \cap A_2| < \mathfrak{n}$ for A_1 and A_2 distinct members of $\mathscr{A}$,
then there exists a uniform ultrafilter A^p on D such that for each Z in A^p,

$$|\{A \in \mathscr{A} : |Z \cap A| = \mathfrak{n}\}| = |\mathscr{A}| .$$

Proof. For each A in $\mathscr{A}$, choose x_A in $\mathrm{cl}_{\beta D} A \cap uD$ and let $B = \{x_A : A \in \mathscr{A}\}$. If $A_1 \neq A_2$, then $x_{A_1} \neq x_{A_2}$ since x_{A_1} and x_{A_2} are in uD and if x_{A_1} were the same point as x_{A_2}, we would have x_{A_1} in $\mathrm{cl}_{\beta D} A_1 \cap \mathrm{cl}_{\beta D} A_2 = \mathrm{cl}_{\beta D}(A_1 \cap A_2)$ and $|A_1 \cap A_2| < \mathfrak{n}$, which contradicts the defining property of uD. Hence, B and $\mathscr{A}$ are of the same cardinality.

Now note that the subspace uD is compact since we saw in the proof of Theorem 5.13 that uD is closed in βD. We will use the compactness of uD to show that

(a) There is a point p in uD such that every neighborhood of p contains $|\mathscr{A}|$ points of B: If not, there is a finite open covering of uD each member of which contains fewer than $|\mathscr{A}|$ points of B. But this is impossible since B is infinite.

(b) A^p is the required ultrafilter: If Z is in A^p and $|Z \cap A| < \mathfrak{n}$, then $x_A \notin \mathrm{cl}_{\beta D} Z \cap uD$. However, since $\mathrm{cl}_{\beta D} Z$ is a neighborhood of p, there must be $|\mathscr{A}|$ members of $\mathscr{A}$ for which this does not happen and for these members, $|Z \cap A| = \mathfrak{n}$. $\square$

5.16. The next result appears in W. Sierpinski's 1928 paper and shows that for an infinite cardinal $\mathfrak{n}$ such that $2^{\mathfrak{m}} \leqslant \mathfrak{n}$ for all $\mathfrak{m} < \mathfrak{n}$, a set of cardinality $\mathfrak{n}$ will admit a family of $2^{\mathfrak{n}}$ subsets satisfying the hypothesis of the preceding lemma. Note that $\aleph_0$ satisfies the requirement on the cardinal number and that in this case, the theorem shows the existence of an almost disjoint family of $\mathfrak{c}$ infinite subsets of $\mathbb{N}$. Thus, Proposition 3.21 is a special case of the theorem.

Theorem (Sierpinski). *If $\mathfrak{n}$ is an infinite cardinal number such that $2^{\mathfrak{m}} \leqslant \mathfrak{n}$ for all $\mathfrak{m} < \mathfrak{n}$, then any set of cardinality $\mathfrak{n}$ admits a family of $2^{\mathfrak{n}}$ subsets each having cardinality $\mathfrak{n}$ such that the cardinality of the intersection of any two members of the family is strictly smaller than $\mathfrak{n}$.*

Proof. Let $\mathscr{F} = \{2^{\mathfrak{m}} : \mathfrak{m} < \mathfrak{n}\}$, i.e. $\mathscr{F}$ is the set of all $\mathfrak{m}$-sequences of 0's and 1's where $\mathfrak{m}$ runs through the ordinals $<\mathfrak{n}$. Then we have

$$|\mathscr{F}| = \sum_{\mathfrak{m}<\mathfrak{n}} 2^{\mathfrak{m}} \leqslant \mathfrak{n}\cdot\mathfrak{n} = \mathfrak{n}$$

and it is clear that the cardinality of $\mathscr{F}$ is at least $\mathfrak{n}$. Thus, the cardinality of $\mathscr{F}$ is $\mathfrak{n}$.

Now let $\mathscr{G}$ be the set of all $\mathfrak{n}$-sequences of 0's and 1's and for each y in $\mathscr{G}$, let E_y be the elements of $\mathscr{F}$ which are initial segments of y. Then each E_y has cardinality $\mathfrak{n}$. If x and y are distinct elements of $\mathscr{G}$, then there is a least ordinal $\mathfrak{m}<\mathfrak{n}$ at which x and y differ. Hence, E_x and E_y have only $\mathfrak{m}$ elements in common. Thus, the set $\mathscr{F}$ of cardinality $\mathfrak{n}$ has a family of subsets of the required type and the proof is complete. $\square$

It is easy to see that a cofinal class of cardinal numbers will satisfy the theorem. For any cardinal $\mathfrak{n}_0$, define inductively $\mathfrak{n}_{i+1}=2^{\mathfrak{n}_i}$. Then $\mathfrak{n}=\sup\{\mathfrak{n}_i : i\in\omega_0\}$ is such a cardinal.

5.17. The preceding two results show that for some cardinal numbers $\mathfrak{n}$, uD contains a collection of $2^{\mathfrak{n}}$ disjoint open subsets and a point p such that every neighborhood of p meets $2^{\mathfrak{n}}$ of the open subsets. However, this does not yet establish p as a $2^{\mathfrak{n}}$-point of uD since each neighborhood may meet a different subfamily of $2^{\mathfrak{n}}$ open subsets. The next theorem shows that the open sets may be chosen so that each neighborhood of p meets the same $2^{\mathfrak{n}}$ subsets.

Theorem (Hindman). *If D is the discrete space of cardinality $\mathfrak{n}$ and $2^{\mathfrak{m}}\leqslant\mathfrak{n}$ for all $\mathfrak{m}<\mathfrak{n}$, there is a $2^{\mathfrak{n}}$ point in uD.*

Proof. By the preceding theorem, there is a family $\mathscr{A}$ of $2^{\mathfrak{n}}$ subsets of D which satisfy the conditions of Lemma 5.15. By the same lemma, there exists a uniform ultrafilter $A^p=\{Z_\alpha : \alpha<2^{\mathfrak{n}}\}$ on D such that every member of A^p meets $2^{\mathfrak{n}}$ sets of $\mathscr{A}$ in a set of cardinality $\mathfrak{n}$.

(a) It is possible to choose a subset X_α of Z_α for each α less than $2^{\mathfrak{n}}$ such that $|X_\alpha|=\mathfrak{n}$, $|X_\alpha\cap X_\gamma|<\mathfrak{n}$ if $\alpha\neq\gamma$, and each X_α is the intersection of Z_α with a member of $\mathscr{A}$: For Z_1, choose A_1 in $\mathscr{A}$ such that $|Z_1\cap A_1|=\mathfrak{n}$. Now assume that for each $\sigma<\alpha$, we have chosen A_σ in $\mathscr{A}$ such that $|Z_\sigma\cap A_\sigma|=\mathfrak{n}$ and $A_\sigma\neq A_\gamma$ for all $\gamma<\sigma$. Now since the lemma shows that there are $2^{\mathfrak{n}}$ such members of $\mathscr{A}$ for each Z_α, there is a member A_α of $\mathscr{A}$ such that $|Z_\alpha\cap A_\alpha|=\mathfrak{n}$ and $A_\alpha\neq A_\sigma$ for all $\sigma<\alpha$. Since the cardinality of the intersection of distinct members of $\mathscr{A}$ is less than $\mathfrak{n}$, setting $X_\alpha=Z_\alpha\cap A_\alpha$ completes the proof of (a).

Since $|X_\alpha|=\mathfrak{n}$ for each α, there is a collection of subsets $\{X_{\tau\alpha} : \tau<2^{\mathfrak{n}}\}$ of X_α such that each $X_{\tau\alpha}$ has cardinality $\mathfrak{n}$ and $|X_{\tau\alpha}\cap X_{\sigma\alpha}|<\mathfrak{n}$ if $\tau\neq\sigma$. For each $\tau<2^{\mathfrak{n}}$, set

$$U_\tau=\left(\bigcup\{\mathrm{cl}_{\beta D}X_{\tau\alpha} : \alpha<2^{\mathfrak{n}}\}\right)\cap uD.$$

It is clear that each U_τ is open in uD. We will show that $\{U_\tau\}$ is a pairwise disjoint family and that each U_τ contains p in its closure.

(b) $\{U_\tau\}$ is pairwise disjoint: If q is in $U_\sigma \cap U_\tau$, then there is a pair α and γ such that

$$q \in \mathrm{cl}_{\beta D} X_{\sigma\alpha} \cap \mathrm{cl}_{\beta D} X_{\tau\gamma} = \mathrm{cl}_{\beta D}(X_{\sigma\alpha} \cap X_{\tau\gamma}).$$

However, since $|X_{\sigma\alpha} \cap X_{\tau\gamma}| < \mathfrak{n}$, this cannot occur since q belongs to uD.

(c) Every uD neighborhood of p meets each U_τ: Each such neighborhood is of the form $\mathrm{cl}_{\beta D} Z_\alpha \cap uD$ for some Z_α in A^p so that it is sufficient to show that $\mathrm{cl}_{\beta D} Z_\alpha \cap uD \cap U_\tau \neq \emptyset$ for all $\tau < 2^\mathfrak{n}$. Since $|X_{\tau\alpha}| = \mathfrak{n}$ and $X_{\tau\alpha} \subset X_\alpha \subset Z_\alpha$, both $X_{\tau\alpha}$ and Z_α belong to some uniform ultrafilter A^q and

$$q \in \mathrm{cl}_{\beta D} Z_\alpha \cap \mathrm{cl}_{\beta D} X_{\tau\alpha} \cap uD \subset \mathrm{cl}_{\beta D} Z_\alpha \cap \mathrm{cl}_{\beta D} U_\tau \cap uD.$$

Hence, p is a $2^\mathfrak{n}$-point of uD. $\square$

5.18. Since $u\mathbb{N}$ is $\mathbb{N}^*$, the following corollary is immediate.

Corollary. *There is a $\mathfrak{c}$-point in $\mathbb{N}^*$.*

This result is particularly interesting since we saw from Proposition 3.24 that two disjoint cozero-sets in $\mathbb{N}^*$ have disjoint closures.

5.19. In the case of the countable discrete space $\mathbb{N}$, Lemma 5.15 shows that with every almost disjoint family of $\mathfrak{n}$ infinite subsets of $\mathbb{N}$ we can associate a free ultrafilter A^p on $\mathbb{N}$ such that p is a potential $\mathfrak{n}$-point of $\mathbb{N}^*$. By assuming the Continuum Hypothesis, we will be able to reverse the steps of the lemma. By assuming that the elements of any free ultrafilter A^p are indexed by $\omega_1 = \mathfrak{c}$, we can construct an almost disjoint family $\mathscr{A}$ of infinite subsets of $\mathbb{N}$ such that each member of A^p contains a member of $\mathscr{A}$. The proof of Theorem 5.17 can then be adapted to A^p and the family $\mathscr{A}$ to show that p is a $\mathfrak{c}$-point of $\mathbb{N}^*$.

Theorem [CH] (Hindman). *Every point of $\mathbb{N}^*$ is a $\mathfrak{c}$-point of $\mathbb{N}^*$.*

Proof. Using the Continuum Hypothesis, let the members of any free ultrafilter A^p be indexed by ω_1, i.e. let $A^p = \{Z_\alpha : \alpha < \omega_1\}$. We show that infinite sets X_α can be chosen such that X_α is a subset of Z_α and $X_\alpha \cap X_\gamma$ is finite if $\alpha \neq \gamma$ as in (a) of the proof of Theorem 5.17. The remainder of the proof is the same as in Theorem 5.17.

Choose any infinite subset X_1 of Z_1 such that X_1 is not in A^p. For $\alpha < \omega_1$, assume that an infinite subset X_σ of Z_σ has been chosen for all $\sigma < \alpha$ such that X_σ is not in A^p and $\{X_\sigma : \sigma < \alpha\}$ is an almost disjoint family. Let $B = \{\sigma < \alpha : X_\sigma \cap Z_\alpha \text{ is infinite}\}$. If B is finite, then $\bigcup\{X_\sigma : \sigma \in B\}$ cannot be in A^p since no single set X_σ is in A^p. Thus, any infinite subset of $Z_\alpha \setminus \bigcup \{X_\sigma : \sigma \in B\}$ which is not in A^p may be chosen to be X_α. If B is infinite, we may index B by the natural numbers and write $B = \{\sigma_n\}$. Since $X_{\sigma_k} \cap Z_\alpha$ is infinite, it is possible to choose

$$x_k \in (X_{\sigma_k} \cap Z_\alpha) \setminus \bigcup_{j<k} X_{\sigma_j}.$$

Let X_α be any infinite subset of $\{x_k\}$ which does not belong to A^p. Then $|X_\alpha \cap X_{\sigma_k}| \leqslant k < \aleph_0$ and if σ is not in B, $|X_\alpha \cap X_\sigma| \leqslant |X_\sigma \cap Z_\alpha| < \aleph_0$ by the definition of B. $\quad\square$

π-Points and Compactifications of $\mathbb{N}$

5.20. Recall from Chapter 2 that an extremally disconnected space is one in which the closure of every open set is open. It can easily be shown that a space is extremally disconnected if and only if disjoint open sets of the space have disjoint closures, i.e. if and only if the space contains no 2-points. It is easy to see that $\beta\mathbb{N}$ is extremally disconnected.

In his 1967 paper, R.S. Pierce was particularly interested in 2- and 3-points of compact totally disconnected spaces. He showed that in the presence of the Continuum Hypothesis, the P-points of $\mathbb{N}^*$ are 3-points. He then utilized this result to show that any totally disconnected compact space contains a closed subspace which admits a 3-point. In the process, Pierce characterized those compactifications of $\mathbb{N}$ which do not admit a 3-point and also showed that every compactification other than $\beta\mathbb{N}$ does contain a 2-point. Recall that any compactification K of $\mathbb{N}$ is the image of $\beta\mathbb{N}$ under a mapping φ which sends $\mathbb{N}^*$ onto $K\setminus\mathbb{N}$ and leaves points of $\mathbb{N}$ fixed. We will call φ the canonical mapping of $\beta\mathbb{N}$ onto K. A *fiber* of a mapping is the inverse image of a point.

Proposition (R.S. Pierce). *A compactification K of $\mathbb{N}$ contains no 3-points exactly when no fiber of the canonical map φ of $\beta\mathbb{N}$ onto K contains more than two points. Further, any point of K which has two points in its fiber is a 2-point of K.*

Proof. Let K be a compactification of $\mathbb{N}$ which contains no 3-points. We will show that no fiber of the canonical map contains more than two points. By Corollary 1.12, any point of K whose fiber contains more than a single point must lie in $K\setminus\mathbb{N}$ and the fiber must be contained in $\mathbb{N}^*$. Suppose that the fiber of a point x contains distinct points p_1, p_2, and p_3. Choose disjoint subsets Z_i for $i=1,2,3$ such that Z_i belongs to A^{p_i}. Then $\varphi[Z_i]$ is open in K and by the continuity of φ,

x belongs to $\operatorname{cl}\varphi[Z_i]$ for each i, contradicting the assumption that K has no 3-point.

To prove the converse, assume that $\varphi^{\leftarrow}(x)=\{p,q\}$ where p and q are not necessarily distinct.

(a) If V and W are open neighborhoods of p and q, respectively, then x belongs to $\operatorname{int}(\varphi[V\cup W])$: The set $F=\varphi^{\leftarrow}(\varphi[\beta\mathbb{N}\setminus(V\cup W)])$ is closed because φ is closed and continuous. Clearly, F cannot contain p or q since x is in $\varphi[V\cup W]$. Then $\varphi[F]=\varphi[\beta\mathbb{N}\setminus(V\cup W)]$ is closed in K and contains $K\setminus\varphi[V\cup W]$ while missing x. Hence, $K\setminus\varphi[F]$ is a neighborhood of x contained in $\varphi[V\cup W]$.

(b) K has no 3-points: Assume the contrary. Let U_1, U_2, and U_3 be disjoint open sets of K such that a point x of K belongs to none of the U_i but does belong to the closures of all three. The sets $\varphi^{\leftarrow}(U_i\cap\mathbb{N})$ are disjoint open sets of $\mathbb{N}$ and therefore have disjoint closures in $\beta\mathbb{N}$. Thus, if $\{p,q\}$ is the fiber of x, there is some $i=1,2$ or 3 such that p and q both fail to belong to $\operatorname{cl}\varphi^{\leftarrow}(U_i\cap\mathbb{N})$. Let V and W be neighborhoods of p and q, respectively, such that V and W both miss $\varphi^{\leftarrow}(U_i\cap\mathbb{N})$. Then $\varphi[V\cup W]\cap U_i\cap\mathbb{N}=\emptyset$. By (a), x belongs to $\operatorname{int}\varphi[U\cup V]$, and hence x cannot belong to $\operatorname{cl}(U_i\cap\mathbb{N})=\operatorname{cl}(U_i)$. Thus, x is not a 3-point.

(c) If $\varphi^{\leftarrow}(x)=\{p,q\}$ and p and q are distinct, then x is a 2-point: Choose disjoint subsets Z and W of $\mathbb{N}$ such that Z belongs to A^p and W to A^q. Then by the continuity of φ, x belongs to $\operatorname{cl}\varphi[U]\cap\operatorname{cl}\varphi[W]$ and thus is a 2-point. □

5.21. If K is any compactification of $\mathbb{N}$ other than $\beta\mathbb{N}$, it is clear that the canonical map φ of $\beta\mathbb{N}$ onto K identifies some pair of points and their common image is a 2-point of K. Thus, we have verified the following

Corollary. *$\beta\mathbb{N}$ is the unique extremally disconnected compactification of $\mathbb{N}$.*

5.22. Example. Proposition 5.20 provides a way to construct interesting compactifications of $\mathbb{N}$. For example, define a permutation σ of $\mathbb{N}$ by

$$\sigma(n) = \begin{cases} n-1 & \text{if } n \text{ is even,} \\ n+1 & \text{if } n \text{ is odd.} \end{cases}$$

Then $\sigma^2=1_{\mathbb{N}}$ and if $\beta(\sigma)$ is the homeomorphism of $\beta\mathbb{N}$ which extends σ, $\beta(\sigma)^2=1_{\beta\mathbb{N}}$. The orbit of any point of $\mathbb{N}^*$ under $\beta(\sigma)$ consists of exactly two points. If K is the compactification of $\mathbb{N}$ which results from identifying each orbit contained in $\mathbb{N}^*$ to a single point, then every point of $K\setminus\mathbb{N}$ is a 2-point of K, but K contains no 3-points. K is easily seen to be totally disconnected by showing that distinct points of $K\setminus\mathbb{N}$ have complementary neighborhoods.

5.23. Pierce proved the next theorem under the assumption of the Continuum Hypothesis because the proof uses the existence of a 3-point in $\mathbb{N}^*$ which he had earlier demonstrated by using that assumption. We can now establish the result without the Continuum Hypothesis by virtue of Corollary 5.18.

Theorem. *Any infinite totally disconnected compact space contains a closed subspace which admits a 3-point in the subspace topology.*

Proof. Since X is an infinite Hausdorff space, it contains a copy N of $\mathbb{N}$. Let $K = \mathrm{cl}\, N$. If K contains a 3-point, we are done. Otherwise, we will show that K contains a copy of $\mathbb{N}^*$ which will be the required subspace. Because K is a compactification of $\mathbb{N}$ without 3-points, each fiber of the canonical mapping φ of $\beta\mathbb{N}$ onto K contains at most two points. Let W be the subspace of $\beta\mathbb{N}$ consisting of singleton fibers, i.e.

$$W = \{p \in \beta\mathbb{N} : \varphi^{\leftarrow}(\{\varphi(p)\}) = \{p\}\} .$$

W clearly contains $\mathbb{N}$. If $W \setminus \mathbb{N}$ contains an infinite closed subspace of $\mathbb{N}^*$, then Theorem 3.3 shows that $W \setminus \mathbb{N}$ contains a copy of $\mathbb{N}^*$ which is embedded into K by φ. Since $\mathbb{N}^*$ contains a 3-point, the copy of $\mathbb{N}^*$ is the required closed subspace of X.

If $W \setminus \mathbb{N}$ fails to contain an infinite closed subspace of $\mathbb{N}^*$, then $S = \beta\mathbb{N} \setminus W$ is dense in $\mathbb{N}^*$. For each point p of S, let $\hat{p}$ denote the point of S such that $\varphi^{\leftarrow}(\{\varphi(p)\}) = \{p, \hat{p}\}$. Note that $\hat{\hat{p}} = p$. Choose p_0 in S and let U_0 and V_0 be disjoint clopen neighborhoods in $\beta\mathbb{N}$ of p_0 and $\hat{p}_0$, respectively, such that $U_0 \cup V_0$ does not contain $\mathbb{N}^*$. Suppose that points $\{p_0, \ldots, p_n\}$ have been chosen in S such that $\varphi(p_i) \neq \varphi(p_j)$ for $i \neq j$ and that disjoint clopen sets U_n and V_n have been found such that U_n contains $\{p_0, \ldots, p_n\}$, V_n contains $\{\hat{p}_0, \ldots, \hat{p}_n\}$, and $U_n \cup V_n$ does not contain $\mathbb{N}^*$. Now since S is dense in $\mathbb{N}^*$, there exists a point p in $S \setminus (U_n \cup V_n)$ and it is clear that $\varphi(p)$ is distinct from all the $\varphi(p_i)$. There are three distinct possibilities for the location of $\hat{p}$: $\hat{p}$ belongs to U_n, $\hat{p}$ belongs to V_n, or $\hat{p}$ fails to belong to $U_n \cup V_n$. In the first case, put $p_{n+1} = \hat{p}$. In the other cases, put $p_{n+1} = p$. The clopen sets U_n and V_n can be enlarged, if necessary, to disjoint clopen sets U_{n+1} and V_{n+1} to include p_{n+1} and $\hat{p}_{n+1}$ such that $U_{n+1} \cup V_{n+1}$ does not contain $\mathbb{N}^*$. Thus, by induction, we can obtain sequences of points $\{p_i\}$ and $\{\hat{p}_i\}$ such that

$$\{p_i : i < \omega_0\} \subset U = \bigcup U_i$$

and

$$\{\hat{p}_i : i < \omega_0\} \subset V = \bigcup V_i ,$$

$\varphi(p_i) \neq \varphi(p_j)$ for $i \neq j$, and U and V are disjoint. Now put

$$C = \mathrm{cl}_K \varphi(U \cap \mathbb{N}) \cap \mathrm{cl}_K \varphi(V \cap \mathbb{N}) .$$

Then C is an infinite closed subspace of $K\setminus N$ and the fiber of each point of C consists of two points—one in $\mathrm{cl}\,U\setminus N$ and one in $\mathrm{cl}\,V\setminus N$. Hence, C is homeomorphic to an infinite closed subspace of N^* and thus contains a copy of N^* which contains a 3-point. $\quad\square$

Exercises

5A. *n-points*

Let x be a non-isolated point of a totally disconnected space X such that there is a well-ordered family $\{U_\alpha : \alpha < \lambda\}$ of clopen sets of X satisfying:

(i) If $\alpha < \beta < \lambda$, then $U_\beta \subset U_\alpha$.

(ii) $\bigcap\limits_{\alpha < \lambda} U_\alpha = \{x\}$.

1. λ is a limit ordinal.
2. For each positive integer n, x is an n-point. [For $\alpha < \lambda$, put $P_\alpha = U_\alpha \setminus U_{\alpha+1}$. For $i = 0, 1, \ldots, n-1$, put $W_i = \bigcup \{P_\alpha : \alpha < \lambda$ and $\alpha \equiv i \,(\mathrm{mod}\,n)\}$. Then $x \in \mathrm{cl}\,W_i$.]
3. Assume the Continuum Hypothesis. Any P-point of N^* is a n-point. Reference: R. S. Pierce, 1967.

5B. *Finiteness at infinity*

A space X is said to be *finite at infinity* if X^* contains only finitely many points. A family $\mathscr{C}$ of subsets of X is said to be *pairwise completely separated* if distinct members of $\mathscr{C}$ are completely separated.

1. A space which is finite at infinity is locally compact. [Theorem 1.59.]
2. If $\mathfrak{m}$ is any cardinal number and $|X^*| = \mathfrak{m}$, then any pairwise completely separated family $\mathscr{C}$ of closed subsets of X having $|\mathscr{C}| > \mathfrak{m}$ must contain a compact member. [Theorem 1.14.]
3. If $n < \omega_0$, then $|X^*| \leqslant n$ if and only if any pairwise completely separated family of $n+1$ closed subsets of X contains a compact member.
4. If $n < \omega_0$, then $|X^*| = n$ if and only if there exists a pairwise completely separated family of n closed, non-compact subsets of X but no such family having $n+1$ sets exists.
5. $|X^*| \leqslant 1$ if and only if of any two completely separated subsets of X, at least one is compact.
6. Any infinite sequence of distinct points of X^* contains an infinite subsequence such that pairwise disjoint closed βX-neighborhoods can be found for the points of the subsequence.

7. X is finite at infinity if and only if every infinite pairwise completely separated family of subsets of X contains a compact member.
8. A space which is the union of a strictly increasing sequence of open subsets contains an infinite pairwise disjoint family of closed countable subsets. [If $X = \bigcup U_n$, $U_n \subset U_{n+1}$, and $x_n \in U_{n+1} \setminus U_n$, consider $A_i = \{x_n : n = 2^i(2m+1), m = 1, 2, \ldots\}$ for each i.]
9. A normal space which is finite at infinity is countably compact.
10. A countably compact normal space need not be finite at infinity. [See Example 8.23. The space $\omega_1 \times \omega_1$ is normal since disjoint closed subspaces of $\omega_1 \times \omega_1$ have disjoint closures in $\beta(\omega_1 \times \omega_1)$.]

References: Based on the 1971 paper of P.A. Firby in which finiteness at infinity is also characterized in terms of subalgebras of $C^*(X)$. The condition of (5) was considered in the 1949 paper of R. Doss and in [GJ, Ex. 6J, 15R.] where it is shown to be equivalent to the existence of a unique uniform structure on X.

Chapter 6. Mappings of βX to X^*

6.1. This chapter is devoted to the investigation of various types of mappings of βX to X^*. The chapter will be in four parts. We will first consider a sequence of preliminary results giving sufficient conditions to ensure that the continuous image of a dense, C^*-embedded subspace will be C^*-embedded in the image of the whole space. These results will then be applied to study mappings of βX onto X^* and in particular to obtain necessary conditions for the existence of a retraction of βX onto X^*. Next, the growths of the compactifications of a locally compact space X will be characterized as the continuous images of X^*. Finally we will investigate mappings of extremally disconnected spaces and apply our results to mappings of Stone-Čech compactifications of discrete spaces. As a corollary we will obtain our third proof that $\mathbb{N}^*$ is not homogeneous.

C^*-Embedding of Images

6.2. The results of this and the succeeding section on retractive spaces appear in the 1965 paper of W. W. Comfort.

Proposition. *If r is an open or closed mapping of X onto Z and Y is a dense subspace of X such that any fiber of r containing distinct points meets Y, then $r[Y]$ is C-embedded (resp. C^*-embedded) in Z if Y is C-embedded (resp. C^*-embedded) in X.*

As we will see in Chapter 10, the diagram which illustrates the proposition indicates that this result has some aspects of an adjunction or pushout. The proof will be given for C-embedding and is similar for C^*-embedding.

Proof. To simplify the notation, write r_1 for $r|Y$. Let f be in $C(r[Y])$ so that $f \circ r_1$ belongs to $C(Y)$. Since Y is C-embedded in X, $f \circ r_1$ has

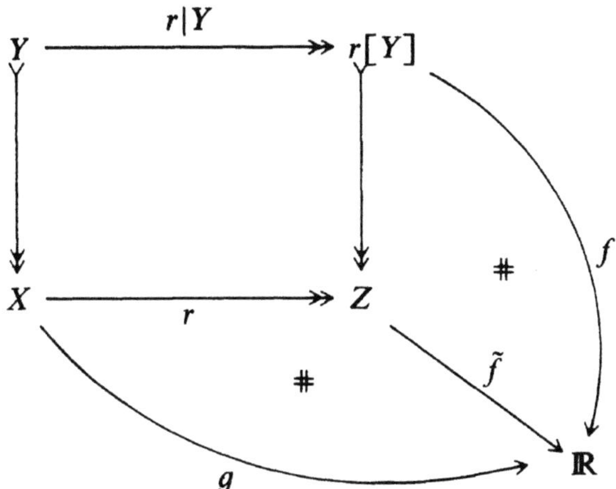

an extension g in $C(X)$. The first step toward defining an extension of f to Z is to show that:

(a) g is constant on any fiber of r: For any fiber containing more than one point, we choose a point y of Y lying in the fiber and show that $g(x)=g(y)$ for any other point x in the fiber. Since Y is dense in X, there is a net $\{y_\alpha\}$ in Y such that

$$x = \lim_\alpha y_\alpha .$$

Then

$$g(x) = \lim_\alpha g(y_\alpha) = \lim_\alpha (f \circ r_1)(y_\alpha) .$$

Since r is continuous on X, we have that

$$\lim_\alpha r(y_\alpha) = r(x) = r(y) \in r[Y] .$$

Since f is continuous on $r[Y]$,

$$g(y) = (f \circ r_1)(y) = f(\lim_\alpha r(y_\alpha)) = \lim_\alpha f(r(y_\alpha))$$
$$= \lim_\alpha (f \circ r_1)(y_\alpha) = g(x) .$$

Since r is onto Z, we can use (a) to define $\tilde{f}$ on Z by

$$\tilde{f}(r(x)) = g(x) .$$

(b) $\tilde{f}$ extends f: For $r(y)$ in $r[Y]$,

$$\tilde{f}(r(y)) = g(y) = (f \circ r_1)(y) = f(r_1(y)) = f(r(y)) .$$

(c) $\tilde{f}$ is continuous: From (a), it follows that for any subset S of $\mathbb{R}$, $\tilde{f}^{\leftarrow}(S)=r[g^{\leftarrow}(S)]$. If r is open, taking S to be any open subset of $\mathbb{R}$ shows that $\tilde{f}$ is continuous. Taking S to be closed if r is closed completes the proof. $\square$

6.3. The assumption that r is a closed or open mapping in the proposition will not be overly restrictive since all of our applications will be to cases where X is compact and a mapping of a compact space into a Hausdorff space is always closed.

Corollary. *If r maps a compact space K onto L and every fiber of r which meets the closure of a C^*-embedded subspace Y of K in more than a single point meets Y, then $r[Y]$ is C^*-embedded in L.*

Proof. Since a mapping defined on a compact space is closed, applying the proposition with $X=\mathrm{cl}_K Y$ and $Z=r[X]$ yields $r[Y]$ C^*-embedded in $r[X]$. It then follows that $r[Y]$ is C^*-embedded in L since the compact space $r[X]$ is C^*-embedded in L by Proposition 1.47. $\square$

6.4. The hypotheses of the preceding corollary are satisfied when K is chosen to be βX and the restriction of r to X^* is required to be one-to-one.

Corollary. *If r maps βX onto Z such that r restricted to X^* is one-to-one, then the image of a closed (in X) C^*-embedded subspace Y of X is C^*-embedded in Z and $\beta(r[Y])=r[\beta Y]$.*

Proof. It follows from the previous corollary that $r[Y]$ is C^*-embedded in Z. Since Z is compact and r is closed,

$$\beta(r[Y]) = \mathrm{cl}_Z r[Y] = r[\mathrm{cl}_{\beta X} Y] = r[\beta Y] . \quad \square$$

Retractive Spaces

6.5. A mapping of a space Y onto a subspace S of Y is called a *retraction* if the mapping leaves points of S fixed. The subspace S is then called a *retract* of Y and it is easily seen that S is closed whenever Y is a Hausdorff space [D, p. 322]. We will call a space X *retractive* if there is a retraction of βX onto X^*. Since a retraction of βX onto X^* is one-to-one on X^*, we can restate Corollary 6.4 for retractive spaces.

Corollary. *If r is a retraction of βX onto X^* and Y is a closed (in X) and C^*-embedded subspace of X, then $r[Y]$ is C^*-embedded in X^* and $\beta(r[Y]) = r[\beta Y]$. In particular, $\beta(r[X]) = X^*$.*

6.6. In the preceding Corollary, note that if Y is both closed and C^*-embedded, then $Y^* = \mathrm{cl}_{\beta X} Y \backslash X$ is the growth of Y in βY and is in the closure of both Y and $r[Y]$. In particular, this will be the case if Y is a C-embedded copy of $\mathbb{N}$ since it follows from Corollary 4.20 that a C-embedded copy of $\mathbb{N}$ is closed. In the next theorem, we will apply the previous corollary to show that in the presence of the Continuum Hypothesis, a retractive space cannot contain a C-embedded copy of $\mathbb{N}$, and thus must be pseudocompact. We will use Corollary 4.30 to show that a P-point exists in $\mathbb{N}^*$ for any C-embedded copy of $\mathbb{N}$. Then by Lemma 4.32, such a point must also be a P-point of X^* if X is retractive since a retractive space is easily seen to be locally compact. We will then see that this together with the fact that $\beta(r[\mathbb{N}]) = r[\beta \mathbb{N}]$ for any retraction $r : \beta X \to X^*$ yields a contradiction.

Theorem [CH] (Comfort). *A retractive space is locally compact and pseudocompact.*

Proof. Consider a retractive space X and let r be a retraction of βX onto X^*. Then X^* is closed in βX and X is therefore locally compact. Suppose that X is not pseudocompact. Then X contains a C-embedded copy M of $\mathbb{N}$ and Corollary 6.5 shows that

$$\beta(r[M]) = r[\beta M] = r[M] \cup M^*.$$

$\beta(r[M])$ is the closure in βX of $r[M]$ and contains M^* so that $r[M]$ cannot be finite. Moreover, infinitely many points of $r[M]$ lie outside of M^* since otherwise $r[M] \cap M^*$ would be a countable dense subspace of M^*, which is impossible since the cellularity of M^* is c by Corollary 5.10. Choose a countably infinite discrete subspace $K = \{p_1, p_2, \ldots\}$ of $r[M]$ which misses M^*. Choose x_n in $r^{\leftarrow}(p_n) \cap M$ for each n and define $N = \{x_n\}$. Since N is a subspace of M, it is clear that N is a countably infinite C^*-embedded closed subspace of X. Because r is a retraction, $K = r[N]$ is C^*-embedded in X^* and $N^* = K^*$.

We now use the fact that $N^* = K^*$ with N a subspace of X and K a subspace of X^* to reach a contradiction. If the Continuum Hypothesis is assumed, Corollary 4.30 shows that there exists a P-point of N^*. By Lemma 4.32, every P-point of N^* is a P-point of X^* since X is locally compact. However, the mapping f defined on K by $f(p_n) = 1/n$ for all n has a continuous extension g to X^* and g vanishes on K^*. But g cannot vanish on any X^* neighborhood of a point of K^* since every such neighborhood meets K. Hence, no point of K^* is a P-point of X^*, which is a contradiction showing that X can contain no C-embedded copy of $\mathbb{N}$ and is therefore pseudocompact by Corollary 4.6. □

6.7. Since any pseudocompact metrizable or realcompact space is compact, we have the

Corollary [CH]. *No realcompact or metrizable space is retractive.*

6.8. The converse of Theorem 6.6 is easily seen to be false.

Example. *There exists a locally compact pseudocompact space Λ such that Λ^* has cardinality 2^c but the only mappings of $\beta\Lambda$ onto Λ^* are the constant mappings.*

If $\mathbb{N}$ is considered as a subspace of $\mathbb{R}$, then $\beta\mathbb{N} = \mathrm{cl}_{\beta\mathbb{R}}\mathbb{N}$ and $\mathbb{N}^*$ is a subspace of $\mathbb{R}^*$. Put $\Lambda = \beta\mathbb{R}\backslash\mathbb{N}^*$. Clearly, Λ is locally compact and the cardinality of Λ^* is 2^c. If f is an unbounded real-valued mapping on Λ, then f must be unbounded on some closed subspace S of $\mathbb{R}$ which misses $\mathbb{N}$. Thus, f must be unbounded on $\mathrm{cl}_{\beta\mathbb{R}}S$, which is impossible. Hence, Λ is pseudocompact. Since $\beta\Lambda$ contains $\mathbb{R}$ as a dense, connected subspace, $\beta\Lambda$ must be connected. However, $\Lambda^* = \mathbb{N}^*$ is totally disconnected, and therefore any mapping of $\beta\Lambda$ to Λ^* must be constant. The space Λ was introduced by M. Katětov in 1953 and is also described in [GJ, ex. 6 P, 9 E].

6.9. If we assume in addition that X is an F-space, we can prove a variation of Theorem 6.6 without assuming the Continuum Hypothesis. Since P-points will not enter into the proof we can replace pseudocompact with the weaker condition that X not contain a closed C^*-embedded copy of $\mathbb{N}$.

Theorem (Comfort). *A retractive space X is locally compact, and if X is also an F-space, then X contains no closed, C^*-embedded copy of $\mathbb{N}$.*

Proof. Assume that X is retractive. As in the proof of Theorem 6.6, X is locally compact. Assume that X contains a closed, C^*-embedded copy M of $\mathbb{N}$ and construct the subspaces K and N as in the proof of Theorem 6.6. Because K misses the compact set $N \cup N^*$ and N misses the compact set $K \cup N^*$, K and N are both open in the subspace $N \cup K$. Thus, the function which is constantly equal to 1 on N and to 0 on K is continuous on $N \cup K$. Therefore, it has a continuous extension to all of βX, since βX is an F-space whenever X is (Theorem

1.60), and therefore any countable subspace of βX is C^*-embedded (Proposition 1.64). But this yields a contradiction since such a function clearly has no extension to any point of $K^* = N^*$. Thus, X can contain no closed, C^*-embedded copy of $\mathbb{N}$. ☐

6.10. Since any P-space is an F-space and any countable subspace of such a space is closed and C^*-embedded, we have the following

Corollary. *No non-empty P-space is retractive, and in particular, no infinite discrete space is retractive.*

6.11. Although no retractive F-space can contain a closed, C^*-embedded copy of $\mathbb{N}$, a retractive F-space can contain a non-closed C^*-embedded copy of $\mathbb{N}$.

Example. *The complement in $\beta\mathbb{N}$ of any finite subspace of $\mathbb{N}^*$ is a retractive F-space which contains $\mathbb{N}$ as a non-closed C^*-embedded subspace.* Such a space is an F-space since it is C^*-embedded in $\beta\mathbb{N}$. It is evident that such a subspace of $\beta\mathbb{N}$ is retractive since $\beta\mathbb{N}$ can be written as the pairwise disjoint union of neighborhoods of finitely many points of $\mathbb{N}^*$ and the retraction consists of shrinking each neighborhood to the respective point.

Growths of Compactifications

6.12. Corollary 1.12 showed that βX is the maximal compactification of the space X in the sense that if K is another compactification of X, then the embedding of X into K extends to a mapping of βX onto K. We will now see that X^* can be thought of as the maximal growth of X since all the points of X^* are sent to points of $K \setminus X$ by the extension, i. e. every growth is the continuous image of X^*.

Recall from Section 1.44 that if f is a mapping of X to Y and $\mathscr{F}$ is a z-filter on X, then the collection of zero-sets in Y whose inverse images belong to $\mathscr{F}$ is a z-filter on Y and is denoted by $f^{\#}\mathscr{F}$.

Proposition. *The growth of any compactification of a space X is a continuous image of X^*.*

Proof. Let K be any compactification of X. Then the embedding h of X into K has an extension $\beta(h)$ of βX onto K. Let p be a point of βX and suppose that $\beta(h)(p) = x$, where x is a point of X. Then $(\beta(h))^{\#} A^p = h^{\#} A^p = A^p$ converges to x as does A^x. Since each point of X is the limit of a unique z-ultrafilter, $x = p$. Hence, p belongs to X and the points of X^* must all be mapped to $K \setminus X$ by $\beta(h)$. ☐

6.13. An immediate corollary occurs in the case of a retractive space.

Corollary. *The growth of any compactification of a retractive space X is a continuous image of βX.*

6.14. The following theorem appears in the 1966B paper of K. D. Magill and shows that the converse of the Proposition 6.12 holds for a locally compact space X, i.e. that every continuous image of X^* is a growth of X in some compactification. A map f of X to Y is a *quotient map* whenever a subset U of Y is open if and only if $f^{\leftarrow}(U)$ is open in X. In the proof of Magill's result we will construct a compactification K of X by identifying the fibers of a function on βX to points and defining K to be the resulting quotient space.

Theorem (Magill). *The growths of the compactifications of a locally compact space X are the continuous images of X^*.*

Proof. Proposition 6.12 showed that every growth of X is a continuous image of X^*. To obtain the converse, let f be a mapping of X^* onto Y and assume that X and Y are disjoint. Define a function g from βX to $X \cup Y$ by

$$\begin{cases} g(p) = p & \text{if } p \in X, \\ g(p) = f(p) & \text{if } p \in X^*. \end{cases}$$

The function g induces an equivalence relation on βX by defining two points of βX to be equivalent if g takes the same value on the two points. Let K be the quotient space obtained by identifying the equivalence classes to points and let φ be the quotient map sending βX onto K. We will show that K is a compactification of X in which the growth of X is homeomorphic to Y.

Call a subset H of βX saturated if H contains the equivalence class of any point belonging to H. Since the locally compact space X is an open saturated subspace of βX and the restriction of g to X is a homeomorphism, $\varphi[X]$ is homeomorphic to X. Because X^* is a closed saturated subspace of βX and the restriction of g to X^* is a closed mapping onto Y, $\varphi[X^*]$ is homeomorphic to Y [D, p. 130]. Thus, the image of X is a copy of X which is easily seen to be dense in K and the complement of the image of X in K is homeomorphic to Y.

Since a continuous image of a compact space is compact, it remains to show that K is Hausdorff. There are three cases to be considered to show that distinct points of K may be separated by disjoint open sets. In each case, we will use that the image under φ of a saturated open subset of βX is open in K. In the first case, consider distinct points x_1 and x_2 belonging to $\varphi[X]$. The fibers $\varphi^{\leftarrow}(x_1)$ and $\varphi^{\leftarrow}(x_2)$ are singletons and have disjoint open neighborhoods U_1 and U_2 in X. Because X is open in βX, U_1 and U_2 are also open in βX and since both sets are saturated, their images are disjoint neighborhoods of x_1 and x_2 in K.

In the second case, if y_1 and y_2 are distinct points of $\varphi[X^*]$, $\varphi^{\leftarrow}(y_1)$ and $\varphi^{\leftarrow}(y_2)$ are disjoint closed subsets of the compact space X^* and are therefore contained in disjoint open sets U_1 and U_2 of X^*. Because the restriction of φ to X^* is a closed map onto the regular space $\varphi[Y]$, there exist closed neighborhoods V_1 and V_2 of y_1 and y_2 in $\varphi[X^*]$ such that

$$\varphi^{\leftarrow}(y_i) \subset \varphi^{\leftarrow}(\operatorname{int} V_i) \subset \varphi^{\leftarrow}(V_i) \subset U_i$$

for each value of i [D, p. 86]. Since the closed sets $\varphi^{\leftarrow}(V_1)$ and $\varphi^{\leftarrow}(V_2)$ are contained in disjoint open subsets of X^*, there exists a pair of disjoint open sets W_1 and W_2 of βX such that

$$W_i \cap X^* = \varphi^{\leftarrow}(\operatorname{int} V_i)$$

for both values of i. But then W_1 and W_2 are saturated open sets of βX and their images under φ are the required disjoint neighborhoods of y_1 and y_2.

In the final case, let x be a point of $\varphi[X]$ and y be a point of $\varphi[X^*]$. Then $\varphi^{\leftarrow}(x)$ and $\varphi^{\leftarrow}(y)$ are contained in disjoint open sets U and V, respectively, and as in the preceding case, we can obtain a saturated X^* neighborhood W of $\varphi^{\leftarrow}(y)$ such that

$$\varphi^{\leftarrow}(y) \subset W \subset V \cap X^*.$$

Let G be an open set of βX such that $G \cap X^* = W$. Then $U \cap X$ and $G \cap V$ are disjoint open saturated neighborhoods of $\varphi^{\leftarrow}(x)$ and $\varphi^{\leftarrow}(y)$, respectively, and their images under φ are disjoint neighborhoods of x and y. $\quad\square$

6.15. Magill's Theorem gives an added importance to any technique which will describe mappings having as their domain the growth of a locally compact space. One such method using Boolean algebras was used in Theorem 3.31 to establish Parovičenko's result that every compact space having weight at most $\aleph_1$ is a continuous image of $\mathbb{N}^*$. That result together with Magill's Theorem yields the

Corollary. *Every compact space of weight at most $\aleph_1$ is the growth of* $\mathbb{N}$ *in some compactification.*

A second technique was illustrated in Example 5.22 where we constructed a compactification of $\mathbb{N}$ by identifying the orbits of points of $\mathbb{N}^*$ under an automorphism of $\beta \mathbb{N}$.

6.16. Corollary 6.15 implies that there exists a compactification of $\mathbb{N}$ whose growth is a copy of the ordinal space $\omega_1 + 1$. We will describe such a compactification by using properties of the clopen subsets of $\mathbb{N}^*$ to construct a mapping of $\mathbb{N}^*$ onto $\omega_1 + 1$. The construction will utilize a modification of the technique used to prove Urysohn's Lemma. Recall that a normal space is one in which every neighborhood of a closed set contains a closed neighborhood of the set. Urysohn's Lemma is the statement that disjoint closed subsets of a normal space are completely separated. If F and H are the disjoint closed subsets of the normal space X, then the definition of normality is applied successively to the neighborhood $X\backslash H$ of F to obtain a sequence of open neighborhoods $\{U_r\}$ where the indexing set is the set of rationals of $[0,1]$ and

$$F \subset U_0 \subset \operatorname{cl} U_0 \subset \cdots \subset U_r \subset \operatorname{cl} U_r \subset U_s \subset \cdots \subset X\backslash H = U_1$$

whenever $r < s$. Then putting

$$f(x) = \inf\{r : x \in U_r\}$$

yields a mapping which completely separates F from H. The details of the proof can be found in [GJ, 3.12].

The following application of Magill's Theorem appears in the 1971 paper of S. P. Franklin and M. Rajagopalan and uses a similar technique where the sets involved are clopen subsets of $\mathbb{N}^*$ and are indexed by ω_1. Recall from Corollary 3.27 that every non-empty G_δ in $\mathbb{N}^*$ has non-empty interior and therefore contains a non-empty clopen set.

Corollary (Franklin and Rajagopalan). *There exists a compactification* $\gamma\mathbb{N}$ *of* $\mathbb{N}$ *such that* $\gamma\mathbb{N}\backslash\mathbb{N}$ *is homeomorphic to* $\omega_1 + 1$.

Proof. We first construct a strictly increasing ω_1-sequence $\{U_\alpha : \alpha < \omega_1\}$ of non-empty clopen subsets of $\mathbb{N}^*$. Choose any proper non-empty clopen set of $\mathbb{N}^*$ as U_0. If U_α has been defined for each $\alpha < \beta$, the complements $\mathbb{N}^*\backslash U_\alpha$ form a decreasing sequence of clopen subsets of $\mathbb{N}^*$. Therefore their intersection contains a non-empty clopen set A (Corollary 3.27). Write $A = B \cup C$ where B and C are disjoint non-empty clopen sets. Choose the complement of B to be U_β. Then C is contained in $U_\beta\backslash U_\alpha$ for all $\alpha < \beta$ so that U_α is properly contained in U_β for all $\alpha < \beta$. Further, the complement of U_β is B, which is non-empty allowing the process to continue on to ω_1.

We now use this ω_1-sequence to define a mapping f of $\mathbb{N}^*$ onto $\omega_1 + 1$. Imitating the Urysohn Lemma technique, define

$$f(p) = \sup\{\alpha < \omega_1 : p \notin U_\alpha\}.$$

Points in U_0 are thus sent to 0. Since the complements of the U_α form a decreasing sequence of compact sets, there exists at least one point outside the union of the U_α's so that f takes on the value ω_1. Moreover, f is onto since the sequence is strictly increasing. To prove that f is continuous, note that intervals of the form $[0,\beta)$ and $(\beta,\omega_1]$ for $0 \leqslant \beta < \omega_1$ form a subbase for the topology of $\omega_1 + 1$. Thus, it is sufficient to show that $f^\leftarrow([0,\beta))$ and $f^\leftarrow((\beta,\omega_1])$ are open. Note that $f(p) < \beta$ if and only if p is in U_α for some $\alpha < \beta$. Then $f^\leftarrow([0,\beta)) = \bigcup\{U_\alpha : \alpha < \beta\}$ and is open. Similarly, $f(p) > \beta$ if and only if p is not in U_α for some $\alpha > \beta$ so that $f^\leftarrow((\beta,\omega_1]) = \bigcup\{\mathbb{N}^* \setminus U_\alpha : \beta < \alpha < \omega_1\}$, and f is continuous. $\square$

6.17. The compactification $\gamma\mathbb{N}$ of $\mathbb{N}$ is useful as a source of examples. If $\mathscr{U}$ and $\mathscr{V}$ are covers of a space, then $\mathscr{V}$ is a *refinement* of $\mathscr{U}$ if every member of $\mathscr{V}$ is a subset of some member of $\mathscr{U}$. A space is said to be *paracompact* if every open cover has a locally finite open refinement. Every paracompact space is also normal [D, p. 163], however, the ordered space ω_1 is a normal space which fails to be paracompact since the open covering of ω_1 consisting of initial segments fails to have a locally finite open refinement.

A second countable normal space is metrizable (Exercise 6 H), and hence is paracompact [D, p. 186]. It is natural to ask if second countable can be replaced by separable in this statement, i.e. is there a separable normal space which fails to be paracompact? In 1956, M. E. Rudin and L. F. McAuley each gave fairly complicated examples of such a space. The compactification $\gamma\mathbb{N}$ was used by Franklin and Rajagopalan to give a simpler example. Since ω_1 is embedded as a closed subspace of $\gamma\mathbb{N}\setminus\{\omega_1\}$ and a closed subspace of a paracompact space must be paracompact, we can use the fact that ω_1 is not paracompact to show that $\gamma\mathbb{N}\setminus\{\omega_1\}$ is a separable normal space which fails to be paracompact.

Example. $\gamma\mathbb{N}\setminus\{\omega_1\}$ *is a separable, totally disconnected locally compact space which is normal but which is not paracompact.* It is immediate that $\gamma\mathbb{N}\setminus\{\omega_1\}$ is separable and locally compact and not paracompact.
(a) $\gamma\mathbb{N}$ is totally disconnected: Since the points of $\mathbb{N}$ are isolated, we need only show that points of $\omega_1 + 1$ can be separated by clopen sets. Because $\omega_1 + 1$ is zero-dimensional, if α and β are distinct points of $\omega_1 + 1$, then there is a clopen set U of $\omega_1 + 1$ which contains α but not β. Thus, if φ is the canonical map of $\beta\mathbb{N}$ onto $\gamma\mathbb{N}$, then $\varphi^\leftarrow(U)$ is clopen in $\mathbb{N}^*$ and $\varphi^\leftarrow(U) = Z^*$ for some subset Z of $\mathbb{N}$. The map φ is closed since $\beta\mathbb{N}$ is compact and therefore φ is a quotient map. Since $\mathrm{cl}\,Z$ is saturated with respect to φ and is clopen, $\varphi[\mathrm{cl}\,Z]$ is a clopen subset of $\gamma\mathbb{N}$ which contains α but not β.
(b) Of any pair of disjoint closed subsets of ω_1, one of the sets must be

bounded: Let R and S be closed subsets of ω_1 and assume that both are unbounded. Then we can choose unbounded sequences $\{r_n\}$ and $\{s_n\}$ contained in R and S respectively and such that $r_n < s_n < r_{n+1}$ for every n. Then the two sequences have the same supremum which must then belong to both R and S.

(c) $\gamma \mathbb{N} \setminus \{\omega_1\}$ is normal: Let F and H be disjoint closed subsets of $\gamma \mathbb{N} \setminus \{\omega_1\}$. Then one of them, say F, meets ω_1 in a closed and bounded set. Since a closed and bounded subspace of ω_1 is compact, there is a clopen subset U of $\gamma \mathbb{N} \setminus \{\omega_1\}$ which contains $F \cap \omega_1$ and misses H. Then $V = U \cup (F \cap \mathbb{N})$ is open since only isolated points have been added and is also closed since any cluster points of $F \cap \mathbb{N}$ are contained in $F \cap \omega_1$. Thus, V is clopen, contains F, and misses H.

6.18. Example. The construction of $\gamma \mathbb{N}$ shows that the space $\Lambda = \beta \mathbb{R} \setminus \mathbb{N}^*$ described in Example 6.8 also has a compactification with growth homeomorphic to $\omega_1 + 1$.

Mappings of βD and other Extremally Disconnected Spaces

6.19. In the remainder of the chapter we will consider mappings of βD into itself and into the growth D^* when D is discrete. We will first deduce two results that depend on the cardinality of D and follow from our consideration of density characters in Chapter 5. Then we will examine mappings of extremally disconnected spaces and apply these results to βD. One result that will be obtained is another proof of the fact that $\mathbb{N}^*$ is not homogeneous.

Although no infinite discrete space is retractive, it is possible to map βD onto D^* whenever the cardinality $\mathfrak{n}$ of D satisfies $\mathfrak{n} = \mathfrak{n}^{\aleph_0}$.

Theorem (Comfort). *If D is the discrete space of infinite cardinality $\mathfrak{n}$, then D^* is a continuous image of βD if and only if $\mathfrak{n} = \mathfrak{n}^{\aleph_0}$.*

Proof. Recall from Theorem 5.12 that the density character of D^* is $\mathfrak{n}^{\aleph_0}$. Then if f is a mapping of βD onto D^*, $f[D]$ is dense in D^* and $\mathfrak{n}^{\aleph_0} \leqslant |f[D]| \leqslant \mathfrak{n}$.

Conversely, map D onto a dense subset of D^* having cardinality $\mathfrak{n} = \mathfrak{n}^{\aleph_0}$. Since D^* is compact, the mapping extends to βD and the extended mapping is onto D^* since the image of D is dense in D^*. $\quad\square$

6.20. Examples. The theorem shows that $\mathbb{N}^*$ *is not a continuous image of* $\beta \mathbb{N}$. However, *if D is the discrete space of cardinality $\mathfrak{c}$, D^* is a continuous image of βD since* $\mathfrak{c}^{\aleph_0} = \mathfrak{c}$.

6.21. By observing that $\aleph_1^{\aleph_0} = \mathfrak{c}$, the following equivalent form of the Continuum Hypothesis follows from Theorem 6.19.

Corollary. *If D is the discrete space of cardinality $\aleph_1$, the Continuum Hypothesis holds if and only if D^* is a continuous image of βD.*

6.22. A *fixed point* of a function f is a point x such that $f(x)=x$. In his 1968A paper, Z. Frolík studied fixed points of mappings of βD into D^*. In the succeeding paper, 1968B, he generalized these results by adapting his proofs to mappings defined on compact extremally disconnected spaces. The remainder of the chapter is drawn from these two papers. A survey of these results and of Frolík's work on non-homogeneity also appears in his 1971 paper.

We begin with an example from his 1968A paper. As we saw in Example 3.43, the theory of types of ultrafilters can be applied to construct mappings of $\beta\mathbb{N}$ into $\mathbb{N}^*$. If two points p and q of $\mathbb{N}^*$ have the same relative types with respect to countable discrete subspaces X and Y, then there is a bijection of X onto Y whose extension to a mapping of cl X onto cl Y sends p to q. Using this fact, we can describe a mapping of $\beta\mathbb{N}$ into $\mathbb{N}^*$ which has exactly one fixed point. The construction will be similar to that of Example 3.43 except for modifications which assure that the map is one-to-one on $\mathbb{N}$.

Example. *There exists a mapping f of $\beta\mathbb{N}$ into $\mathbb{N}^*$ such that*
   (a) *f restricted to $\mathbb{N}$ is one-to-one,*
   (b) *$f[\mathbb{N}]$ is a countable union of discrete subspaces, and*
   (c) *$f^2[\beta\mathbb{N}]$ is a singleton implying that f has a unique fixed point.*

To construct f, choose a discrete countable set Y_n of $\mathbb{N}^*$ for each $n \geqslant 1$ such that Y_{n+1} is a subspace of Y_n^* and choose a point p which belongs to $\bigcap_n Y_n^*$. Let $\{N_n\}$ be any decomposition of $\mathbb{N}$ into infinite sets and choose a sequence $X = \{x_n\}$ in $\mathbb{N}^*$ such that the type of x_n relative to N_n is the same as the type of p relative to Y_n. The sequence X is a discrete subspace of $\mathbb{N}^*$ so that there exists a homeomorphism h of $\beta\mathbb{N}$ onto cl X. Put $q = h(p)$. Then because h is a homeomorphism, the relative type of q with respect to the discrete subspace $h[Y_n]$ is the same as that of x_n with respect to N_n. Therefore, there exists a bijection of N_n with $h[Y_n]$ which sends the traces of neighborhoods of x_n on N_n to traces of neighborhoods of q on $h[Y_n]$ for each n. Thus, we have defined a bijection of $\mathbb{N}$ onto $\bigcup h[Y_n]$ such that if f is the Stone-Čech extension of the bijection, then $f(x_n)=q$ for each n. Because q is in cl X, this implies that $f(q)=q$. Further, since $f[\beta\mathbb{N}]$ is contained in the image of h, $f[\beta\mathbb{N}]$ is contained in cl X. Thus, by continuity,

$$f^2[\beta\mathbb{N}] \subset f[\mathrm{cl}\, X] \subset \{q\},$$

and q is the unique fixed point of f.

Although we have seen two examples of mappings of $\beta\mathbb{N}$ into $\mathbb{N}^*$ which have fixed points, neither of the examples was an embedding. The following sequence of results will show that no embedding of $\beta\mathbb{N}$ into $\mathbb{N}^*$ can have a fixed point.

6.23. Let f map a space X into itself. If x is a fixed point of f, then any neighborhood U of x must meet its image, i.e. $U \cap f[U] \neq \emptyset$. This fact will enable us to describe certain subspaces of X which cannot contain a fixed point.

A subset B of X is said to be *f-invariant* if $f[B]$ is contained in B and *f-coinvariant* if $f^{\leftarrow}(B)$ is contained in B. Call a subset B of X *3-decomposable* (with respect to f) if B is f-invariant and is the pairwise disjoint union of three clopen subsets B_1, B_2, and B_3 which satisfy $B_i \cap f[B_i] = \emptyset$ for each i. Note that if a 3-decomposable set is non-empty, then at least two of the sets forming the decomposition must be non-empty. Clearly, a 3-decomposable subset cannot contain a fixed point.

Recall that a space is said to be *extremally disconnected* if the closure of every open subset is open. The next two lemmas will show in turn the existence of a largest 3-decomposable subset for any map f if X is extremally disconnected and then of a non-empty 3-decomposable subset if X is also compact and f is a homeomorphism other than the identity. The two lemmas are then combined to obtain a theorem which asserts that the set of fixed points of an embedding of a compact extremally disconnected space into itself is clopen.

Lemma. *For each mapping f of an extremally disconnected space X into itself, X has a largest 3-decomposable subset which contains any other 3-decomposable subset. Moreover, the largest 3-decomposable subset is also f-coinvariant.*

Proof. Consider the collection $\mathscr{D}$ of all subsets of X which are 3-decomposable with respect to f. Observe that $\mathscr{D}$ may contain only the empty set, e.g. this will be the case if f is the identity.

We first show that:

(a) $\mathscr{D}$ contains an element S such that any element of $\mathscr{D}$ that misses S must be empty: It is easily seen that the closure of a union of an arbitrary pairwise disjoint family of members of $\mathscr{D}$ belongs to $\mathscr{D}$. Thus, to obtain an element of the required type it is sufficient to choose any element of $\mathscr{D}$ and successively enlarge it by adding disjoint members of $\mathscr{D}$ and taking closures.

However, the 3-decomposable set described in (a) need not be the largest. To obtain the largest set, we need a further property of $\mathscr{D}$:

(b) Any element D_0 of $\mathscr{D}$ is contained in an f-coinvariant element of $\mathscr{D}$:

Let $\mathscr{B}_0 = \{B_1^0, B_2^0, B_3^0\}$ be a 3-decomposition of D_0 and consider the family $\{D_\alpha\}$ of members of $\mathscr{D}$ and their 3-decompositions $\{\mathscr{B}_\alpha\}$ defined as follows: For $\alpha = \sigma + 1$ a non-limit ordinal, define $D_\alpha = f^\leftarrow(D_\sigma)$ and define $\mathscr{B}_\alpha$ by:

$$B_1^\alpha = B_1^\sigma \cup [f^\leftarrow(B_2^\sigma \cup B_3^\sigma)\setminus D_\sigma],$$

$$B_2^\alpha = B_2^\sigma \cup [f^\leftarrow(B_1^\sigma)\setminus D_\sigma], \quad \text{and}$$

$$B_3^\alpha = B_3^\sigma.$$

For α a limit ordinal, define $D_\alpha = \mathrm{cl}(\bigcup\{D_\gamma : \gamma < \alpha\})$ and define $\mathscr{B}_\alpha$ by $B_i^\alpha = \mathrm{cl}(\bigcup\{B_i^\gamma : \gamma < \alpha\})$ for each $i = 1, 2, 3$. The continuity of f together with the fact that disjoint open subsets of an extremally disconnected space have disjoint closures shows that $\mathscr{B}_\alpha$ is a 3-decomposition of D_α. Because $\{D_{\alpha+1}\setminus D_\alpha\}$ is a family of pairwise disjoint clopen subsets, we must have $D_{\alpha+1}\setminus D_\alpha = \emptyset$ for some α at most equal to the cellularity of X. When this occurs, D_α is an f-coinvariant member of $\mathscr{D}$ which contains D_0.

Now we combine properties (a) and (b) of the collection $\mathscr{D}$ to obtain the largest element. Let B be an f-coinvariant member of $\mathscr{D}$ containing a member S satisfying (a). Given any D in $\mathscr{D}$, $D\setminus B$ must be f-invariant since B is f-coinvariant. Thus, by restricting the 3-decomposition of D, $D\setminus B$ is seen to belong to $\mathscr{D}$. But then $D\setminus B$ must be empty since it misses S. Thus, B contains every member of $\mathscr{D}$. $\quad\square$

6.24. Lemma. *If f is a homeomorphism of an extremally disconnected compact space K into itself which is not the identity, then K contains a non-empty set which is 3-decomposable with respect to f.*

Proof. Because f is not the identity, there is some x in K such that $f(x) \neq x$. Therefore, there exists a clopen neighborhood B_3 of x such that $f[B_3] \cap B_3 = \emptyset$. The set $f[B_3]$ is clopen in $f[K]$, and hence is compact. Thus, there are finitely many clopen subsets $U_1, \ldots, U_n$ of K such that $f[B_3] = (\bigcup U_i) \cap f[K]$ and $U_i \cap B_3 = \emptyset$ for each i. By putting $Z_1 = \bigcup U_i$, we obtain a clopen subset Z_1 of K such that $Z_1 \cap f[K] = f[B_3]$ and $Z_1 \cap B_3 = \emptyset$. Further, $f[Z_1] \cap Z_1 = \emptyset$ since f is one-to-one, Z_1 and B_3 are disjoint, and $f^\leftarrow(Z_1)$ is contained in B_3. Therefore, there exists a clopen subset Z_2 of K such that $Z_2 \cap (B_3 \cup Z_1) = \emptyset$ and $Z_2 \cap f[K] = f[Z_1]\setminus B_3$. Continuing by induction, we obtain a sequence $\{Z_n\}$ of clopen subsets of K such that for $n \geq 1$,

$$Z_{n+1} \cap (B_3 \cup Z_1 \cup \cdots \cup Z_n) = \emptyset, \quad \text{and}$$

$$Z_{n+1} \cap f[K] = f[Z_n]\setminus B_3.$$

Such a Z_{n+1} exists since

$$f^{\leftarrow}(Z_n) \subset Z_{n-1} \cup B_3$$

and this implies that

$$f[Z_n] \cap (\bigcup \{Z_i : i \leqslant n\}) = \emptyset.$$

Now let B_1 be the closure of the union of all Z_n with n odd and B_2 be the closure of the union of all Z_n with n even. Because K is extremally disconnected, the sets B_1 and B_2 are clopen and disjoint. From the continuity of f, we observe that

$$f[B_3] \subset B_1 ,$$
$$f[B_2] \subset B_1 \cup B_3 , \quad \text{and}$$
$$f[B_1] \subset B_2 \cup B_3 .$$

Hence, $B = B_1 \cup B_2 \cup B_3$ is a non-empty 3-decomposable subset of K. $\square$

6.25. The combination of the previous lemmas yields the main result.

Theorem (Frolík). *If f is a one-to-one mapping of an extremally disconnected compact space K into itself, then the set F of fixed points of f is clopen and $K \setminus F$ is 3-decomposable with respect to f.*

Proof. By Lemma 6.23, there exists a largest 3-decomposable subset B. We will show that $F = K \setminus B$ is the set of fixed points of f. By the Lemma, B is f-coinvariant so that F is f-invariant. Because the 3-decomposable subset B is clopen, F is a clopen subset of K and hence is compact. Thus, $f|F$ is a closed one-to-one mapping and therefore is a homeomorphism of F into itself. The restriction $f|F$ must therefore be the identity since otherwise $F = K \setminus B$ would contain a non-empty 3-decomposable set by Lemma 6.24, which would allow B to be enlarged to form a larger 3-decomposable subset, which would be a contradiction. $\square$

6.26. Recall from Exercise 2K that a space in which the closure of every cozero-set is open is called a *basically disconnected space*. Hence, every extremally disconnected space is also basically disconnected. The following example shows that the hypothesis in the theorem cannot be weakened by replacing extremally disconnected with basically disconnected.

Example. *There exists a basically disconnected compact space Y and a homeomorphism f of Y into itself such that f has exactly one fixed point and that point is not isolated.*

Let D be an uncountable discrete space and consider βD and the subset C of D^* consisting of those points which belong to the closure of no countable subset of D. The set C is clearly non-empty since it contains the set uD of points corresponding to uniform ultrafilters on D (see Theorem 5.13). Let π be the quotient map which identifies C to a point, say y_0, and let Y be the image of π. The quotient space Y is easily seen to be Hausdorff by using the observation that a neighborhood of y_0 can fail to contain only countably many points of D since its preimage must be a neighborhood of every point of C.

(a) Y is not extremally disconnected: If D is written as the disjoint union of two uncountable subsets, then y_0 belongs to the closure of both of them. Thus, Y cannot be extremally disconnected in light of Exercise 2J.

To demonstrate that Y is basically disconnected, we first show that:

(b) The intersection of a cozero-set of Y with D is either countable or has a countable complement in D: Let U be a cozero-set of Y. If y_0 is in U, then U is a neighborhood of y_0 so that $(Y \setminus U) \cap D$ is countable. Now suppose that y_0 is in $Y \setminus U$. The zero-set $Y \setminus U$ is a G_δ and therefore can be written $Y \setminus U = \bigcap \{U_n : n \geqslant 1\}$ where each U_n is a neighborhood of y_0. Thus, $(Y \setminus U_n) \cap D$ is countable for each n, and therefore, the intersection $U \cap D = \bigcup \{(Y \setminus U_n) \cap D : n \geqslant 1\}$ is countable.

(c) Y is basically disconnected: We must show that the closure of any cozero-set of Y is open. If the intersection of the cozero-set with D is countable, then its closure misses y_0 and hence is the one-to-one image of a clopen subset of βD under the quotient mapping π. Otherwise, from (b) the complement of the closure of U is a one-to-one image of a clopen subset of βD.

To define an embedding f of Y into itself which has a single fixed point, we begin by letting g be any embedding of βD into D^* such that $g[D] \cap C = \emptyset$. Then if we put $X = g[D]$, the image of g can be written $g[\beta D] = \mathrm{cl}_{D^*} X$. Now observe that the ultrafilters corresponding to the points of $X^* = \mathrm{cl}_{D^*} X \setminus X$ can be described as sums of ultrafilters as in Exercise 3F. An examination of this description shows that the points not lying in the closure of any countable subset of X are precisely those points belonging to $g[\beta D] \cap C$ and that points belonging to the closure of a countable subset of X lie outside of C.

Now by defining f on Y by $f(y) = g(y)$ for $y \neq y_0$ and $f(y_0) = y_0$, we obtain an embedding of Y into itself such that $f \circ \pi = \pi \circ g$. Further, Corollary 6.29 to follow shows that y_0 is the only fixed point of f. Since y_0 is certainly not an isolated point, the set of fixed points of f cannot be open.

This same example will be useful following Theorem 6.34 to show that the hypothesis of that theorem cannot be weakened either. There we will need only to make an additional requirement in the choice of g.

6.27. Because the class of extremally disconnected compact spaces corresponds to the class of complete Boolean algebras under the duality described in Chapter 2 (see Proposition 2.5), Theorem 6.25 has an interpretation in terms of homomorphisms of complete Boolean algebras (Exercise 6J).

6.28. The applications of Theorem 6.25 to homogeneity problems are described in the subsequent sequence of corollaries.

Corollary. *A one-to-one mapping f of an extremally disconnected compact space K into itself has no fixed point if either of the following conditions is fulfilled:*
   (a) *There exists a dense subset of K containing no fixed point.*
   (b) *$f[K]$ is nowhere dense in K.*

Proof. If (b) is satisfied, then $K \setminus f[K]$ is a dense subset which can contain no fixed point so that it suffices to show that (a) implies that f has no fixed point. The set F consisting of the fixed points of f is clopen and hence must meet every dense set. Thus, if (a) is satisfied, then f can have no fixed points. $\square$

6.29. Since the Stone-Čech compactification of an extremally disconnected space is extremally disconnected and a one-to-one mapping defined on a compact space is an embedding, the following corollary is immediate:

Corollary. *If X is extremally disconnected, in particular if X is discrete, then an embedding of βX into X^* has no fixed point.*

6.30. In the particular case of $\beta \mathbb{N}$, the fact that no embedding of $\beta \mathbb{N}$ into $\mathbb{N}^*$ has a fixed point can be interpreted in terms of relative types of ultrafilters and the producing relation which was introduced in 3.44. This interpretation will then yield a quick proof of the non-homogeneity of $\mathbb{N}^*$.

Corollary. *The type of a point of $\mathbb{N}^*$ is not also a relative type of the point, i.e. no type produces itself.*

Proof. Suppose that p in $\mathbb{N}^*$ is of type t and that the type of p relative to a discrete sequence X contained in $\mathbb{N}^*$ is also t. Then there exists a bijection of $\mathbb{N}$ onto X which sends the ultrafilter A^p onto p_X, the neighborhood traces of p on X. But then the Stone-Čech extension of the bijection is a homeomorphism of $\beta \mathbb{N}$ into $\mathbb{N}^*$ having p as a fixed point. $\square$

6.31. Corollary. $\mathbb{N}^*$ *is not homogeneous.*

Proof. If h is an automorphism of $\mathbb{N}^*$ and $h(p)=q$, then the relative types of p and q coincide. Thus, if p is of type t and q has t as a relative type, then we cannot have $h(p)=q$ for any h. The existence of a point q having t as a relative type follows from (b) of the proof of Theorem 3.45. $\square$

6.32. By abstracting the method just used to show that $\mathbb{N}^*$ is not homogeneous, we can obtain a similar result in the setting of extremally disconnected spaces.

Corollary. *If F is a closed nowhere dense subspace of an extremally disconnected compact space K and if F contains a copy of K, then F is not homogeneous.*

Proof. Let g be a homeomorphism of K into F. Choose any point x of F and put $y=g(x)$. If $h(y)=x$ for some homeomorphism of F into F, then x would be a fixed point of the homeomorphism $f=h\circ g$. But this is impossible by Corollary 6.28 since F is nowhere dense in K. $\square$

6.33. From Corollary 5.10 it follows that for any infinite discrete space D, D^* contains a copy of βD. In the special case of $\mathbb{N}$, any infinite closed subspace of $\mathbb{N}^*$ contains a copy of $\beta\mathbb{N}$. Coupled with the previous result, these two observations yield the

Corollary. *The growth of any infinite discrete space fails to be homogeneous. In the case of $\mathbb{N}$, no infinite closed subspace of $\mathbb{N}^*$ is homogeneous.*

6.34. Theorem 6.25 implies that an embedding of an extremally disconnected compact space into itself is the identity on a neighborhood of each of its fixed points. Since the mapping of Example 6.22 has just one fixed point and that one is not isolated, we see that the statement of Theorem 6.25 fails for more general mappings. However, the following result shows that such a fixed point does at least have an invariant neighborhood.

Theorem (Frolík). *If f is a mapping of an extremally disconnected space into itself, then each fixed point of f has a neighborhood base consisting of f-invariant clopen neighborhoods.*

Before undertaking the proof, which will involve a transfinite induction argument, it is necessary to define odd and even ordinal numbers. It is shown in E. Kamke's book, *Theory of Sets,* that every ordinal number α can be written in the form $\alpha=\omega_0\beta+n$ where n is a natural number, i.e. $n<\omega_0$. We shall say that the ordinal α is *even* or *odd* according as the natural number n is even or odd.

Proof. Let X be an extremally disconnected space and let x be a fixed point of a mapping f of X into itself. Any neighborhood U of x contains a clopen neighborhood V of x since X is regular (see Exercise 6M). Suppose that V is not f-invariant. Then the set defined by $S_0 = \{y \in V : f(y) \notin V\}$ is non-empty. Since $S_0 = f^{\leftarrow}(X \setminus V) \cap V$, S_0 is clopen and therefore $V \setminus S_0$ is a clopen neighborhood of x. However, $V \setminus S_0$ is not necessarily an f-invariant neighborhood since it may contain points y such that $f(y)$ belongs to S_0. In order to obtain the required f-invariant neighborhood, we will eliminate this possibility by subtracting such points from V.

Using S_0 to begin an inductive construction, for each ordinal $\alpha > 0$ put

$$P_\alpha = \mathrm{cl}(\bigcup\{S_\beta : \beta < \alpha, \ \beta \text{ non-limit}\}), \quad \text{and}$$

$$Q_\alpha = \mathrm{cl}(\bigcup\{S_\beta : \beta < \alpha\}).$$

Now define

$$S_\alpha = \{y \in V \setminus Q_\alpha : f(y) \in Q_\alpha\}$$

if α is a non-limit ordinal, and define

$$S_\alpha = \{y \in V \setminus Q_\alpha : f(y) \in P_\alpha\}$$

if α is a limit ordinal. By construction, $\{S_\alpha\}$ is a pairwise disjoint family of subsets of V. Further, the construction shows that if S_α is empty, then S_σ must be empty for all $\sigma > \alpha$. Thus, if the induction is continued only until the cardinality of X is exceeded by that of the index, we are certain to obtain a non-limit ordinal γ such that S_γ is empty. We will show that $W = V \setminus Q_\gamma$ is the required neighborhood of x. W is f-invariant since if y is a point of W such that $f(y)$ is not in W, then y must belong to $S_{\gamma+1}$, but $S_{\gamma+1} = \emptyset$.

To show that W is a neighborhood of x, we first define the following sets, each of which is clopen and misses the other two since X is extremally disconnected:

$$Z_0 = \mathrm{cl}(\bigcup\{S_\alpha : \alpha \text{ is a limit ordinal}\}),$$

$$Z_1 = \mathrm{cl}(\bigcup\{S_\alpha : \alpha \text{ is odd}\}),$$

$$Z_2 = \mathrm{cl}(\bigcup\{S_\alpha : \alpha \text{ is an even, non-limit ordinal}\}).$$

Then $W = V \setminus Q_\gamma$ can also be written as

$$W = V \setminus (Z_0 \cup Z_1 \cup Z_2)$$

since $S_\alpha = \emptyset$ for $\alpha > \gamma$. To show that W is a neighborhood of x, we need only demonstrate that no Z_i is a neighborhood of x. Since any neighborhood of a fixed point must meet its image, it suffices to show that $Z_i \cap f[Z_i] = \emptyset$ for each $i = 1, 2, 3$. But this follows easily since the Z_i's are pairwise disjoint and by construction they satisfy:

$$f[Z_0] \subset Z_1 \cup Z_2,$$

$$f[Z_1] \subset Z_0 \cup Z_2, \quad \text{and}$$

$$f[Z_2] \subset (X \setminus V) \cup Z_1.$$

Finally, W is clopen since each Z_i is clopen. $\square$

It follows immediately from the theorem that for any infinite discrete space D, a fixed point of a mapping of βD into itself has a neighborhood base of invariant neighborhoods. Further, since the neighborhoods are clopen, they are closures of subsets of D.

The dual statement of the preceding theorem in terms of complete Boolean algebras is given in Exercise 6J.

6.35. We now return to Example 6.26 and adapt it to show that the extremally disconnected hypothesis cannot be weakened in the theorem.

Example. *There exists a compact basically disconnected space Y and an embedding f of Y into itself such that f has exactly one fixed point y_0 and y_0 does not have a neighborhood base of f-invariant neighborhoods.*

Let Y and f be as described in Example 6.26. Recall that the construction of f utilized an embedding g of βD in D^*. By further specifying the properties of g, we will show that f can be constructed so that y_0 has a neighborhood which contains no f-invariant neighborhood. Our description of g is motivated by the proof of the previous theorem in that we will describe a neighborhood V of y_0 such that the method of subtracting points from V used in the proof to obtain an f-invariant neighborhood will fail.

Let $\{M_\sigma : \sigma < \beta\}$ be an uncountable family of pairwise disjoint countable subsets of D. Define g so that:

$$g[M_\sigma] \subset \mathrm{cl}_{\beta D} M_{\sigma - 1} \text{ if } \sigma \text{ is non-limit, and}$$

$$g[M_\sigma] \subset \mathrm{cl}_{\beta D}(\bigcup\{M_\gamma : \gamma < \sigma\}) \setminus \bigcup\{\mathrm{cl}_{\beta D} M_\gamma : \gamma < \sigma\} \text{ otherwise.}$$

Put $V = \pi[\mathrm{cl}_{\beta D}(D \setminus M_0)]$. Then in the successive elimination of points from V as in the proof of the theorem, we find that $S_\alpha = M_{\alpha + 1}$. Thus, by the construction of g, an uncountable number of points of D must be removed from V before an f-invariant set is obtained. However, since a neighborhood of y_0 can exclude only countably many points of D, the f-invariant set obtained is not a neighborhood of y_0.

Exercises

6A. *Retractions of βD*

If D is an infinite discrete space and $f:\beta D\to D^*$ is continuous, then f is a retraction if and only if $f^2(d)=f(d)$ for every d in D.
Reference: Frolík, 1968A.

6B. *An application of cellularity*

1. A space X whose cellularity is at most $\aleph_0$ and which admits a mapping of βX onto X^* is pseudocompact. [Consider the inverse image under such a mapping of a family of disjoint open sets provided by Corollary 5.6.]
2. If X is realcompact and separable, then there is no mapping of βX onto X^*. [Corollary 1.58.]
Reference: Comfort, 1965.

6C. *The converse of Magill's Theorem*

If a space X is such that every continuous image of X^* is homeomorphic to the growth of X in some Hausdorff compactification of X, then X is locally compact. [Shrink X^* to a point.]

6D. *Growths and the unit interval*

Let X be a locally compact normal space which contains a closed, infinite, discrete subspace.
1. X contains a closed, C^*-embedded copy of $\mathbb{N}$.
2. X^* contains a copy of $\mathbb{N}^*$.
3. There is a countably infinite discrete space Z such that βZ is contained in X^*. [$\mathbb{N}^*$ contains a copy of $\mathbb{N}$.]
4. $I=[0,1]$ is a continuous image of X^*. [Map Z to the rationals of I and use Proposition 1.47.]
5. Any continuous image of I is a growth of X in some compactification of X. [Theorem 6.14.]
6. All the hypotheses are needed. [Deleted Tychonoff Plank, ω_1, any non-locally compact normal space.]
References: Magill, 1966B. Hahn, in 1914, and Mazurkiewicz, in 1920, characterized the continuous images of I as *Peano spaces*, i.e. as compact, connected, locally connected metric spaces. For a proof of this result, see Willard, 1970.

6E. *Non-pseudocompactness and growths*

If X is a locally compact space which fails to be pseudocompact, then any continuous image of the unit interval is homeomorphic to the growth of X in some compactification. [Do the preceding Exercise first.]

6F. *Growths as graph closures*

Let X be a locally compact but non-compact space and let K be a compact space. Let $\alpha X = X \cup \{\infty\}$ be the one point compactification of X and let f map X into K such that the image of the trace on X of every neighborhood of ∞ is dense in K.

1. X is homeomorphic to the graph of f in $X \times K$. $[h(x) = (x, f(x)).]$
2. If the graph of f is considered as a subspace of $\alpha X \times K$, then no point of the form (x, k) where x belongs to X and $f(x) \neq k$ is in the closure of the graph. [K is Hausdorff and f is continuous.]
3. Points of the form (∞, k) are in the closure of the graph.
4. X has a compactification with growth homeomorphic to K.
5. The half-open interval $(0, 1]$ has a compactification with growth homeomorphic to $[-1, 1]$. $[f(x) = \sin(1/x).]$
6. If X is an infinite discrete space and K is compact with density at most the cardinality of X, then X has a compactification with growth homeomorphic to K. [Write X as a union of ω_0 disjoint subsets each having the same cardinality as X. Map each of these subsets to a dense subspace of K.]
7. Not all compactifications of $\mathbb{N}$ can be obtained as in 6. [$\mathbb{N}^*$ has density character $\mathfrak{c}$.]
8. The result 6D.5 can be obtained as a corollary of 4 above.

Reference: A. K. Steiner and E. F. Steiner, 1968A. For some examples in the special case $X = [0, \infty)$, particularly with K the torus $S^1 \times S^1$, see B. Simon, 1969.

6G. *Extremally disconnected compactifications*

If X is locally compact and extremally disconnected, then βX is the unique extremally disconnected compactification of X.

6H. *Second countable normal spaces*

1. A second countable normal space is metrizable and hence is para-compact. [Apply the fact that disjoint closed subsets are completely separated to pairs (U, V) of basic open sets such that cl $U \subset V$. Then use the Embedding Lemma, 1.5.]
2. A separable metrizable space is second countable.
3. Any separable metric space can be embedded in the Hilbert cube $I^{\aleph_0}$.

Reference: In 1925A, P. Urysohn proved a more general version of 1, showing that a second countable regular T_1-space is metrizable. See also [D, p. 195].

6I. *Decomposition of a non-compact space*

If f is a homeomorphism of an extremally disconnected space X into itself such that $f[X]$ is C^*-embedded, then X contains a clopen set F such that f is the identity on F and $X\backslash F$ is 3-decomposable with respect to f. [Apply Theorem 6.25 to $\beta(f)$.]
Reference: Frolík, 1968 B.

6J. *Boolean algebra formulations*

1. Let h be a homomorphism of a complete Boolean algebra L onto itself. Then there exist B_0, B_1, B_2, and B_3 in L with $\bigvee B_i = 1$ and $B_i \wedge B_j = 0$ for $i \neq j$ such that h is the identity on the principal ideal generated by B_0 and $B_i \wedge h(B_i) = 0$ for $i = 1, 2, 3$. [Theorem 6.25.]
2. Let h be a homomorphism of a complete Boolean algebra L into itself. If a maximal ideal $\mathscr{I}$ in L is h-invariant, then each principal ideal contained in $\mathscr{I}$ is contained in an h-invariant principal ideal contained in $\mathscr{I}$. [Theorem 6.34.]
Reference: Frolík, 1968 B.

6K. *Decomposable subsets*

Let f be a function of a set D into itself.
1. If B is the largest 3-decomposable subset of D, then some iteration of every point in $D\backslash B$ is a fixed point. [$D\backslash B$ cannot contain a sequence $\{x_n\}$ of distinct points such that $f(x_n) = x_{n+1}$ or a periodic point.]
2. Let F_0 denote the set of fixed points of f and denote the iterated inverse images of f by $F_n = (f^{\leftarrow})^n(F)$. Then the largest 3-decomposable subset of D is given by

$$B = D\backslash(\bigcup\{F_i : i = 0, 1, 2, \ldots\}).$$

3. If f has no fixed points, then D is 3-decomposable with respect to f.
References: (2) appears in Frolík, 1968 B. (3) appears in Katětov, 1967, and a proof was given by I. N. Baker in Amer. Math. Monthly, 1964, pp. 219—220, in response to a problem posed by H. Kenyon.

6L. *Fixed points of an extension*

If D is an infinite discrete space and f maps D into itself, then the set of fixed points of the extension mapping $\beta(f)$ of βD into itself is the closure of the set of fixed points of f. [Write D as a disjoint union $D = F \cup B_1 \cup B_2 \cup B_3$] where F is the set of fixed points of f and $f[B_i] \cap B_i = \emptyset$ for each i.]
Reference: Katětov, 1967.

6 M. *Extremally disconnected spaces and regularity*

In Section 1.8 we made the assumption that all spaces mentioned would be assumed to be completely regular. In the proof of Theorem 6.34, this assumption has been used in an essential way to assert that an extremally disconnected space is zero-dimensional. Consider the following example. Let X have the same underlying set as $\beta\mathbb{N}$ with the following topology: points of $\mathbb{N}$ are isolated and a basic neighborhood of a point p of $\mathbb{N}^*$ is a set of the form $Z \cup \{p\}$ where Z belongs to p.
1. X is Hausdorff.
2. The topology of X is larger than the usual topology on $\beta\mathbb{N}$.
3. The closure of an open subspace of X is identical with the closure of the subset as a subset of $\beta\mathbb{N}$ with the usual topology. Thus, X is extremally disconnected.
4. Points of $\mathbb{N}^*$ do not have a neighborhood basis of closed sets in X, i.e. X is not regular.
5. X is not zero-dimensional.
Reference: This example is also described in Steen and Seebach, Ex. 113.

Chapter 7. $\beta\mathbb{N}$ Revisited

7.1. We have obtained two results since leaving Chapter 3 which make this return to $\beta\mathbb{N}$ possible. The first of these was Corollary 4.30 in which we found, under the assumption of the Continuum Hypothesis, that $\mathbb{N}^*$ contains a dense subset of 2^c P-points. The second was Corollary 6.30 in which we saw that no point of $\mathbb{N}^*$ has its own type as a relative type with respect to any copy of $\mathbb{N}$ contained in $\mathbb{N}^*$. Here we will make use of these two properties of $\mathbb{N}^*$ to continue the investigation of $\beta\mathbb{N}$ and especially of $\mathbb{N}^*$.

We will begin by considering three examples. First we will show that the subspace $\mathbb{N}^*\setminus\{p\}$ is not normal for any point p of $\mathbb{N}^*$. Separate arguments will be required depending upon whether or not p is a P-point. The argument in the P-point case will be based on the construction used in Corollary 6.16. Next, we will consider an example of a compact space which is unchanged by the addition of two isolated points but which is altered by adding a single isolated point. This example will depend on properties of $\beta\mathbb{N}$ and was developed to demonstrate the difficulties involved in seeking generalizations of the Banach-Stone Theorem. In the third example, we will see that points exist in $\mathbb{N}^*$ which have c distinct relative types. The purpose of this example is to show that Theorem 3.45 gives the best possible upper bound for the cardinality of the set of relative types of a point in $\mathbb{N}^*$.

Next, we will consider the set of P-points in the context of the theory of types as developed in Chapter 3. We will show that if p and q are any two P-points, then there exists an automorphism of $\mathbb{N}^*$ which sends p to q.

Finally, we will show that the producing relation which was introduced in Chapter 3 defines a partial order on the set of types of points of $\mathbb{N}^*$. The P-points will be seen to be minimal in this partial order. The chapter will conclude by using the minimal types to show that points of $\mathbb{N}^*$ exist which have only finitely many relative types.

$\mathbb{N}^*\backslash\{p\}$ is not Normal

7.2. If p is any point of $\mathbb{N}^*$ and the Continuum Hypothesis is assumed, then $\mathbb{N}^*\backslash\{p\}$ is not normal. The proof of this result was given by L. Gillman for the case where p is a non-P-point of $\mathbb{N}^*$ and independently by M. Rajagopalan and N. Warren where p is a P-point. We will consider Gillman's argument as it appeared in the 1968 paper of W.W. Comfort and S. Negrepontis.

Recall from Section 1.65 that a non-P-point must belong to the boundary of a zero-set. That fact together with the following result form the key to demonstrating the non-normality of $\mathbb{N}^*\backslash\{p\}$ when p is not a P-point.

Proposition [CH] (Gillman). *The boundary of a non-open zero-set of $\mathbb{N}^*$ is homeomorphic to $\mathbb{N}^*$.*

Proof. Let Z be a non-open zero-set of $\mathbb{N}^*$ so that its boundary ∂Z is non-empty. We will show that ∂Z is a copy of $\mathbb{N}^*$ by showing that it satisfies the hypotheses of Theorem 3.32. It is clear that ∂Z is compact and totally disconnected. Because ∂Z is compact, it is C^*-embedded in $\mathbb{N}^*$ and therefore has a base of $\mathfrak{c}$ clopen sets. Further, since $\mathbb{N}^*$ is an F-space (Proposition 1.62 or 3.24), the C^*-embedded subspace ∂Z is also an F-space. To show that the zero-sets of ∂Z are regular closed and that ∂Z has no isolated points, we will show that ∂Z is the growth of a locally compact, realcompact space (Proposition 4.21). The cozero-set $C = \mathbb{N}^*\backslash Z$ is C^*-embedded in $\mathbb{N}^*$ since $\mathbb{N}^*$ is an F-space. Hence, $\beta C = \mathrm{cl}_{\mathbb{N}^*} C$ and $C^* = \partial Z$. A cozero-set in a realcompact space is realcompact [GJ, 8.14], so that C is realcompact. C is locally compact since it is open in $\mathbb{N}^*$. Proposition 4.21 now implies that the zero-sets of ∂Z are regular closed. Finally, in the proof of Theorem 4.23 we showed that the growth of a realcompact locally compact space has no isolated points. Hence, with the assumption of the Continuum Hypothesis, Theorem 3.32 implies that ∂Z is homeomorphic to $\mathbb{N}^*$. $\quad\square$

7.3. Theorem [CH] (Gillman). $\mathbb{N}^*\backslash\{p\}$ *is not normal if p is a non-P-point of $\mathbb{N}^*$.*

Proof. Since p is a non-P-point, there is a zero-set Z in $\mathbb{N}^*$ such that p belongs to ∂Z. Since ∂Z is a copy of $\mathbb{N}^*$ and the Continuum Hypothesis implies that dense subspaces of $\mathbb{N}^*$ are not C^*-embedded (Proposition 3.30), $\partial Z\backslash\{p\}$ is not C^*-embedded in ∂Z. We will show that the assumption that $\mathbb{N}^*\backslash\{p\}$ is normal will contradict this fact. $\partial Z\backslash\{p\}$ is a closed subspace of $\mathbb{N}^*\backslash\{p\}$. Hence, if $\mathbb{N}^*\backslash\{p\}$ is normal, a bounded, real-valued mapping f on $\partial Z\backslash\{p\}$ will extend to a mapping g on $\mathbb{N}^*\backslash\{p\}$. As in the previous proposition, if $C = \mathbb{N}^*\backslash Z$, then $C^* = \partial Z$.

Hence, the restriction $g|C$ extends to βC and $\beta(g|C)|\partial Z$ extends f to ∂Z. But this is a contradiction. $\square$

7.4. The case where p is a P-point is more complicated. Rajagopalan and Warren each give a proof in their respective 1972 papers. Our proof will be the one given by Rajogopalan. We will show that $\mathbb{N}^*\backslash\{p\}$ can be mapped onto the space $\omega_1 \times (\omega_1 + 1)$ by a closed mapping. Since $\omega_1 \times (\omega_1 + 1)$ is not normal [GJ, ex. 8M] and a closed continuous image of a normal space must be normal [D, p. 145], this will show that $\mathbb{N}^*\backslash\{p\}$ is not normal. (A proof that $\omega_1 \times (\omega_1 + 1)$ is not normal will be provided in Example 8.37.) The construction of the closed mapping will be based on the construction of the mapping of $\mathbb{N}^*$ onto $\omega_1 + 1$ as described in the proof of Corollary 6.16.

Theorem [CH] (Rajagopalan, Warren). *$\mathbb{N}^*\backslash\{p\}$ is not normal if p is a P-point.*

Proof. We begin by modifying the construction of the mapping f of $\mathbb{N}^*$ onto $\omega_1 + 1$ which was described in the proof of Corollary 6.16. Recall that we constructed a strictly increasing ω_1-sequence $\{U_\alpha : \alpha < \omega_1\}$ of proper clopen subsets of $\mathbb{N}^*$. The definition $f(p) = \sup\{\alpha < \omega_1 : p \notin U_\alpha\}$ yields a mapping of $\mathbb{N}^*$ onto $\omega_1 + 1$ such that

$$f^{\leftarrow}(\{\omega_1\}) = \bigcap\{\mathbb{N}^*\backslash U_\alpha : \alpha < \omega_1\} .$$

Assuming the Continuum Hypothesis, there exists a P-point p in $\mathbb{N}^*$ and the base of $\mathfrak{c}$ clopen neighborhoods of p can be indexed by ω_1. Begin the construction of f as before by choosing U_0 to be a non-empty clopen set which misses p. Then because every G_δ containing p is a neighborhood of p, at each stage of the construction U_α can be chosen so that the complement of U_α is a basic neighborhood of p. Hence, $\bigcap\{\mathbb{N}^*\backslash U_\alpha : \alpha < \omega_1\} = \{p\}$ and if f is defined as above, $f^{\leftarrow}(\{\omega_1\}) = \{p\}$.

Using the mapping f, we will describe a partition of $\mathbb{N}^*\backslash\{p\}$ which induces a quotient mapping of $\mathbb{N}^*\backslash\{p\}$ onto $\omega_1 \times (\omega_1 + 1)$. For each α, put $W_\alpha = f^{\leftarrow}([0,\alpha])$ and $V_\alpha = W_\alpha \backslash (\bigcup\{W_\beta : \beta < \alpha\})$ so that $f^{\leftarrow}(\{\alpha\}) = V_\alpha$. Now we will define a family of subsets $\{V_{\alpha\beta} : \alpha < \omega_1, \beta \leqslant \omega_1\}$ so that:

(a) For every $\alpha < \omega_1$, the subcollection $\{V_{ij} : i \leqslant \alpha, j \leqslant \omega_1\}$ partitions W_α, and

(b) The partition of W_α induces a quotient mapping g_α of W_α onto $\alpha \times (\omega_1 + 1)$ and $g_{\alpha+1}|W_\alpha = g_\alpha$.

The family will be constructed by transfinite induction. We will construct the partition for the non-limit ordinals first. If α is non-limit, $\bigcup\{W_\beta : \beta < \alpha\}$ is clopen so that $V_\alpha = f^{\leftarrow}(\{\alpha\})$ is the difference of clopen sets and is therefore clopen. Hence, there exists a P-point p_α in V_α since the P-points are dense in $\mathbb{N}^*$ (Corollary 4.30). Since the clopen set V_α is a copy of $\mathbb{N}^*$ (Proposition 3.16), there exists a mapping $f_\alpha : V_\alpha \to \omega_1 + 1$ such that $f_\alpha^\leftarrow(\{\omega_1\}) = \{p_\alpha\}$. Put $V_{\alpha\beta} = f_\alpha^\leftarrow(\{\beta\})$ for $\beta \leqslant \omega_1$. This completes the construction for non-limit ordinals.

The situation is somewhat more complicated for the limit ordinal case. Let α be a limit ordinal and assume that the partition of W_δ has been defined for all $\delta < \alpha$. To define $V_{\alpha\beta}$ for each $\beta < \omega_1$ we will construct an increasing sequence of clopen sets dominated by a decreasing sequence and use DuBois-Reymond separability to define $V_{\alpha\beta}$ (Section 2.20). Let $\{\alpha_n\}$ be a sequence of non-limit ordinals whose supremum is α. To obtain $V_{\alpha 0}$, begin by defining the following two sequences:

$$A_1 = \bigcup \{V_{i0} : i \leqslant \alpha_1\}, \ldots, A_n = \bigcup \{V_{i0} : i \leqslant \alpha_n\}, \ldots$$

Thus, $\bigcup A_n$ includes all points of $\bigcup \{W_\beta : \beta < \alpha\}$ which are mapped to points having second coordinate equal to 0.

$$B_1 = (W_\alpha \backslash (\bigcup \{W_i : i \leqslant \alpha_1\})) \cup A_1, \ldots, B_n = (W_\alpha \backslash (\bigcup \{W_i : i \leqslant \alpha_n\})) \cup A_n, \ldots$$

B_{n+1} is contained in B_n since the family $\{W_i\}$ is increasing and for any index i, $\alpha_n < i \leqslant \alpha_{n+1}$, the points of A_i which are included in B_{n+1} also belong to B_n. Hence, $\bigcap B_n$ excludes all points of $\bigcup \{W_\beta : \beta < \alpha\}$ which are sent to points having second coordinate greater than 0. Since $A_n = g_{\alpha_n}^\leftarrow(\alpha_n \times \{0\})$, each A_n is clopen. Similarly,

$$B_n = (f^\leftarrow([0,\alpha]) \backslash f^\leftarrow([0,\alpha_n])) \cup A_n$$

and is also clopen. Thus, we have two sequences of clopen sets such that

$$A_1 \subset \cdots \subset A_n \subset \cdots \subset B_n \subset \cdots \subset B_1 .$$

Since $CO(\mathbb{N}^*)$ is DuBois-Reymond separable (Corollary 3.25), there exists a clopen set H such that $A_n \subset H \subset B_n$ for all n. Put $V_{\alpha 0} = H \cap V_\alpha$. If $\{y_n\}$ is a sequence of points such that y_n is in $V_{\alpha_n 0}$ for each n, then the cluster points of $\{y_n\}$ must belong to $H \cap V_\alpha$ so that $V_{\alpha 0}$ is non-empty.

Now suppose that for some $\beta < \omega_1$, $V_{\alpha\gamma}$ has been defined for all $\gamma < \beta$. We define our sequences $\{A_n\}$ and $\{B_n\}$ somewhat as before. However, now we want A_n to include those points which g_{α_n} maps to points with second coordinate at most β. B_n will exclude those points which are mapped by g_{α_n} to points having second coordinate greater than β. Thus, define

$$A_n = \bigcup \{V_{ij} : i \leqslant \alpha_n, j \leqslant \beta\}$$

and

$$B_n = (W_\alpha \backslash W_{\alpha_n}) \cup A_n .$$

Then again we have sequences of clopen sets such that

$$A_1 \subset \cdots \subset A_n \subset \cdots \subset B_n \subset \cdots \subset B_1$$

and there exists a clopen set H such that $A_n \subset H \subset B_n$ for all n. Put $V_{\alpha\beta} = (H \cap V_\alpha) \backslash (\bigcup \{V_{\alpha j} : j < \beta\})$. The set $V_{\alpha\beta}$ is seen to be non-empty by choosing a sequence $\{y_n\}$ as before, this time with y_n in $V_{\alpha_n \beta}$.

Finally, put $V_{\alpha\omega_1} = V_\alpha \backslash (\bigcup \{V_{\alpha j} : j < \omega_1\})$. Recall that for each of the indices α_n, $V_{\alpha_n \omega_1} = \{p_{\alpha_n}\}$ where p_{α_n} is a P-point. $V_{\alpha\omega_1}$ is non-empty since it must contain the cluster points of the sequence $\{p_{\alpha_n}\}$. Thus, we have defined a partition of W_α and it is clear that the quotient map g_α induced by the partition maps W_α onto $\alpha \times (\omega_1 + 1)$.

Now consider the partition of $\mathbb{N}^*$ given by

$$\Pi = \{V_{\alpha\beta} : \alpha < \omega_1, \beta \leqslant \omega_1\} \cup \{p\} \, .$$

It remains to show that the quotient map $g_{\omega_1} : \mathbb{N}^* \backslash \{p\} \to \omega_1 \times (\omega_1 + 1)$ is closed. To do this, we show that the partition Π of $\mathbb{N}^*$ induces a Hausdorff quotient space Y. Thus, the quotient map on $\mathbb{N}^*$ is closed and g_{ω_1} is just the restriction to a saturated set and hence is also closed.

Since each W_α is clopen and saturated, $g_{\omega_1}[W_\alpha]$ is clopen in Y. Hence, because the images of any two points other than p are contained in some $g_{\omega_1}[W_\alpha]$, they can be separated by open sets in Y. On the other hand, any point of Y other than the image of p is separated from the image of p by some $g_{\omega_1}[W_\alpha]$ and its complement. $\square$

An Example Concerning the Banach-Stone Theorem

7.5. In Corollary 1.25, we saw that two compact spaces X and Y are homeomorphic if and only if $C^*(X)$ and $C^*(Y)$ are isomorphic. If a metric structure is imposed on the linear spaces by means of the supremum norm, then the result becomes the

Banach-Stone Theorem. *Two compact spaces X and Y are homeomorphic if and only if $C^*(X)$ and $C^*(Y)$ are (linearly) isometric.*

This result was proved by S. Banach in his book, *Theorie des operations lineaires*, in 1932, for compact metric spaces and by M. H. Stone for compact spaces in his 1937 paper. It is sufficient to require only that $C^*(X)$ and $C^*(Y)$ be isometric, since Banach also showed that if two Banach spaces over the real field are isometric, then they are also linearly isometric. For a discussion of this theorem, see M. M. Day's book, *Normed Linear Spaces*.

The example to be presented here is one that developed out of efforts to generalize the Banach-Stone Theorem by replacing the Banach space $C^*(X)$ of real-valued mappings by a Banach space $C^*(X,E)$ of E-valued mappings where E is a Banach space. Investigations of the possibility of such a generalization are carried out in the 1950 paper of M. Jerison and the 1973 paper of K. Sundaresan. Each of these papers presents a Banach-Stone type of theorem and also a counterexample to show that such a theorem must place some restriction on the compact space X or on the Banach space E.

7.6. The present example is taken from K. Sundaresan's 1971 paper and uses a construction based on properties of $\beta\mathbb{N}$ to show that $\mathbb{R}$ cannot be replaced by the 2-dimensional Banach space $\mathbb{R}^2$ in the Banach-Stone Theorem. The following lemma will be used to relate the example to be given to Banach spaces of $\mathbb{R}^2$-valued mappings. Let $S \oplus T$ denote the topological sum of two spaces S and T.

Lemma. *If X and Y are compact spaces, then $C^*(X,\mathbb{R}^2)$ is (linearly) isometric with $C^*(Y,\mathbb{R}^2)$ if and only if $X \oplus X$ is homeomorphic with $Y \oplus Y$.*

Proof. Let **2** denote the discrete space having two points. We see in the natural way that $C^*(X,\mathbb{R}^2)$ is isometric with $C^*(Y,\mathbb{R}^2)$ if and only if $C^*(X \times 2, \mathbb{R})$ is isometric with $C^*(Y \times 2, \mathbb{R})$. Then the Banach-Stone Theorem implies that this isometry occurs exactly when $X \times 2$ and $Y \times 2$ are homeomorphic. The proof is then completed by noting that $X \times 2 \approx X \oplus X$ and $Y \times 2 \approx Y \oplus Y$. ◻

7.7. Example (Sundaresan). *There exists a compact space X such that X is homeomorphic with $X \oplus \{p\} \oplus \{q\}$ while X is not homeomorphic with $X \oplus \{p\}$.*

Before describing the example, we first see the relevance of such a space in the setting of the Banach-Stone Theorem.

Proposition. *There exist compact spaces X and Y such that $C^*(X,\mathbb{R}^2)$ and $C^*(Y,\mathbb{R}^2)$ are linearly isometric but X and Y are not homeomorphic.*

Proof. Let X be a space having the properties described in the example and let $Y = X \oplus \{p\}$. Then X and Y are not homeomorphic. However,

$$X \oplus X \approx X \oplus X \oplus \{p\} \oplus \{q\} \approx Y \oplus Y,$$

and thus it follows from Lemma 7.6 that $C^*(X,\mathbb{R}^2)$ and $C^*(Y,\mathbb{R}^2)$ are linearly isometric. ◻

To begin the description of X, let $N_1 = \{a_i : i \geqslant 1\}$ and $N_2 = \{b_i : i \geqslant 1\}$ be two countable discrete spaces which are disjoint. Let f be the function on $N_1 \cup N_2$ defined by $f(a_i) = b_i$ and $f(b_i) = a_i$ for all i. The restriction $f | N_1$ extends to a homeomorphism $\beta(f)$ of βN_1 onto βN_2. Let X be the adjunction space $\beta N_1 \cup_{\beta(f)|N_1^*} \beta N_2$ obtained by attaching βN_1 to βN_2 by the restriction $\beta(f) | N_1^*$. We will show that X is the required space.

A copy of $N_1 \cup N_2$ is embedded in X as an open subspace and we will continue to denote this copy by $N_1 \cup N_2$. X is compact because it is a quotient of a compact space and X is Hausdorff since N_1^* is closed in βN_1 and both βN_1 and βN_2 are normal [D, p. 145].

Now consider the space $X \oplus \{p\} \oplus \{q\}$. Define a mapping $g : N_1 \cup N_2 \oplus \{p\} \oplus \{q\} \to N_1 \cup N_2$ by $g(p) = a_1$, $g(q) = b_1$, $g(a_i) = a_{i+1}$, and $g(b_i) = b_{i+1}$. Clearly, g extends to a homeomorphism of $X \oplus \{p\} \oplus \{q\}$ onto X.

It remains to show that X is not homeomorphic to $X \oplus \{p\}$. We will demonstrate this by exhibiting a particular type of automorphism of $X \oplus \{p\}$ and then showing that X does not admit such an automorphism. If we define a mapping σ of $X \oplus \{p\}$ onto itself by $\sigma(p) = p, \sigma | N_1 \cup N_2 = f$, and $\sigma(x) = x$ for x in $X \setminus (N_1 \cup N_2)$, then σ is an automorphism of $X \oplus \{p\}$ such that σ^2 is the identity, i.e. σ is an involution. Further, p is the only isolated point which is invariant under σ. We will complete the proof by showing that X does not admit an involutory automorphism with exactly one invariant isolated point, and hence cannot be homeomorphic with $X \oplus \{p\}$.

Suppose that h is such an involutory automorphism of X. Without loss of generality, we can assume that a_1 is the isolated point which is invariant under h. Define a sequence $C = \{c_n : n \geqslant 1\}$ of points of X as follows:

$$c_1 = a_1 , \qquad c_2 = f(c_1) = b_1 , \qquad c_3 = h(c_2) ,$$

and in general,

$$c_{n+1} = f(c_n) \quad \text{if } n \text{ is odd, and}$$

$$c_{n+1} = h(c_n) \quad \text{if } n \text{ is even.}$$

The points of C are distinct: Suppose that $c_1, \ldots, c_{2n}$ are distinct and consider $c_{2n+1} = h(c_{2n})$. If $c_{2n+1} = c_{2i+1}$ for some $i < n$, then we would have $h(c_{2n}) = c_{2n+1} = h(c_{2i})$ which is impossible since h is one-to-one. On the other hand, if $c_{2n+1} = c_{2i}$ for some $i < n$, then we would have

$$c_{2n} = h \circ h(c_{2n}) = h(c_{2n+1}) = h(c_{2i}) = c_{2i+1}$$

which is also impossible. Now suppose that $c_1, \ldots, c_{2n}, c_{2n+1}$ are distinct.

By using the facts that f is one-to-one and $f \circ f$ is the identity, we see in a similar manner that c_{2n+2} is not equal to any previous term of the sequence.

In order to relate the points of the sequence to N_1 and N_2, it is convenient to relabel C in the following manner: $c_{2n} = \eta_n$ and $c_{2n+1} = \xi_{n+1}$ for $n = 1, 2, \dots$. Note that $f(\xi_n) = \eta_n$ and $f(\eta_n) = f \circ f(\xi_n) = \xi_n$ for each n and if $n \geqslant 2$, then $h(\eta_n) = \xi_{n+1}$ and $h(\xi_n) = h \circ h(\eta_{n-1}) = \eta_{n-1}$. Thus, ξ_n belongs to N_1 (resp. N_2) if and only if η_n belongs to N_2 (resp. N_1) and for each n, $\{\xi_n, \eta_n\} = \{a_i, b_i\}$ for some i depending on n. Now consider the subsequences $S_1 = \{\xi_{3k+2} : k \geqslant 1\}$ and $S_2 = \{\eta_{3k+2} : k \geqslant 1\}$ of C. Note that $f[S_1] = S_2$ and $f[S_2] = S_1$. Since C is one-to-one, $N_1 \cap (S_1 \cup S_2)$ is infinite so that at least one of the sets $S_1 \cap N_1$ and $S_2 \cap N_1$ is infinite. Without loss of generality, assume that $S_1 \cap N_1$ is infinite. For any subset M of X, let $M^* = \mathrm{cl}_X M \setminus M$. Then it follows from the definition of X that

$$(S_1 \cap N_1)^* = (f[S_1 \cap N_1])^* = (S_2 \cap N_2)^*.$$

Put $D_i = h[S_i \cap N_i]$ for $i = 1, 2$. Since h is a homeomorphism, we must have $D_1^* = D_2^* \neq \emptyset$. However, we will show that $\mathrm{cl}_X D_1 \cap \mathrm{cl}_X D_2 = \emptyset$, thereby reaching a contradiction which will complete the proof.

First note that if x belongs to D_1 (resp. D_2), then x is of the form η_{3k+1} (resp. ξ_{3k}) for some k since $h(\xi_{3k+2}) = \eta_{3k+1}$ and $h(\eta_{3k+2}) = \xi_{3(k+1)}$. Thus,

$$\mathrm{cl}_X D_1 \cap \mathrm{cl}_X D_2 \subset \mathrm{cl}_X \{\xi_{3k} : k \geqslant 1\} \cap \mathrm{cl}_X \{\eta_{3k+1} : k \geqslant 1\}.$$

Now let us define sequences $T_1 = \{t_{3k} : k \geqslant 1\}$ and $T_2 = \{t_{3k+1} : k \geqslant 1\}$ in the following manner:

$$t_{3k} = \begin{cases} \xi_{3k} & \text{if } \xi_{3k} \in N_1, \\ f^{\leftarrow}(\xi_{3k}) & \text{if } \xi_{3k} \in N_2; \end{cases}$$

$$t_{3k+1} = \begin{cases} \eta_{3k+1} & \text{if } \eta_{3k+1} \in N_1, \\ f^{\leftarrow}(\eta_{3k+1}) & \text{if } \eta_{3k+1} \in N_2. \end{cases}$$

By noting that the sequence C is one-to-one, that $f(\xi_n) = \eta_n$, and that $f \circ f$ is the identity on $N_1 \cup N_2$, we see that T_1 and T_2 are disjoint subsets of N_1 and thus have disjoint closures both in βN_1 and in X. From the construction of X, note that if we identify βN_1 with its image in X, we obtain

$$\mathrm{cl}_{\beta N_1} T_1 = \mathrm{cl}_X \{\xi_{3k} : k \geqslant 1\} \quad \text{and} \quad \mathrm{cl}_{\beta N_1} T_2 = \mathrm{cl}_X \{\eta_{3k+1} : k \geqslant 1\}.$$

Hence, we conclude that

$$\mathrm{cl}_X D_1 \cap \mathrm{cl}_X D_2 \subset \mathrm{cl}_{\beta N_1} T_1 \cap \mathrm{cl}_{\beta N_1} T_2 = \emptyset.$$

Thus, a contradiction is reached and X does not admit such an automorphism so that X is not homeomorphic with $X \oplus \{p\}$.

Similar examples are known for the n-dimensional case. In his 1957 paper, W. Hanf used Boolean algebra techniques to obtain, for any $n \geqslant 2$, non-homeomorphic compact spaces X and Y such that $X \times n \approx Y \times n$. In her 1971 paper, N. Kroonenberg also used the properties of βN to obtain, for $n \geqslant 2$, a pair of compact spaces X and Y such that for $k = 1, 2, \ldots, n-1$, $X \times k \not\approx Y \times k$ and yet $X \times n \approx Y \times n$.

A Point of $\mathbb{N}^*$ with $\mathfrak{c}$ Relative Types

7.8. Recall from Chapter 3 that a point p of $\mathbb{N}^*$ has t as a relative type if there is a countable discrete subset X of βN such that p belongs to X^* and $\tau(p, X) = t$. We showed in Theorem 3.45, due to Z. Frolík, that there can be no more than $\mathfrak{c}$ relative types for any point p. Here we will show that this upper bound is the best possible, i.e. that there exists a point having $\mathfrak{c}$ relative types. This example of such a point appeared in the 1971 paper of A. K. and E. F. Steiner.

For S a subset of βN and p a point of $\mathbb{N}^*$, recall that

$$\tau[p, S] = \{\tau(p, X) : X \subset S, X \approx \mathbb{N}\} .$$

Thus, $\tau[p, S]$ is the set of relative types of p which occur relative to countable discrete subspaces of S. The only new result on relative types since Chapter 3 is Corollary 6.30 which states that no type produces itself, i.e. that $\tau(p)$ is not in $\tau[p, \mathbb{N}^*]$. The result which will be used in the description of a point having $\mathfrak{c}$ relative types is the following lemma, which is merely a restatement of Corollary 6.30 since $\mathrm{cl}\, X$ is a copy of βN for any countable discrete subspace X of βN.

Lemma. *If X is a countable discrete subspace of βN and Y is a countable discrete subspace of X^* with p in Y^*, then $\tau(p, X) \neq \tau(p, Y)$.*

7.9. We can now describe a point having $\mathfrak{c}$ relative types.

Example (A. K. and E. F. Steiner). *There exists a point in $\mathbb{N}^*$ which has exactly $\mathfrak{c}$ relative types.*

Our description of the required point will be aided by using a diagram to illustrate the relationships among certain countable discrete subspaces of βN. A diagram such as

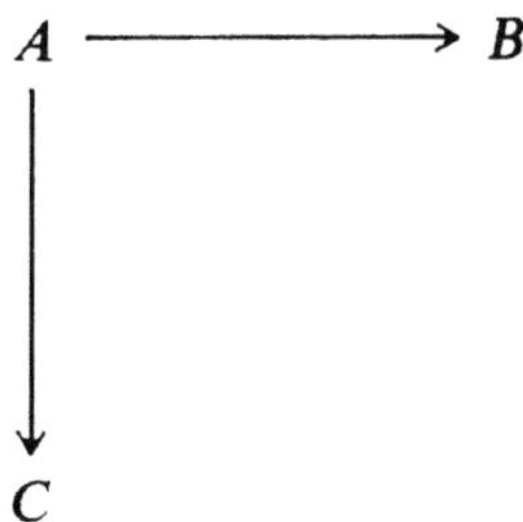

will mean that A is a countable discrete subspace of $\beta\mathbb{N}$, that B and C are countable discrete subspaces of A^*, and that $\operatorname{cl} B \cap \operatorname{cl} C = \emptyset$. When $C = \emptyset$, we will simply write $A \to B$.

Begin by letting $A(1)$ be a countable discrete subset of $\mathbb{N}^*$. Write $A(1)$ as the disjoint union of infinite subsets K_1 and K_2. Let $A(2)$ be a countable discrete subset of K_1^* and let $B(0)$ be a countable discrete subset of K_2^*. Let $B(1)$ be a countable discrete subset of $B(0)^*$ and choose a point x_1 in $B(1)^*$. Then we have described the construction of the first row of the following diagram:

$$\mathbb{N} \to A(1) \to B(0) \to B(1) \to \{x_1\}$$
$$\downarrow$$
$$A(2) \to B(0,0) \to B(0,1) \to B(1,0) \to B(1,1) \to \{x_2\}$$
$$\downarrow$$
$$\vdots$$
$$\downarrow$$
$$A(n) \to B(0,0,\ldots,0) \to B(0,0,\ldots,1) \to \cdots \to B(1,1,\ldots,1) \to \{x_n\}.$$
$$\downarrow$$
$$\vdots$$

In order to complete the construction by induction, we will need to consider an ordering on the set of sequences of k zeros and ones for each positive integer k. If $(\alpha_1, \ldots, \alpha_k)$ and $(\beta_1, \ldots, \beta_k)$ are two such sequences, define $(\alpha_1, \ldots, \alpha_k) > (\beta_1, \ldots, \beta_k)$ if for the first i where $\alpha_i \neq \beta_i$, $\alpha_i > \beta_i$. Note that this definition yields a total order on the 2^k sequences. Now for some integer $n > 1$, assume that $A(n)$ has been defined and that the k-th row has been defined for all $k < n$. Write $A(n)$ as the disjoint union of two infinite sets K_1 and K_2. Choose $A(n+1)$ to be a countable discrete subspace of K_1^* and $B(0,0,\ldots,0)$ to be a countable discrete subspace of K_2^*. Now for each sequence $(\alpha_1, \ldots, \alpha_n)$, choose $B(\alpha_1, \ldots, \alpha_n)$ to be a countable discrete subspace of $B(\beta_1, \ldots, \beta_n)^*$ where $(\beta_1, \ldots, \beta_n)$ is the greatest sequence preceding $(\alpha_1, \ldots, \alpha_n)$, i.e. the immediate predecessor of $(\alpha_1, \ldots, \alpha_n)$. Finally, choose x_n to be a point of $B(1,1,\ldots,1)^*$. The induction is now complete.

Now let $X = \{x_n : n \geqslant 1\}$ and for each infinite sequence $\{\alpha_n\}$ of zeros and ones, let $B\{\alpha_n\} = \bigcup \{B(\alpha_1, \ldots, \alpha_i) : i \geqslant 1\}$. It follows from the construction that X and $B\{\alpha_n\}$ are countable discrete subspaces of $\mathbb{N}^*$. Since x_i belongs to $B(\alpha_1, \ldots, \alpha_i)^*$ for each finite sequence of length i, it follows that for each q in X^*, q belongs to $B\{\alpha_n\}^*$ for every sequence $\{\alpha_n\}$. Choose a point p in X^*. We will show that p has c distinct relative types by showing that $\tau(p, B\{\alpha_i\}) \neq \tau(p, B\{\beta_i\})$ if $\{\alpha_i\} \neq \{\beta_i\}$. Let k be the first index such that $\alpha_k \neq \beta_k$ and assume that $\alpha_k > \beta_k$. Let

$$Y = \bigcup \{B(\alpha_1, \ldots, \alpha_n) : n \geqslant k\} \subset B\{\alpha_i\}.$$

Since x_n is in $B(\alpha_1, \ldots, \alpha_n)^*$ for all $n \geqslant k$, it follows that p is in Y^* and from Lemma 3.42(a), we have that $\tau(p, Y) = \tau(p, B\{\alpha_i\})$. From the construction, Y is a countable discrete subspace of $B\{\beta_i\}^*$ since $(\alpha_1, \ldots, \alpha_n) > (\beta_1, \ldots, \beta_n)$ for all $n \geqslant k$. Hence, from Lemma 7.8, $\tau(p, Y) \neq \tau(p, B\{\beta_i\})$. Thus, p has at least c distinct relative types, and since it can have no more by Frolík's result, p has exactly c relative types.

Types, $\mathbb{N}^*$-Types, and P-Points

7.10. In this section, we will introduce the notion of an $\mathbb{N}^*$-type and relate it to types of ultrafilters as discussed in Chapter 3 and to the P-points of $\mathbb{N}^*$. Recall that two points of $\mathbb{N}^*$ are said to be of the same type if there is an automorphism of $\beta\mathbb{N}$ which maps one to the other. We will define two points of $\mathbb{N}^*$ to be of the same $\mathbb{N}^*$-*type* if there is an automorphism of $\mathbb{N}^*$ which maps one to the other. Since an automorphism of $\beta\mathbb{N}$ restricts to one of $\mathbb{N}^*$, any points having the same type must also have the same $\mathbb{N}^*$-type. However, because an automorphism of $\mathbb{N}^*$ is not determined by its behavior on a countable set as are those of $\beta\mathbb{N}$, we should not be surprised if we discover that there are more automorphisms of $\mathbb{N}^*$ than there are of $\beta\mathbb{N}$. Should this be the case, we would expect to find more points of each $\mathbb{N}^*$-type than of each type and perhaps as a consequence that there are fewer $\mathbb{N}^*$-types than there are types. Actually, we will find that the set of $\mathbb{N}^*$-types has the same cardinality as the set of types, but that under the assumption of the Continuum Hypothesis, there are 2^c automorphisms of $\mathbb{N}^*$ while we have seen that there are only c of $\beta\mathbb{N}$.

All of the tools necessary to demonstrate that there are 2^c $\mathbb{N}^*$-types were developed in the proof of Corollary 3.46 where we showed that $\mathbb{N}^*$ is not homogeneous. As was the discussion of types of ultrafilters in Chapter 3, this result is taken from Z. Frolík's 1967B paper. Note that it does not depend on the Continuum Hypothesis.

Proposition (Frolík). *There are 2^c $\mathbb{N}^*$-types of points in $\mathbb{N}^*$ and $\mathbb{N}^*$ contains a dense subset of each type.*

Proof. If p is a point of $\mathbb{N}^*$, let $\tau[p,\mathbb{N}^*]$ denote the set of relative types of p as in Corollary 3.46. Consider the family $\mathscr{G} = \{\tau[p,\mathbb{N}^*] : p \in \mathbb{N}^*\}$ of all such sets. Theorem 3.45 implies that each set $\tau[p,\mathbb{N}^*]$ contains at most c types. Since $\mathscr{G}$ is a cover of the set of all types and there are 2^c types, $\mathscr{G}$ must contain 2^c distinct sets.

Since the set of relative types is an invariant under automorphisms of $\mathbb{N}^*$, $\tau[p,\mathbb{N}^*]$ must equal $\tau[q,\mathbb{N}^*]$ if p and q are of the same $\mathbb{N}^*$-type. Hence, because $\mathscr{G}$ contains 2^c distinct members, there must be at least 2^c $\mathbb{N}^*$-types. There can be no more since $\mathbb{N}^*$ has just 2^c points. The density of the points of any given $\mathbb{N}^*$-type is clear since the orbit of any point is dense in $\mathbb{N}^*$ (Corollary 3.20). $\square$

7.11. Thus, we see that the sets of $\mathbb{N}^*$-types and of types have the same cardinality. However, by assuming the Continuum Hypothesis, we can show that there are 2^c automorphisms of $\mathbb{N}^*$ versus c for $\beta\mathbb{N}$. If a point is a P-point of $\mathbb{N}^*$, then every point of the same type or $\mathbb{N}^*$-type must also be a P-point. Since there are 2^c P-points and only c points of each type, there are 2^c distinct types of P-points. However, in 1956 W. Rudin showed that all P-points are of the same $\mathbb{N}^*$-type. The proof is accomplished by showing that for any two P-points there is an automorphism of the Boolean algebra $CO(\mathbb{N}^*)$ which interchanges the filters determined by the two P-points. The resultant automorphism of $\mathbb{N}^*$ therefore interchanges the two P-points. The construction of the Boolean algebra automorphism is similar to that given in the proof of Theorem 3.31.

Theorem [CH] (W. Rudin). *All P-points of $\mathbb{N}^*$ are of the same $\mathbb{N}^*$-type. Hence, $\mathbb{N}^*$ admits precisely 2^c automorphisms.*

Proof. There can be no more than 2^c automorphisms since distinct automorphisms of $\mathbb{N}^*$ induce distinct permutations of the clopen sets of $\mathbb{N}^*$ and there are only 2^c such permutations. Because $\mathbb{N}^*$ has 2^c P-points, the first statement of the theorem implies the second.

Now let p and q be P-points of $\mathbb{N}^*$. Each of the points is contained in c clopen subsets of $\mathbb{N}^*$. Using the Continuum Hypothesis, let the families $\{S_\alpha : \alpha < \omega_1\}$ and $\{T_\alpha : \alpha < \omega_1\}$ of all clopen subsets containing p and q, respectively, be indexed by the countable ordinals. Further, assume that $S_0 = T_0 = \mathbb{N}^*$. We now construct an automorphism σ of $CO(\mathbb{N}^*)$ so that $\sigma[\{S_\alpha\}] = \{T_\alpha\}$. Note that it is sufficient to define σ for the family $\{S_\alpha\}$ since the requirement that σ preserve the Boolean operations dictates the value of σ on the other clopen sets which are merely complements of the S_α's. The construction will be carried out by induction. We will show that if σ has been defined for a countable field of sets not containing a set S_α, then σ can be extended to a countable field containing S_α. As in Theorem 3.31, Lemma 2.24 will be used to define the extension.

Begin by putting $\sigma(S_0) = T_0$ and $\sigma(\emptyset) = \emptyset$ so that the family for which σ is initially defined forms a field. Now assume that σ has been defined for a countable field $\mathscr{C}$ and that α is the least ordinal for which $\sigma(S_\alpha)$ has not been defined. Consider the intersection $S_\alpha \cap (\bigcap \{ S_\beta : \sigma(S_\beta)$ has been defined$\})$. Since p is a P-point, there exists an S_γ with $\gamma > \alpha$ such that S_γ is contained in this intersection. Similarly, since q is a P-point and each $\sigma(S_\beta)$ is a clopen set containing q, there exists a set T_γ with $T_\gamma \subset \bigcap \{ \sigma(S_\beta) \}$. Define $\sigma(S_\gamma) = T_\gamma$ and $\sigma(\mathbb{N}^* \backslash S_\gamma) = \mathbb{N}^* \backslash T_\gamma$. Let $\mathscr{C}'$ be the field generated by S_γ and $\mathscr{C}$. Extend σ to $\mathscr{C}'$ by the requirement that σ preserve the Boolean operations. Thus, σ is now defined on a countable field which includes a subset of S_α. Write $\mathscr{C}'$ as the union of three sub-families $\{F_i\}$, $\{G_i\}$, and $\{C_i\}$ such that $F_i \subset S_\alpha$, $S_\alpha \subset G_i$, and no inclusion relation holds between S_α and C_i for all values of i. Put $A_n = F_1 \cup \cdots \cup F_n$ and $B_n = G_1 \cap \cdots \cap G_n$ so that

$$A_1 \subset \cdots \subset A_n \subset \cdots \subset S_\alpha \subset \cdots \subset B_n \subset \cdots \subset B_1$$

and no C_i is contained in any A_n nor contains any B_n. Since σ preserves the Boolean operations, we have

$$\sigma(A_1) \subset \cdots \subset \sigma(A_n) \subset \cdots \subset \sigma(B_n) \subset \cdots \subset \sigma(B_1)$$

and no $\sigma(C_i)$ is contained in any $\sigma(A_n)$ and no $\sigma(C_i)$ contains any $\sigma(B_n)$. Since $CO(\mathbb{N}^*)$ is dense in itself and Cantor and DuBois-Reymond separable, Lemma 2.24 implies that there exists a set T_α such that

$$\sigma(A_1) \subset \cdots \subset \sigma(A_n) \subset \cdots \subset T_\alpha \subset \cdots \subset \sigma(B_n) \subset \cdots \subset \sigma(B_1)$$

and T_α is unrelated by inclusion to any $\sigma(C_n)$. Define $\sigma(S_\alpha) = T_\alpha$ and $\sigma(\mathbb{N}^* \backslash S_\alpha) = \mathbb{N}^* \backslash T_\alpha$. Let $\mathscr{C}''$ be the field generated by $\mathscr{C}' \cup \{ S_\alpha \}$. The requirement that σ preserve the Boolean operations dictates a unique extension of σ to $\mathscr{C}''$.

Now having defined $\sigma(S_\alpha)$, $\sigma^\leftarrow$ satisfies the same induction hypothesis and by the same procedure, we can define $\sigma^\leftarrow(T_\delta)$ where δ is the least ordinal for which T_δ does not belong to the range of σ. The process is continued to yield the automorphism σ.

Finally, the automorphism σ induces an automorphism of $\mathbb{N}^*$ as in Theorem 3.31. Since the induced automorphism exchanges the points of $\mathbb{N}^*$ corresponding to the maximal filters exchanged by σ, the automorphism sends p to q. □ ∴

7.12. The question of whether two points of $\mathbb{N}^*$ are of the same type is really a question of the existence of a permutation of $\mathbb{N}$ whose extension to an automorphism of $\beta\mathbb{N}$ will interchange the two points. A similar type of condition for $\mathbb{N}^*$-types will be seen to replace the permutation of $\mathbb{N}$ with the existence of a rather special bijection between two countable subsets of $\mathbb{N}^*$. As in the proof of the preceding theorem, the required automorphism of $\mathbb{N}^*$ will be obtained by first describing an automorphism of the Boolean algebra $CO(\mathbb{N}^*)$ and it is here that the Continuum Hypothesis will be assumed.

The next two results appear in the 1966 paper of M. E. Rudin. The first of these can be thought of as an extension theorem since it provides an automorphism which simultaneously extends certain restrictions of a countable family of automorphisms.

Theorem [CH]. *For each $n<\omega_0$, let h_n be an automorphism of $\mathbb{N}^*$ and let $\{U_n:n<\omega_0\}$ be a family of clopen subsets of $\mathbb{N}^*$ such that both $\{U_n:n<\omega_0\}$ and $\{h_n[U_n]:n<\omega_0\}$ are pairwise disjoint families. Then there exists an automorphism h of $\mathbb{N}^*$ such that the restrictions $h|U_n$ and $h_n|U_n$ are equal for each n.*

Proof. Let $\mathscr{C}$ be the field generated by $\{U_n\}\cup\{\emptyset\}$. Because the family $\{U_n\}$ is pairwise disjoint, a member of $CO(\mathbb{N}^*)$ belongs to $\mathscr{C}$ if and only if it is a union of finitely many of these sets or the complement of such a union. By requiring the preservation of unions and complements, a function $f:\mathscr{C}\to CO(\mathbb{N}^*)$ can be defined which has these properties and also satisfies $f(\emptyset)=\emptyset$ and $f(U_n)=h[U_n]$.

Using the Continuum Hypothesis, index $CO(\mathbb{N}^*)=\{W_\alpha:\alpha<\omega_1\}$ by the countable ordinals. We will employ transfinite induction to extend f to an automorphism σ of $CO(\mathbb{N}^*)$.

Assume that $\mathscr{F}$ is a countable field containing $\mathscr{C}$ and that the extension $\sigma:\mathscr{F}\to CO(\mathbb{N}^*)$ has been defined so that:

(1) σ is a Boolean algebra monomorphism, and

(2) if W_γ in $\mathscr{F}$ is contained in U_n for some n, then $\sigma(W_\gamma)=h_n[W_\gamma]$.

Let α be the least ordinal for which $\sigma(W_\alpha)$ has not been defined. We will show that σ can be extended to a countable field containing W_α in such a way that (1) and (2) are preserved.

Begin by letting $V_n=W_\alpha\cap U_n$ for each $n<\omega_0$ and let $\mathscr{F}'$ be the field generated by $\mathscr{F}\cup\{V_n\}$. Then $\mathscr{F}'$ is countable. To satisfy (2), we define $\sigma(V_n)=h[V_n]$ for each $n<\omega_0$, and σ can then be extended to all of $\mathscr{F}'$ by requiring that (1) be satisfied. Let $\mathscr{F}''$ be the field generated by $\mathscr{F}'\cup\{W_\alpha\}$. Now as in the proof of Theorem 7.11, we can extend σ to $\mathscr{F}''$ by first writing $\mathscr{F}''$ as the union of three subfamilies: $\{F_i\}$ consisting of sets contained in W_α, $\{G_i\}$ consisting of sets containing W_α, and $\{C_i\}$

consisting of sets bearing no inclusion relation to W_α, and then using Lemma 2.24 to define $\sigma(W_\alpha)$.

Now having defined $\sigma(W_\alpha)$, $\sigma^\leftarrow$ satisfies the same induction hypothesis with h_n replaced by $h_n^\leftarrow$ and we can repeat the argument to define $\sigma^\leftarrow(W_\beta)$ where β is the least ordinal such that W_β is not yet in the range of σ.

The automorphism σ of $CO(\mathbb{N}^*)$ now induces an automorphism h of $\mathbb{N}^*$ and condition (2) of the induction hypothesis guarantees that $h|U_n = h_n|U_n$ for each n. $\quad\square$

7.13. The next theorem provides a necessary and sufficient condition for two points of $\mathbb{N}^*$ to be of the same $\mathbb{N}^*$-type provided that one of them is a limit point of a countable discrete subspace of $\mathbb{N}^*$. Recall that if a point p belongs to X^* for some countable discrete subspace X, then p_X denotes the trace of the neighborhood ultrafilter of p on X.

Theorem [CH] (M. E. Rudin). *Let p and q belong to $\mathbb{N}^*$ and let X be a countable discrete subspace of $\mathbb{N}^*$ with p in X^*. Then p and q are of the same $\mathbb{N}^*$-type if and only if there exists a countable discrete subspace Y of $\mathbb{N}^*$ with q in Y^* and a one-to-one correspondence f of X onto Y such that $f[p_X] = q_Y$ and x and $f(x)$ are of the same $\mathbb{N}^*$-type for each x in X.*

Proof. First suppose that p and q are of the same $\mathbb{N}^*$-type. Then there exist an automorphism h of $\mathbb{N}^*$ such that $h(p) = q$. Put $Y = h[X]$. Then h clearly induces a one-to-one correspondence between X and Y, $h[p_X] = q_Y$, and x and $h(x)$ are of the same $\mathbb{N}^*$-type for all x in X.

Now suppose that p and q belong to X^* and Y^*, respectively, and that there exists a one-to-one correspondence f of $X = \{x_n : n < \omega_0\}$ onto Y such that $f[p_X] = q_Y$ and x_n and $f(x_n)$ are of the same $\mathbb{N}^*$-type for each n. Thus there is an automorphism h_n of $\mathbb{N}^*$ such that $h_n(x_n) = f(x_n)$ for each n. Since X and Y are discrete, there are pairwise disjoint families $\{W_n\}$ and $\{V_n\}$ of clopen subsets of $\mathbb{N}^*$ such that x_n belongs to W_n and $f(x_n)$ belongs to V_n for each n. Now define $U_n = W_n \cap h_n^\leftarrow(V_n)$. Note that x_n belongs to U_n and that $\{U_n\}$ and $\{h_n[U_n]\}$ are each pairwise disjoint families of clopen subsets of $\mathbb{N}^*$. Hence, by the previous theorem, there exists an automorphism h of $\mathbb{N}^*$ such that $h|U_n = h_n|U_n$. Finally, if U is a clopen set containing p, then $h[U]$ must contain q since $h[p_X] = f[p_X] = q_Y$, so that $h(p) = q$. Hence, p and q are of the same $\mathbb{N}^*$-type. $\quad\square$

7.14. The following corollary is a special case of the theorem where all points of the countable discrete subspaces are of the same $\mathbb{N}^*$-type.

Corollary [CH]. *Let X and Y be countable discrete subspaces of $\mathbb{N}^*$ such that all points of X and Y are of the same $\mathbb{N}^*$-type. If p and q are limit points of X and Y, respectively, then p and q are of the same $\mathbb{N}^*$-type if and only if $\tau(p, X) = \tau(q, Y)$.*

7.15. Observe that from Theorem 7.13 we see again, this time with the assumption of the Continuum Hypothesis, that there are 2^c $\mathbb{N}^*$-types. The countable discrete subspace X has 2^c limit points and because there can be only c permutations of X, the theorem shows that only c of these limit points can be of the same $\mathbb{N}^*$-type. Thus, the limit points of X must include 2^c different $\mathbb{N}^*$-types.

In the same manner, Corollary 7.14 shows that if the points of X are all of the same $\mathbb{N}^*$-type, then the limit points of X also contain 2^c $\mathbb{N}^*$-types. Of course, the P-points cannot be included among these types since a P-point cannot be the limit point of a countable set. On the other hand, not all $\mathbb{N}^*$-types can be found among the limit points of countable sets of P-points, as the following example shows.

Example (M. E. Rudin). *There exists a non-P-point of $\mathbb{N}^*$ which is not a limit point of any countable set of P-points of $\mathbb{N}^*$.*

For each subset E of $\mathbb{N}$ and each positive integer n, let $\varphi(E,n)=|\{i\in E: i\leqslant n\}|$. Define the density of E to be $D(E)=\lim\limits_{n\to\prime}\varphi(E,n)/n$ if the limit exists. Let $\mathscr{I}$ be the set of all subsets of $\mathbb{N}$ having density equal to 1.

(a) $\mathscr{I}$ is closed under finite intersections: Let E and F belong to $\mathscr{I}$ and let $\varepsilon>0$ be given. Then there exists an integer N such that

$$\frac{\varphi(E,n)}{n} > 1 - \frac{\varepsilon}{2} \quad \text{and} \quad \frac{\varphi(F,n)}{n} > 1 - \frac{\varepsilon}{2}$$

for all $n>N$. Thus,

$$\frac{\varphi(E\cap F,n)}{n} \geqslant 1 - \frac{\varphi(\mathbb{N}\setminus E)}{n} - \frac{\varphi(\mathbb{N}\setminus F)}{n} > 1-\varepsilon,$$

so that $D(E\cap F)=1$.

Thus, the set $S=\{q\in\mathbb{N}^*:\mathscr{I}\subset A^q\}$ is non-empty. We will show that no point of S can be a P-point of $\mathbb{N}^*$ nor lie in the closure of any countable set of P-points. First, we show that:

(b) Any P-point p of $\mathbb{N}^*$ includes a set of zero density in A^p: For any ultrafilter $\mathscr{U}$ on $\mathbb{N}$ and any positive integer n, there is exactly one integer m, $0\leqslant m\leqslant n-1$, such that the set of positive integers congruent to m modulo n belongs to $\mathscr{U}$. Hence, every ultrafilter contains a set Z_n of density $1/n$ for each n. However, if p is a P-point of $\mathbb{N}^*$, A^p contains a set Z such that Z^* is contained within Z_n^* for each n (Exercise 4A). Hence, $D(Z)=0$.

Thus, S can contain no P-point since every set of density zero is the complement of a set having density one. The key to showing that the closure of any countable discrete subspace of $\mathbb{N}^* \backslash S$ must miss S is the following:

(c) A set of zero-density can be obtained from any countable family $\{E_i : i \geqslant 1\}$ of sets such that $D(E_i) = 0$ for all i by removing finitely many points from each E_i and forming the union of the resulting sets: For each positive integer n, there exists an integer j_n such that for all $k > j_n$,

$$\sum_{i=1}^{n} \frac{\varphi(E_i, k)}{k} < \frac{1}{n}.$$

We can assume that the sequence $\{j_n\}$ is strictly increasing. For each n, define $D_n = \{x \in E_n : x > j_n\}$ and put $E = \bigcup \{D_n : n \geqslant 1\}$. We must demonstrate that $D(E) = 0$. We will show that for any n, if $k > j_n$ then $\varphi(E, k)/k \leqslant 1/n$. There is an integer m such that $j_m < k \leqslant j_{m+1}$. Clearly, $m \geqslant n$. Then

$$\{x \in E : x \leqslant k\} \subseteq \bigcup_{i=1}^{m} \{x \in E_i : x \leqslant k\}$$

so that $\varphi(E, k)/k < 1/m < 1/n$.

Now let $X = \{p_n\}$ be a countable set of P-points of $\mathbb{N}^*$. X therefore is discrete and lies outside of S. For each n, there exists a set E_n in the ultrafilter A^{p_n} such that $D(E_n) = 0$. If D_n is any set such that $E_n \backslash D_n$ is finite, then it follows that D_n also belongs to A^{p_n}. Thus, sets D_n can be chosen as in (c) above so that $E = \bigcup \{D_n : n \geqslant 1\}$ has density zero and E^* contains X and therefore X^* as well. But then for any point q of X^*, E also belongs to A^q so that A^q can contain no set of density 1. Hence, $X^* \cap S = \emptyset$, and no point of S is a limit point of any countable set of P-points.

Minimal Types and Points with Finitely Many Relative Types

7.16. As we have just observed in the previous section, a P-point cannot be a limit point of any countable set. Thus, it follows from the discussion of the producing relation in Section 3.44 that no type which consists of P-points can be produced by any type. In the concluding sections of the chapter, we will see that the producing relation defines a partial order on the set of types and that as a consequence of this fact, a type consisting of P-points is minimal with respect to this partial order. We will then introduce a multiplication on the set of types and prove that any point which has as its type the product of finitely many minimal types has only finitely many relative types. Most of the results are from the 1971 paper of A. K. and E. F. Steiner.

7.17. Recall that a type s is said to produce a type t if there is a point p in $\mathbb{N}^*$ of type t such that $\tau(p,X)=s$ for some countable discrete subspace X of $\mathbb{N}^*$. We should observe here that $\mathbb{N}$ can be replaced in this definition by any countable discrete subspace Y of $\beta\mathbb{N}$ since $\operatorname{cl} Y$ is a copy of $\beta\mathbb{N}$. Thus, we can say that s produces a type t if and only if there is a countable discrete subspace Y of $\beta\mathbb{N}$, a countable discrete subspace X of Y^*, and a point p in X^* such that $\tau(p,X)=s$ and $\tau(p,Y)=t$.

We will begin by showing that the producing relation induces a total order on the set $\tau[p,\beta\mathbb{N}]$ of relative types of a point p. The following two lemmas are taken from the 1967 D paper of Z. Frolík.

Lemma. *If X and Y are countable discrete subspaces of $\beta\mathbb{N}$, then the set $Z=(X\cap Y)\cup(X^*\cap Y)\cup(X\cap Y^*)$ is discrete, $\operatorname{cl} Z=\operatorname{cl} X\cap\operatorname{cl} Y$, and $Z^*=X^*\cap Y^*$.*

Proof. The set Z is discrete since each of the three sets in the union is discrete and no point belonging to any one of the sets can be an accumulation point of the other two sets. The equalities are evident since a point belonging to both $\operatorname{cl} X$ and $\operatorname{cl} Y$ must belong to the closure of one of the three sets making up Z. $\square$

7.18. If t_1 and t_2 are both in $\tau[p,\beta\mathbb{N}]$, then we will show that as a consequence of the preceding lemma, either the two types are equal, or one produces the other. Write $t_1>_p t_2$ if t_2 produces t_1 and define $t_1\geqslant_p t_2$ if and only if $t_1=t_2$ or $t_1>_p t_2$.

Lemma. *The relation $\geqslant_p$ is a total order on $\tau[p,\beta\mathbb{N}]$. Hence, two relative types of p are either equal or one produces the other.*

Proof. Let t_1 and t_2 belong to $\tau[p,\beta\mathbb{N}]$. Then there exist countable discrete subspaces X_1 and X_2 of $\beta\mathbb{N}$ such that $\tau(p,X_1)=t_1$ and $\tau(p,X_2)=t_2$. Thus, p belongs to $X_1^*\cap X_2^*$. If p belongs to $\operatorname{cl}(X_1\cap X_2)$, then it follows from Lemma 3.42(a) that $t_1=t_2$. Otherwise, it follows from Lemma 7.17 that p belongs either to $(X_1^*\cap X_2)^*$ or to $(X_1\cap X_2^*)^*$. Assume without loss of generality that p is in $(X_1^*\cap X_2)^*$. Then because $\tau(p,X_2)=\tau(p,X_1^*\cap X_2)$, we can assume that X_2 is contained in X_1^*. Thus, t_2 produces t_1 and we can write $t_1>_p t_2$.

If $t_1\geqslant_p t_2$ and $t_2\geqslant_p t_3$ and either is an equality, then we have $t_1\geqslant_p t_3$. Thus, suppose that $t_1>_p t_2$ and $t_2>_p t_3$. Then there exist countable discrete subspaces $X_1,X_2,Y_2,$ and Y_3 of $\mathbb{N}^*$ such that $\tau(p,X_i)=t_i$ for $i=1,2$, $\tau(p,Y_j)=t_j$ for $j=2,3$ and X_2 and Y_3 are subsets of X_1^* and Y_2^*, respectively. Since $\tau(p,X_2)=\tau(p,Y_2)$, by Lemma 3.42(b) there is a homeomorphism h of $\operatorname{cl} Y_2$ onto $\operatorname{cl} X_2$ such that p is mapped to itself. Then $h[Y_3]$ is contained in X_1^* and $\tau(p,h[Y_3])=t_3$ by Lemma 3.42(c), so that we can write $t_1>_p t_3$. Hence, $\geqslant_p$ is transitive.

The relation $\geqslant_p$ is clearly reflexive. The antisymmetry of $\geqslant_p$ follows easily from transitivity and the fact that no type can produce itself (Corollary 6.30). Thus, $\geqslant_p$ is a total order on $\tau[p, \beta\mathbb{N}]$. $\square$

7.19. Each of the total orders $\geqslant_p$ is defined only on the set $\tau[p, \beta\mathbb{N}]$ of relative types of p. We will use the family $\{\geqslant_p : p \in \mathbb{N}^*\}$ to define a partial order $\geqslant$ on the set T of all types in such a way that the restriction of $\geqslant$ to $\tau[p, \beta\mathbb{N}]$ is $\geqslant_p$ for each p. For two types t_1 and t_2, define $t_1 > t_2$ if $t_1 >_p t_2$ for some p, i.e. $t_1 > t_2$ if and only if t_1 is produced by t_2. The following lemma shows that the relation $>$ is well defined.

Lemma. *For any two points p and q of $\mathbb{N}^*$, the relation $>_p$ coincides with $>_q$ on $\tau[p, \beta\mathbb{N}] \cap \tau[q, \beta\mathbb{N}]$.*

Proof. Suppose that t_1 and t_2 belong to $\tau[p, \beta\mathbb{N}] \cap \tau[q, \beta\mathbb{N}]$ and that $t_1 >_p t_2$. Then there are countable discrete subsets X_1 and X_2 of $\beta\mathbb{N}$ such that $\tau(p, X_1) = t_1$, $\tau(p, X_2) = t_2$, and X_2 is contained in X_1^*. Since t_1 belongs to $\tau[q, \beta\mathbb{N}]$, there is a countable discrete subspace Y of $\beta\mathbb{N}$ such that $\tau(q, Y) = t_1$. By Lemma 3.42(b), there is a homeomorphism h of $\mathrm{cl}\, X_1$ onto $\mathrm{cl}\, Y$ such that $h(p) = q$. Then $\tau(q, h[X_2]) = t_2$ and since $h[X_2]$ is contained in Y^*, $t_1 >_q t_2$. $\square$

7.20. If t_1 and t_2 are types, write $t_1 \geqslant t_2$ if and only if $t_1 = t_2$ or $t_1 > t_2$. Then we see that

Lemma. *The relation $\geqslant$ is a partial order on the set of types.*

Proof. It is immediate that $\geqslant$ is reflexive and antisymmetry will follow from transitivity and the fact that no type produces itself. Let t_1, t_2, and t_3 belong to T and assume that $t_1 > t_2$ and $t_2 > t_3$. Then $t_1 >_p t_2$ and $t_2 >_q t_3$ for some p and q in $\mathbb{N}^*$. The proof now continues exactly as that of Lemma 7.18 except that here the homeomorphism h must send q to p. $\square$

This partial order is also considered in the 1970 paper of D. D. Booth. In her 1971 paper, M. E. Rudin considers this partial order and relates it to two other partial orders on the set of types. One of these is discussed in Exercises 7B and 7C.

7.21. If p in $\mathbb{N}^*$ is a P-point, then p is not a limit point of any countable subset of $\mathbb{N}^*$. Thus, the definition of the partial order $\geqslant$ shows that $\tau(p)$ must be minimal in T, i.e. no type consisting of P-points is produced by any type. If the Continuum Hypothesis is assumed, then there are 2^c P-points in $\mathbb{N}^*$ and since there are only c points of each type, there must be 2^c minimal types which consist of P-points. K. Kunen, also assuming the Continuum Hypothesis, has shown in 1970 that there are other minimal types.

7.22. The following proposition characterizing points of minimal type follows from the observation that if p is any point of $\mathbb{N}^*$, then the types preceding $\tau(p)$ are the relative types of p.

Proposition. *A point p of $\mathbb{N}^*$ is of minimal type if and only if $\tau[p,\mathbb{N}^*]=\emptyset$, i.e. if and only if p is not a limit point of any countable discrete subset of $\mathbb{N}^*$.*

We will shortly introduce a multiplication in T and find that if a type can be expressed as the product of finitely many minimal types, then that expression is unique.

7.23. The following lemma will make the definition of multiplication of types possible. If t is a type and S is a subset of $\beta\mathbb{N}$, define

$$t[S]=\{x\in\mathbb{N}^*:t\in\tau[x,S]\}\,.$$

Note that in the lemma we are making use of the fact that a type is actually a member of a partition of $\mathbb{N}^*$ and thus is a subset of $\beta\mathbb{N}$.

Lemma. *If t_1 and t_2 are types, then $t_1[t_2]$ is also a type.*

Proof. We must first show that any two points p and q belonging to $t_1[t_2]$ are of the same type. There are countable discrete subsets X and Y of t_2 such that $\tau(p,X)=t_1=\tau(q,Y)$. Thus, there exists a homeomorphism f of $\operatorname{cl}X$ onto $\operatorname{cl}Y$ such that $f(p)=q$. Since X and Y are discrete, there are pairwise disjoint families of neighborhoods $\{U_x:x\in X\}$ and $\{V_y:y\in Y\}$ such that $\mathbb{N}$ is contained in $\bigcup U_x$ and in $\bigcup V_y$. For each x in X and y in Y, $\tau(x,U_x\cap\mathbb{N})=t_2=\tau(y,V_y\cap\mathbb{N})$ by Lemma 3.42. For each x in X, let g_x be a homeomorphism of $\operatorname{cl}(U_x\cap\mathbb{N})$ onto $\operatorname{cl}(V_{f(x)}\cap\mathbb{N})$ such that $g_x(x)=f(x)$. If σ is the permutation on $\mathbb{N}$ defined by $\sigma(n)=g_x(n)$ for n in $U_x\cap\mathbb{N}$, then the extension $\beta(\sigma)$ of σ to $\beta\mathbb{N}$ is an automorphism of $\beta\mathbb{N}$ such that $\sigma|\operatorname{cl}X=f$. Hence, $\beta(\sigma)(p)=q$ and p and q are of the same type.

Having seen that all points of $t_1[t_2]$ are of the same type, it remains to show that every point of this type belongs to $t_1[t_2]$. Suppose that $\tau(p)=\tau(q)$ and p is in $t_1[t_2]$. There is a countable discrete subspace X contained in t_2 such that $\tau(p,X)=t_1$. Since p and q are of the same type, there is an automorphism h of $\beta\mathbb{N}$ such that $h(p)=q$. Thus, q belongs to $(h[X])^*$, $\tau(q,h[X])=t_1$ by Lemma 3.42(c), and $h[X]$ is a subspace of t_2. Thus, q is also in $t_1[t_2]$. $\square$

7.24. Thus, we may define the product $t_1\circ t_2$ of two types to be the type $t_1[t_2]$. Then from this definition and the definition of relative type, we see that if X is a countable discrete subspace of $\beta\mathbb{N}$ and p is in X^*, then $\tau(p, X) = t_1 \circ t_2$ if and only if there is a countable discrete subspace Y of X^* such that $\tau(p, Y) = t_1$ and $\tau(y, X) = t_2$ for every y in Y. It follows from the next lemma that this multiplication is associative.

Lemma. *For any types t_1 and t_2 and any subset S of $\beta\mathbb{N}$, $t_1[t_2[S]] = t_1 \circ t_2[S]$.*

Before taking up the proof, let us first see that the lemma implies the associativity of the multiplication. Let S in the lemma be a third type t_3. Then we have

$$(t_1 \circ t_2) \circ t_3 = t_1 \circ t_2[t_3] = t_1[t_2[t_3]] = t_1 \circ (t_2 \circ t_3)$$

so that the multiplication is associative.

Proof of Lemma. If p is in $t_1[t_2[S]]$, then there is a countable discrete set X in $t_2[S]$ such that $\tau(p, X) = t_1$. Further, for each x in X, there is a countable discrete set Y_x in S such that $\tau(x, Y_x) = t_2$. Because X is discrete there is a pairwise disjoint family of open sets $\{U_x : x \in X\}$ such that $U_x \cap X = \{x\}$. If we put $Z = \bigcup\{U_x \cap Y_x : x \in X\}$, then Z is a countable discrete subspace of S and X is contained in Z^*. By Lemma 3.42, each point of x has type t_2 relative to Z and thus $\tau(p, Z) = t_1 \circ t_2$ so that p is in $t_1 \circ t_2[S]$.

Conversely, if p is in $t_1 \circ t_2[S]$, then there is a countable discrete subspace X contained in S such that $\tau(p, X) = t_1 \circ t_2$. Then there is a countable discrete subspace Y inside X^* such that $\tau(p, Y) = t_1$ and $\tau(y, X) = t_2$ for every y in Y. But this implies that Y is contained in $t_2[S]$ so that p belongs to $t_1[t_2[S]]$. $\square$

7.25. The following lemma is the last preliminary result needed to show that if a type can be written as the product of finitely many minimal types, then it can be done so in only one way.

Lemma. *If X and Y are countable discrete subsets of $\beta\mathbb{N}$ having a common limit point p such that $\tau(p, X) = t$ and $\tau(p, Y) = s$ where both t and s are minimal, then $t = s$, $X \cap Y$ is infinite, and p is in $(X \cap Y)^*$.*

Proof. Consider the set $Z = (X \cap Y) \cup (X^* \cap Y) \cup (X \cap Y^*)$ as described in Lemma 7.17. The point p must belong to the closure of one of the three sets in the union since it is in $X^* \cap Y^*$. If p belonged to the closure of either $X^* \cap Y$ or $X \cap Y^*$, then either t or s must produce the other. But this is impossible since both t and s are minimal. Hence, $t = s$ by Lemma 3.42, $X \cap Y$ is infinite, and p is in $(X \cap Y)^*$. $\square$

7.26. We can now prove the main result from the Steiner's 1971 paper, the Unique Decomposition Theorem.

Theorem (A. K. and E. F. Steiner). *If $s_1 \circ s_2 \circ \cdots \circ s_n = t_1 \circ t_2 \circ \cdots \circ t_m$ where s_i and t_j are minimal types for $1 \leqslant i \leqslant n$ and $1 \leqslant j \leqslant m$, then $n = m$ and $s_i = t_i$ for $1 \leqslant i \leqslant n$. Thus, each finite sequence of minimal types represents a distinct type in T.*

Proof. The proof will be by induction on n. If $s_1 = t_1 \circ t_2 \circ \cdots \circ t_m$, then $s_1 = \tau(p)$ for some p in $\mathbb{N}^*$ and p is not a limit point of any countable subset of $\mathbb{N}^*$ since s_1 is a minimal type. Thus, $\tau(p) = t_1 \circ t_2 \circ \cdots \circ t_m$ cannot be a composition of types which implies that $m = 1$, and the previous lemma then yields $s_1 = t_1$.

Assume that for all $k \leqslant n$, if $s_1 \circ s_2 \circ \cdots \circ s_k = t_1 \circ t_2 \circ \cdots \circ t_m$ and each s_i and t_j are minimal, then $k = m$ and $s_i = t_i$ for $1 \leqslant i \leqslant k$.

Now suppose that $s_1 \circ s_2 \circ \cdots \circ s_{n+1} = t_1 \circ t_2 \circ \cdots \circ t_m$. Then there is a p in $\mathbb{N}^*$ such that $\tau(p) = s_1 \circ s_2 \circ \cdots \circ s_{n+1} = t_1 \circ t_2 \circ \cdots \circ t_m$, so there are countable discrete subsets X and Y of $s_2 \circ \cdots \circ s_{n+1}$ and $t_2 \circ \cdots \circ t_m$, respectively, such that $\tau(p, X) = s_1$ and $\tau(p, Y) = t_1$. From Lemma 7.25, we then have that $s_1 = t_1$ and $X \cap Y$ is infinite. Thus, $s_2 \circ \cdots \circ s_{n+1}$ and $t_2 \circ \cdots \circ t_m$ are types which are not disjoint so that they must be equal. By the induction hypothesis, $n + 1 = m$ and $s_i = t_i$ for $2 \leqslant i \leqslant n+1$, and the proof is complete. $\square$

7.27. There is a result for $\mathbb{N}^*$-types similar to Lemma 7.23 except that its proof relies on an earlier Corollary which required the assumption of the Continuum Hypothesis.

Proposition [CH]. *If S is an $\mathbb{N}^*$-type and t is any type, then $t[S]$ is an $\mathbb{N}^*$-type.*

Proof. If p and q are in $t[S]$, then there exist countable discrete subsets X and Y of S such that $\tau(p, X) = t = \tau(q, Y)$. Since the points of X and Y are of the same $\mathbb{N}^*$-type, Corollary 7.14 implies that p and q are of the same $\mathbb{N}^*$-type. Conversely, if p is in $t[S]$ and q is of the same $\mathbb{N}^*$-type as p, then there is an automorphism h of $\mathbb{N}^*$ such that $h(p) = q$. Since t is in $\tau[p, S]$, Lemma 3.42 (c) implies that t is also in $\tau[h(p), h[S]] = \tau[q, S]$. Thus, q is in $t[S]$. $\square$

7.28. The Unique Decomposition Theorem also has an analogue for $\mathbb{N}^*$-types.

Theorem [CH]. *If S is an $\mathbb{N}^*$-type such that $\tau[S]$ contains only minimal types, and $s_1 \circ s_2 \circ \cdots \circ s_n[S] = t_1 \circ t_2 \circ \cdots \circ t_m[S]$ where each s_i and t_j is a minimal type, then $n = m$ and $s_i = t_i$ for $1 \leqslant i \leqslant n$. Thus, distinct finite sequences of minimal types represent distinct $\mathbb{N}^*$-types which can be obtained from S.*

Proof. The proof will be by induction on n. If p belongs to $s_1[S] = t_1 \circ t_2 \circ \cdots \circ t_m[S]$, then there are countable discrete subsets X and Y of S and $t_2 \circ \cdots \circ t_m[S]$, respectively, such that $\tau(p, X) = s_1$ and $\tau(p, Y) = t_1$. Since s_1 and t_1 are minimal, Lemma 7.25 implies that $s_1 = t_1$ and that $X \cap Y$ is infinite. If y is a point of $X \cap Y$, then y is in S so that $\tau(y)$ is minimal and y cannot be a limit point of any countable discrete subset of $\mathbb{N}^*$. Since y is also in $t_2 \circ \cdots \circ t_m[S]$, this implies that $m = 1$.

Assume that for all $k \leqslant n$, if $s_1 \circ s_2 \circ \cdots \circ s_k[S] = t_1 \circ t_2 \circ \cdots \circ t_m[S]$, then $k = m$ and $s_i = t_i$ for $1 \leqslant i \leqslant k$.

If p is in $s_1 \circ s_2 \circ \cdots \circ s_{n+1}[S] = t_1 \circ t_2 \circ \cdots \circ t_m[S]$, we use Lemma 7.25 as in the $n = 1$ case to show that $s_1 = t_1$ and that the $\mathbb{N}^*$-types $s_2 \circ \cdots \circ s_{n+1}[S]$ and $t_2 \circ \cdots \circ t_m[S]$ have non-empty intersection and thus are equal. By the induction hypothesis, $n + 1 = m$ and $s_i = t_i$ for $1 \leqslant i \leqslant m$. $\square$

7.29. We are now able to describe completely the set $\tau[p, \beta\mathbb{N}]$ of relative types when the type of p is the product of finitely many minimal types.

Theorem (A. K. and E. F. Steiner). *If* $\tau(p) = t_1 \circ t_2 \circ \cdots \circ t_n$ *where each* t_i *is minimal, then*

$$\tau[p, \beta\mathbb{N}] = \{t_1, t_1 \circ t_2, \ldots, t_1 \circ t_2 \circ \cdots \circ t_n\}\,.$$

Proof. From Lemma 7.24, it follows from the fact that $\tau(p) = t_1 \circ t_2 \circ \cdots \circ t_n$ that there exist countable discrete subsets $X_1, X_2, \ldots, X_n = \mathbb{N}$ of $\beta\mathbb{N}$ such that

$$X_i \subset X_{i+1}^* \qquad \text{for } 1 \leqslant i \leqslant n-1\,,$$

$$\tau(x, X_{i+1}) = t_{i+1} \quad \text{for each } x \in X_i, \quad 1 \leqslant i \leqslant n-1\,, \quad \text{and}$$

$$\tau(p, X_i) = t_1 \circ \cdots \circ t_i \quad \text{for } 1 \leqslant i \leqslant n\,.$$

Thus, the set $\{t_1 \circ \cdots \circ t_i : 1 \leqslant i \leqslant n\}$ is contained in $\tau[p, \beta\mathbb{N}]$.

Now we must show that every relative type of p is one of these products. Let Y be a countable discrete subspace of $\beta\mathbb{N}$ with p in Y^*. Put $Y_1 = Y \cap \operatorname{cl} X_1$ and $Y_i = Y \cap (\operatorname{cl} X_i \setminus \operatorname{cl} X_{i-1})$ for $2 \leqslant i \leqslant n$. Then the Y_i's are pairwise disjoint and $Y = \bigcup \{Y_i : 1 \leqslant i \leqslant n\}$ so that p must belong to Y_i^* for exactly one i. If p belongs to Y_i^*, then by Lemma 7.17 p is in either $(Y_i \cap X_i)^*$ or $(Y_i \cap X_i^*)^*$. In the first case, we have

$$\tau(p, Y) = \tau(p, Y_i) = \tau(p, X_i) = t_1 \circ \cdots \circ t_i\,.$$

The proof will be completed by showing that the second case is impossible.

If $i=1$, then $Y_1 \cap X_1^*$ is a countable discrete subspace of X_1^* and since $\tau(p, X_1) = t_1$ is a minimal type, Lemma 7.22 shows that it is impossible for p to belong to $(Y_1 \cap X_1^*)^*$. If $i \geq 2$, then we will show that $(Y_i \cap X_i^*)^* \cap X_{i-1}^* = \emptyset$, which implies that p cannot be in $(Y_i \cap X_i^*)^*$ since p is in X_{i-1}^*. From the definition of Y_i, $\mathrm{cl}\, X_{i-1}$ misses $Y_i \cap X_i^*$. On the other hand, for each $x \cdot$ in X_{i-1}, $\tau(x, X_i) = t_i$ so that x cannot belong to $(Y_i \cap X_i^*)^*$ since t_i is a minimal type. Hence, the two countable sets X_{i-1} and $Y_i \cap X_i^*$ each miss the closure of the other. Thus, it follows from Lemma 7.17 that their closures are disjoint, so that

$$(Y_i \cap X_i^*)^* \cap X_{i-1}^* = \emptyset . \quad \square$$

Note that if $t_1, \ldots, t_n$ are not necessarily minimal types and $\tau(p) = t_1 \circ \cdots \circ t_n$, then the first part of the proof shows that each of the products $t_1 \circ \cdots \circ t_i$ is a relative type of p. However, if the t_i are not minimal, then these products need not exhaust the set of relative types. For instance, if $\tau(p) = t_1 \circ \cdots \circ t_n$ and t_1 is not minimal, then there exists a type s such that $t_1 > s$. But we also have $\tau(p) > t_1$ since t_1 is a relative type of p, and therefore it follows that $\tau(p) > s$. Thus, s is also a relative type of p.

Note also that not every type is a product of minimal types, since a point of such a type can have only finitely many relative types and in Example 7.9 we saw that a point may have as many as c relative types. Also, R. C. Solomon has recently shown the existence of a point in $\mathbb{N}^*$ having exactly $\aleph_0$ relative types, although unlike Example 7.9, his example depends on the Continuum Hypothesis. The problem of characterizing the points having only finitely many relative types remains open and even the existence of minimal types themselves has not been demonstrated without a set-theoretic assumption such as the Continuum Hypothesis.

7.30. There is an analogue of the preceding theorem which describes an $\mathbb{N}^*$-type having the property that each point of that $\mathbb{N}^*$-type has only finitely many relative types. Note that the set of P-points is an example of an $\mathbb{N}^*$-type containing only minimal types as required by the theorem.

Theorem [CH]. *If S is an $\mathbb{N}^*$-type containing only minimal types and $t_1, \ldots, t_n$ are minimal types, then for each p in $t_1 \circ \cdots \circ t_n[S]$,*

$$\tau[p, \mathbb{N}^*] = \{t_1, t_1 \circ t_2, \ldots, t_1 \circ t_2 \circ \cdots \circ t_n\} .$$

Proof. From Proposition 7.27, which utilized the Continuum Hypothesis, $t_1 \circ \cdots \circ t_n[S]$ is an $\mathbb{N}^*$-type. If p belongs to $t_1 \circ \cdots \circ t_n[S]$, then $\tau(p, X)$

$= t_1 \circ \cdots \circ t_n$ for some countable discrete subspace X of S. Since $\operatorname{cl} X$ is a copy of $\beta \mathbb{N}$, Theorem 7.29 implies that $\tau[p, \operatorname{cl} X] = \{t_1 \circ \cdots \circ t_i : 1 \leqslant i \leqslant n\}$. Since a countable discrete subspace of $\operatorname{cl} X$ is also a countable discrete subspace of $\mathbb{N}^*$, we see that $\tau[p, \operatorname{cl} X]$ is contained in $\tau[p, \mathbb{N}^*]$. The proof will be complete if we can show the opposite containment.

Let Y be a countable discrete subspace of $\mathbb{N}^*$ such that p belongs to Y^*. From Lemma 7.17, either p is in $(X \cap Y)^*$, p is in $(Y^* \cap X)^*$, or p is in $(Y \cap X^*)^*$. In the first case, $\tau(p, Y) = \tau(p, X) = t_1 \circ \cdots \circ t_n$. The second case is impossible since X consists of points having minimal type because it is a subset of S, and therefore no point of X can be a limit point of Y. In the third case, p is in $(Y \cap \operatorname{cl} X)^*$. Since $Y \cap \operatorname{cl} X$ is a countable discrete subset of $\operatorname{cl} X$, $\tau(p, Y) = \tau(p, Y \cap \operatorname{cl} X)$ and this relative type is in $\tau[p, \operatorname{cl} X]$ and must therefore be a product $t_1 \circ \cdots \circ t_i$ for some i. Hence, $\tau[p, \mathbb{N}^*]$ is contained in $\tau[p, \operatorname{cl} X]$. $\quad\square$

7.31. If S in the theorem is chosen to be the set of P-points of $\mathbb{N}^*$, note that the $\mathbb{N}^*$-type $t_1 \circ \cdots \circ t_n[S]$ consists of 2^c different types since there are 2^c types of P-points. However, by the theorem, each of these types has the same set of relative types. Thus, we have verified the

Corollary [CH]. *There exist 2^c distinct types each of which is produced by the same set of types.*

7.32. The fact that the Continuum Hypothesis has been employed several times in the discussion of $\beta \mathbb{N}$ and $\mathbb{N}^*$ only begins to indicate the role which set-theoretic assumptions play in further investigations of these topics. One focus of attention is the question of the existence of P-points in $\mathbb{N}^*$. Our proofs that such points exist and that they are all of the same $\mathbb{N}^*$-type both required the Continuum Hypothesis. Additional investigation has been carried out using another set-theoretic assumption, Martin's Axiom, which like the Continuum Hypothesis has been shown to be independent of the usual axioms of set theory. We state here a topological equivalent.

Martin's Axiom. *If X is a compact space in which every collection of disjoint open subsets is countable, then X is not the union of fewer than $2^{\aleph_0}$ nowhere dense subsets.*

The relationships between Martin's Axiom and the Continuum Hypothesis and the comparative usefulness of these two axioms is discussed in the 1970 paper of Martin and Solovay. Here we will state only one of the results that have been obtained by assuming Martin's Axiom. A point is said to be a P_n-*point* of a space X if it belongs to the interior of the intersection of any family of $\aleph_n$ open sets containing it. With this definition, a P-point is a P_0-point. The following theorem appears in D.D. Booth's thesis. The assumption in the theorem was shown to be consistent with the axioms of Zermelo-Fraenkel set theory by Solovay and Tennenbaum in their 1971 paper.

Theorem. (Booth). *If Martin's Axiom and the inequality $2^{\aleph_0} > \aleph_1$ are assumed, then any non-empty intersection of $\aleph_1$ open subsets of $\mathbb{N}^*$ contains a P_1-point.*

Thus, we have an example of a result which implies the existence of P-points in $\mathbb{N}^*$ and which includes the negation of the Continuum Hypothesis as an assumption. Additional results concerning P-points and using Martin's Axiom can be found in Booth's 1970 paper.

The reader should note that there is another, non-equivalent, form of Martin's Axiom in which the phrase "fewer than $2^{\aleph_0}$" is replaced by "$\aleph_1$". By observing that the closed unit interval is the union of exactly c singletons, we see that in this form, Martin's Axiom implies the negation of the Continuum Hypothesis.

Since we noted in Section 7.21 that the minimal types discovered by K. Kunen also involve the assumption of the Continuum Hypothesis, we see that the existence of minimal types, and therefore of points having only finitely many relative types, seems to require a set-theoretic assumption of some nature. However, Example 7.9 of a point having c relative types and therefore whose type is not the product of finitely many minimal types required no set-theoretic assumption.

Exercises

7A. *The semigroup $(T, \circ)$*

A semigroup is a set with an associative binary relation defined on it. Hence, it follows from Lemma 7.24 that $(T, \circ)$ is a semigroup.
1. $(T, \circ)$ is not commutative. [Theorems 7.26 and 7.29.]
2. $(T, \circ)$ has no identity element.
Reference: These properties of $(T, \circ)$ were noted by the Steiners in their 1971 paper.

7B. *Types and mappings of $\mathbb{N}$ onto $\mathbb{N}$*

Let g and f be mappings of $\mathbb{N}$ onto itself and let $L = \{n \in \mathbb{N} : (g \circ f)(n) > n\}$, $M = \{n \in \mathbb{N} : (g \circ f)(n) = n\}$, and $Q = \{n \in \mathbb{N} : (g \circ f)(n) < n\}$. Let $(g \circ f)^i$ denote $g \circ f$ composed with itself i times, with $(g \circ f)^0$ denoting the identity. Let p and q be points of $\beta\mathbb{N}$ such that $f(A^p) = A^q$ and $g(A^q) = A^p$.

1. For n and k in L, define $n\sim k$ provided that $(g\circ f)^i(n)=(g\circ f)^j(k)$ for some non-negative integers i and j. Then $\sim$ is an equivalence relation.
2. For each equivalence class E of $\sim$, choose a point n_E in E. Then for any n in L, n belongs to exactly one such E so that there exist i and j such that $(g\circ f)^i(n)=(g\circ f)^j(n_E)$. Then the function $\gamma:L\to\{0,1\}$ defined by $\gamma(n)=0$ if $|i-j|$ is even and $\gamma(n)=1$ is $|i-j|$ is odd is well-defined.
3. L is not in A^p. [If L were in A^p, then either $\gamma^{\leftarrow}(0)$ or $\gamma^{\leftarrow}(1)$ would be in A^p. But $(g\circ f)(\gamma^{\leftarrow}(0))\cap L\subset\gamma^{\leftarrow}(1)$.]
4. M belongs to A^p.
5. The points p and q are of the same type. [$f|M$ is one-to-one and can be used to define a permutation of $\mathbb{N}$.]

Reference: M. E. Rudin, 1971.

7C. *Another partial ordering of types*

For two types s and t, define $s\leqslant t$ if there are points p and q in $\beta\mathbb{N}$ of types s and t, respectively, and a mapping of $\mathbb{N}$ onto $\mathbb{N}$ such that $f(A^q)=A^p$, or equivalently, $\beta(f)(q)=p$. .

1. If the definition of type from Section 3.41 is extended to points of $\mathbb{N}$, then every point of $\mathbb{N}$ (or equivalently, every fixed ultrafilter), is of the same type.
2. The relation $\leqslant$ is well defined and is a partial order on the set of types. [Anti-symmetry follows from Exercise 7B.]
3. The type consisting of fixed ultrafilters is less than every other type.
4. If $X=\{x_n\}$ is a countable discrete subspace of $\mathbb{N}^*$ and p is in X^*, then $\tau(p,X)\leqslant\tau(p)$. [Let $\{E_n\}$ be a pairwise disjoint family such that $E_n\in A^{x_n}$ and $\mathbb{N}=\bigcup E_n$. Define f to send E_n to $\{n\}$.]
5. If $s\leqslant t$, then $s\leqslant t$. [(4) above.]
6. In the partial order $\leqslant$, each type is greater than at most $\mathfrak{c}$ types and is smaller than $2^{\mathfrak{c}}$ types. [The latter statement follows by considering a mapping f of $\mathbb{N}$ onto itself such that each fiber is infinite.]

Reference: M. E. Rudin, 1971.

Chapter 8. Product Theorems

8.1. If $\{X_\alpha\}$ is any family of spaces, then $\underset{}{\times}(\beta X_\alpha)$ is a compactification of $\underset{}{\times} X_\alpha$. In Example 1.67 we observed that in the case of $\mathbb{R} \times \mathbb{R}$, $\beta\mathbb{R} \times \beta\mathbb{R}$ is not the Stone-Čech compactification of $\mathbb{R} \times \mathbb{R}$. The present chapter will be largely devoted to determining when it will be true that $\underset{}{\times}(\beta X_\alpha)$ is $\beta(\underset{}{\times} X_\alpha)$, i.e. when $\underset{}{\times} X_\alpha$ will be C^*-embedded in $\underset{}{\times}(\beta X_\alpha)$. I. Glicksberg showed in 1959 that for infinite spaces, this will be the case exactly when the product $\underset{}{\times} X_\alpha$ is pseudocompact.

Our program will be to first investigate conditions under which the product of two spaces will be pseudocompact and to obtain the Glicksberg Theorem for finite products. We will then find that the proof for a pseudocompact product of infinitely many spaces can be reduced to a situation involving the product of two spaces. At the end of the chapter we will consider some assorted product theorems which are somewhat related to the Glicksberg Theorem.

8.2. The following proposition eliminates the trivial case by showing that β will always distribute over the product of two spaces if one of them is finite.

Proposition. *If X is finite, then $X \times Y$ is C^*-embedded in $X \times \beta Y$*, i.e. $\beta(X \times Y) = X \times \beta Y$.

Proof. Let $X = \{x_1, \ldots, x_n\}$ and let f belong to $C^*(X \times Y)$. For each i, $_{x_i}f(y) \equiv f(x_i, y)$ belongs to $C^*(Y)$ and therefore has a continuous extension g_i to βY. Define $\beta(f)$ on $X \times \beta Y$ by $\beta(f)(x_i, p) = g_i(p)$ for each x_i in X and p in βY. The function $\beta(f)$ is continuous since its restriction to each set $\{x_i\} \times \beta Y$ is continuous and the family $\{\{x_i\} \times \beta Y\}$ is a finite closed cover of $X \times \beta Y$. $\quad\square$

Glicksberg's Theorem for Finite Products

8.3. Glicksberg proved that $\beta(X \times Y)$ is $\beta X \times \beta Y$ precisely when $X \times Y$ is a pseudocompact product of two infinite spaces. We will first show that the product of two pseudocompact spaces may fail to be pseudo-compact. The next result appeared in the 1962 paper of Fine and Gillman and describes a useful family of pseudocompact spaces.

Lemma. *Let Y be a realcompact space, let H be a subspace of Y^*, and put $X = Y \cup H$. If X is pseudocompact, then H is dense in Y^*. Conversely, if H is dense in Y^* and Y is also locally compact, then X is pseudocompact:*

Proof. We first prove the contrapositive of the first statement. If H is not dense in Y^*, then there exists an open set U of Y^* which misses H and $U = V \cap Y^*$ for some open set V of βY. Choose p in U. Then there exists a zero-set Z_1 of βY such that p belongs to Z_1 and Z_1 is contained in V. Because Y is realcompact, by Theorem 1.53, there exists a zero-set Z_2 of βY which contains p and misses Y. Then $Z_1 \cap Z_2$ is a zero-set which contains p and is contained in $Y^* \backslash H = X^*$. Hence, X fails to be pseudo-compact by Theorem 1.57.

The converse is also demonstrated by showing the contrapositive. If X is not pseudocompact, then there exists a non-empty zero-set Z of βX which is contained in X^* and hence in Y^* as well. But because Y is both realcompact and locally compact, Proposition 4.21 shows that the interior of Z is a non-empty open subset of Y^* which misses H. Hence, H is not dense in Y^*. $\square$

8.4. The pseudocompact spaces described in the lemma make it easy to construct two pseudocompact spaces whose product fails to be pseudocompact.

Examples. We will first describe a simple example and then digress to consider a more complex one which is somewhat related to our present discussion in that it involves the uniform completion of a product space.

For the first example, choose two points p and q of $\mathbb{N}^*$ which are of different types. Let Σ_p and Σ_q denote their orbits under automorphisms of $\beta\mathbb{N}$. Then Σ_p and Σ_q are disjoint dense subspaces of $\mathbb{N}^*$. Hence, $\mathbb{N} \cup \Sigma_p$ and $\mathbb{N} \cup \Sigma_q$ are pseudocompact. However, $(\mathbb{N} \cup \Sigma_p) \times (\mathbb{N} \cup \Sigma_q)$ contains the diagonal $\Delta = \{(n,n) : n \in \mathbb{N}\}$ as a clopen, and therefore C-embedded, copy of $\mathbb{N}$ and so is not pseudocompact.

The second example will consist of two countably compact spaces whose product fails to be pseudocompact. In addition, the example answers in the negative, assuming the Continuum Hypothesis, a question of K. Morita: If X and Y are countably compact and $X \times Y$ is an

M-space, is $X \times Y$ countably compact? An *M-space* is a space which admits a closed mapping onto a metric space such that the point inverses are countably compact. Note that from the definition, any countably compact space is an M-space. Morita also gave an equivalent formulation of his question in terms of uniform completions. Let μX denote the completion of the space X with respect to its finest uniformity. In 1970A, Morita asked if the equality $\mu(X \times Y) = \mu X \times \mu Y$ is satisfied whenever $X \times Y$ is an M-space and showed that this question is equivalent to the previous one. A survey of the theory of M-spaces and the background of these two questions is provided in Morita's 1971 paper. The following example of A. K. Steiner from 1971 provides a negative answer to both questions.

Index the set of copies of $\mathbb{N}$ contained in $\beta\mathbb{N}$ by the least ordinal $\aleph$ having cardinality 2^c and write $\{M_\alpha : \alpha < \aleph\}$ for this set. Assume the Continuum Hypothesis. For each M_α, let p_α be a P-point of M_α^* chosen so that $\tau(p_\alpha, M_\alpha) \neq \tau(p_\beta, M_\beta)$ for any $\beta < \alpha$. This is possible since we saw in Section 7.11 that there are 2^c distinct types of P-points in M_α^* and each α has fewer than 2^c predecessors. Now define an ω_1-sequence of subsets of $\beta\mathbb{N}$ by putting $X_0 = \mathbb{N}$ and for each $\gamma < \omega_1$, put

$$X_\gamma = (\bigcup\{X_\alpha : \alpha < \gamma\}) \cup \{p_\alpha : M_\alpha \subset \bigcup\{X_\alpha : \alpha < \gamma\}\}.$$

Put $X = \bigcup\{X_\gamma : \gamma < \omega_1\}$. Since any countable discrete subspace of X is contained in X_γ for some $\gamma < \omega_1$, it has an accumulation point in $X_{\gamma+1}$. X is therefore countably compact. For each $\gamma < \omega_1$,

$$|X_\gamma| \leqslant c \cdot c + (c \cdot c)^{\aleph_0} = c,$$

so that the cardinality of X is at most c.

Put $Y = \mathbb{N} \cup (\mathbb{N}^* \setminus X)$. Each countable discrete subspace of Y has 2^c accumulation points in $\beta\mathbb{N}$, and since Y fails to contain at most c points of $\beta\mathbb{N}$, Y is countably compact. However,

(a) $X \times Y$ is not pseudocompact, and hence not countably compact, since as in the previous example the diagonal Δ is a C-embedded copy of $\mathbb{N}$.

The key to showing that $X \times Y$ is an M-space is the following:
(b) $(X \times Y) \setminus \Delta$ is countably compact: Let $\{(x_i, y_i)\}$ be a copy of $\mathbb{N}$ in $(X \times Y) \setminus \Delta$. If $x_i = x$ for infinitely many indices, then $\{(x_i, y_i) : x_i = x\}$ is a countable discrete subspace of $\{x\} \times Y$ and therefore has an accumulation point in $(X \times Y) \setminus \Delta$. Similarly, $\{(x_i, y_i)\}$ has an accumulation point if $y_i = y$ for infinitely many indices. Thus, we may assume that $x_i \neq x_j$ and $y_i \neq y_j$ for $i \neq j$. Because $x_i \neq y_i$ for each i, by perhaps choosing a subset we may also assume that $\{x_i\}$ and $\{y_i\}$ are disjoint

subspaces of $\beta\mathbb{N}$. Thus, there exist α and β such that $M_\alpha = \{x_i\}$ and $M_\beta = \{y_i\}$. Let $h: \text{cl}_{\beta\mathbb{N}} M_\alpha \to \text{cl}_{\beta\mathbb{N}} M_\beta$ be the homeomorphism determined by putting $h(x_i) = y_i$ for each i. The point p_α corresponding to M_α is in X and since h is a homeomorphism, $\tau(p_\alpha, M_\alpha) = \tau(h(p_\alpha), M_\beta)$ by Lemma 3.42. Further, $h(p_\alpha)$ is a P-point of M_β^*. Assume that $h(p_\alpha)$ belongs to X. Then there is some M_γ contained in X such that $h(p_\alpha) = p_\gamma$ so that $h(p_\alpha)$ is also a P-point of M_γ^*. But then it follows from Lemma 7.25 that $h(p_\alpha)$ is in $(M_\gamma \cap M_\beta)^*$ and that $\tau(h(p_\alpha), M_\gamma) = \tau(h(p_\alpha), M_\beta)$ since the type of any P-point is minimal. However, now we have

$$\tau(p_\gamma, M_\gamma) = \tau(h(p_\alpha), M_\beta) = \tau(p_\alpha, M_\alpha)$$

and the choice of the points $\{p_\alpha\}$ implies that $M_\gamma = M_\alpha$. But this yields a contradiction, since it implies that $M_\alpha \cap M_\beta \neq \emptyset$. Hence, $h(p_\alpha)$ is not in X, and therefore must belong to Y. Thus, $(p_\alpha, h(p_\alpha))$ is an accumulation point of (x_i, y_i) since h is a homeomorphism, and $(p_\alpha, h(p_\alpha))$ clearly belongs to $(X \times Y) \backslash \Delta$.

Finally, we show that

(c) $X \times Y$ is an M-space: Define a function $f: X \times Y \to \mathbb{R}$ as follows:

$$f(x, y) = \begin{cases} n & \text{if } (x, y) = (n, n) \in \Delta, \\ 0 & \text{if } (x, y) \notin \Delta. \end{cases}$$

Then f is easily seen to be continuous and closed since Δ is clopen in $X \times Y$, and from (b) it follows that the point inverses of f are countably compact. Thus, $X \times Y$ is an M-space.

Several examples similar to this one have appeared in the literature. Perhaps the first two appeared in the 1952 paper of H. Terasaka and the 1953 B paper of J. Novák. Novák's example was simplified by Frolík in his 1959 paper. Frolík's version is described in Engelking's text, *Outline of General Topology*. Terasaka's example is presented in [GJ, 9.15]. The present example of A. K. Steiner is a modification of these earlier examples in which the properties of the set of P-points have been exploited to obtain additional properties of $X \times Y$ in order to answer the questions of Morita.

8.5. The preceding examples show that we must require more than that the factor spaces be pseudocompact in order to ensure that the product will be pseudocompact. In order to describe the proper additional conditions, we will need different characterizations of pseudocompactness than those given in Chapters 1, 4, and 5. A point is said to be a *cluster point of a family of subsets* of a space if every neighborhood of the point meets infinitely many subsets of the family.

Proposition. *The following are equivalent for any space X:*
- (1) *X is pseudocompact.*
- (2) *Each member of $C^*(X)$ assumes its supremum and infimum.*
- (3) *Each sequence of non-empty open subsets of X has a cluster point.*

Proof. $(1)\Leftrightarrow(3)$ is merely a restatement of Proposition 5.5.

$(1)\Rightarrow(2)$: If X is pseudocompact and f in $C^*(X)$ fails to assume it's supremum (or infimum) r, then the mapping defined by $g(x)=1/(f(x)-r)$ is an unbounded member of $C(X)$, which is a contradiction.

$(2)\Rightarrow(1)$: If f belongs to $C(X)$, then the mapping $g=1/(|f|\vee\mathbf{1})$ belongs to $C^*(X)$ and thus obtains its infimum. Hence, f belongs to $C^*(X)$ and X is pseudocompact. $\square$

8.6. The next theorem is a compilation of nine equivalent conditions which have been introduced in order to investigate pseudocompact products and related topics. We will see shortly that any of the nine conditions on $X \times Y$ is both necessary and sufficient for the product space to be pseudocompact whenever both of the factors are pseudocompact. This theorem appears, together with several additional conditions, in the comprehensive 1971 paper of W.W. Comfort and A.W. Hager which forms the basis of the present discussion. Some terminology must be introduced before stating the theorem and we will also take this opportunity to mention the sources of some of the nine conditions.

A *z-closed mapping* is one in which the image of every zero-set is closed. Condition (1) of the theorem states that the projection $\pi_X\colon X \times Y \to X$ is z-closed and was apparently introduced for the first time by H. Tamano in 1960A. Tamano's result will be Theorem 8.8 below. Condition (1) has also been used in the 1968 and 1969 papers of Stephenson and in the 1967 and 1969 papers of T. Isiwata.

Let M be a metric space with metric d. For each y in M and $\varepsilon>0$, let $B(y,\varepsilon)=\{z\in M:d(y,z)<\varepsilon\}$. A family $\mathscr{F}$ of mappings from X into M is said to be *equicontinuous* at a point x_0 of X if for every $\varepsilon>0$, there exists a neighborhood U of x_0 such that $f[U]$ is contained in $B(f(x_0),\varepsilon)$ for every f in $\mathscr{F}$. We will simply say that $\mathscr{F}$ is *equicontinuous* if it is equicontinuous at each point of X.

To each real-valued mapping f on a product space $X \times Y$ we can associate several related mappings. Two such mappings are defined by

$$f_y(x)=f(x,y) \quad \text{for each } y \text{ in } Y,$$

and

$$_xf(y)=f(x,y) \quad \text{for each } x \text{ in } X,$$

and belong to $C(X)$ and $C(Y)$, respectively. With these definitions, (6) of the theorem states that the family $\{f_y : y \in Y\}$ is equicontinuous on X for each f in $C^*(X \times Y)$.

The topology of uniform convergence on $C^*(Y)$ is the topology induced by the sup norm, i.e. the norm defined by $\|g\| = \sup\{|g(y)| : y \in Y\}$ for each g in $C^*(Y)$. Condition (7) of the theorem states that for each f in $C^*(X \times Y)$, the assignment

$$\Phi(f)(x) = {}_x f$$

is a continuous mapping of X into $C^*(Y)$, where $C^*(Y)$ has the topology of uniform convergence.

Glicksberg's proof of the product theorem used lemmas which showed that $X \times Y$ being pseudocompact implies (6) which in turn implies that $X \times Y$ is C^*-embedded in $X \times \beta Y$. This last statement is condition (3) of the theorem.

A *pseudometric* on a set X is a function φ on $X \times X$ into the non-negative reals which satisfies, for all x, y, and z in X:
   (1) $\varphi(x,x) = 0$,
   (2) $\varphi(x,y) = \varphi(y,x)$,
   (3) $\varphi(x,y) \leqslant \varphi(x,z) + \varphi(z,y)$.
If X is a topological space, then a pseudometric φ on X is said to be a *continuous pseudometric* if it is continuous on $X \times X$. Any pseudometric satisfies the inequality:

$$|\varphi(x_1, y_1) - \varphi(x, y)| \leqslant \varphi(x_1, x) + \varphi(y_1, y).$$

Using this inequality, we can show that any pseudometric which is separately continuous on $X \times X$ is continuous. If φ is separately continuous and $\varepsilon > 0$ is given, choose neighborhoods U_1 of x_1 and U_2 of x_2 such that the mappings $y \mapsto \varphi(x_1, y)$ and $y \mapsto \varphi(x_2, y)$ vary by less than $\varepsilon/2$ on U_1 and U_2, respectively. Then if (y,z) belongs to the neighborhood $U_1 \times U_2$ of (x_1, x_2), we have

$$|\varphi(x_1, x_2) - \varphi(y, z)| \leqslant \varphi(x_1, y) + \varphi(x_2, z) < \varepsilon$$

so that φ is jointly continuous at (x_1, x_2). Pseudometrics are discussed in detail in [GJ, Chapter 15].

Other conditions which can be stated in terms of uniformities can be added to the nine which are described here. These conditions are presented in the previously mentioned paper of Comfort and Hager. The interested reader should also consult the 1969B paper of N. Noble, Chapter 17 of J.R. Isbell's book, *Uniform Spaces*, and the 1969A paper of Hager. C- and C^*-embedding have also been treated in terms of extensions of continuous pseudometrics in the 1969 paper of R.A. Alò and the 1970 paper of Alò and H. L. Shapiro.

Condition (5) of the theorem states that any member f of $C^*(X \times Y)$ induces a continuous pseudometric γ on X by means of the definition:

$$\gamma(x_1, x_2) \equiv \sup\{|f(x_1, y) - f(x_2, y)| : y \in Y\}\ .$$

Condition (4) is closely related and states that the function defined by

$$F(x) \equiv \sup\{f(x, y) : y \in Y\}$$

belongs to $C^*(X)$.

In 1960B, Z. Frolík reproved Glicksberg's Theorem and showed that if $X \times Y$ is pseudocompact, then conditions (4), (5), and (7) are satisfied. He also showed that those members of $C^*(X \times Y)$ which extend continuously to $\beta X \times \beta Y$ are characterized by a condition similar to (9). A variant of Frolík's result will be given in Theorem 8.14.

One final bit of notation is needed before stating the theorem. If f belongs to $C^*(X)$ and S is a subspace of Y, then we put

$$\operatorname{osc}_S(f) = \sup\{|f(s) - f(t)| : s, t \in S\}\ .$$

With this definition, we can write that a bounded real-valued function is continuous on X if whenever x in X and $\varepsilon > 0$ are given, then there is a neighborhood U of x such that $\operatorname{osc}_U(f) < \varepsilon$.

We can now state the theorem.

Theorem. *The following conditions on the product space $X \times Y$ are equivalent:*

(1) The projection map π_X from $X \times Y$ onto X is z-closed.

(2) If Z is a zero-set in $X \times Y$, then

$$\operatorname{cl} Z = \bigcup\{\operatorname{cl}(Z \cap (\{x\} \times Y)) : x \in X\}$$

where the closures are taken in $X \times \beta Y$.

(3) $X \times Y$ is C^*-embedded in $X \times \beta Y$.

(4) If f belongs to $C^*(X \times Y)$, then the function defined by

$$F(x) \equiv \sup\{f(x, y) : y \in Y\}$$

is continuous on X (as is $\inf\{f(x, y) : y \in Y\}$).

(5) If f belongs to $C^*(X \times Y)$, then the function defined by

$$\gamma(x_1, x_2) \equiv \sup\{|f(x_1, y) - f(x_2, y)| : y \in Y\}$$

is a continuous pseudometric on X.

(6) *If f belongs to $C^*(X \times Y)$, then $\{f_y : y \in Y\}$ is an equicontinuous family on X.*

(7) *If f belongs to $C^*(X \times Y)$, then $\Phi(f)(x) = {}_x f$ defines a continuous mapping $\Phi(f)$ from X into $C^*(Y)$.*

(8) *If f belongs to $C^*(X \times Y)$, x_0 is a point of X, and $\varepsilon > 0$, then there is a neighborhood U of x_0 and g in $C^*(Y)$ such that $|f(x,y) - g(y)| < \varepsilon$ whenever (x,y) is in $U \times Y$.*

(9) *If f belongs to $C^*(X \times Y)$ and $\varepsilon > 0$, then there is an open cover $\mathcal{U}$ of X, and for each U in $\mathcal{U}$ a finite open cover $\mathcal{V}(U)$ of Y such that $\operatorname{osc}_{U \times V}(f) < \varepsilon$ whenever U is in $\mathcal{U}$ and V is in $\mathcal{V}(U)$.*

Proof. $(1) \Rightarrow (2)$: Denote the set $Z \cap (\{x\} \times Y)$ by Z_x. If (2) fails, then for some (x_0, p) in $X \times \beta Y$,

$$(x_0, p) \in \operatorname{cl} Z \setminus \bigcup \{\operatorname{cl} Z_x : x \in X\},$$

and in particular, (x_0, p) does not belong to $\operatorname{cl} Z_{x_0}$. We will now obtain a contradiction of (1) by constructing a zero-set Z_1 of $X \times Y$ such that (x_0, p) is in $\operatorname{cl} Z_1$ but $\pi_X[Z_1]$ misses x_0. There is a real-valued mapping f on $X \times \beta Y$ such that f is identically equal to 1 on $\operatorname{cl} Z_{x_0}$ and is identically equal to 0 on some neighborhood of (x_0, p). Now let $Z_1 = Z \cap Z(f)$. The zero-set Z_1 contains (x_0, p) in its closure in $X \times \beta Y$, but $\pi_X[Z_1]$ misses x_0. Applying (1) gives a contradiction since

$$x_0 \in \pi_X[\operatorname{cl} Z_1] \subset \operatorname{cl}_X(\pi_X[Z_1]) = \pi_X[Z_1].$$

$(2) \Rightarrow (3)$: From Theorem 1.46 it follows that a dense subspace S of a space T is C^*-embedded in T if and only if disjoint zero-sets of S have disjoint closures in T. We will use (2) to show that this condition is satisfied with $S = X \times Y$ and $T = X \times \beta Y$. If Z_1 and Z_2 are zero-sets of $X \times Y$, then

$$\operatorname{cl} Z_1 \cap \operatorname{cl} Z_2 = \left[\bigcup \{ \operatorname{cl}(Z_1)_x : x \in X \} \right] \cap \left[\bigcup \{ \operatorname{cl}(Z_2)_x : x \in X \} \right]$$
$$= \bigcup \{ \operatorname{cl}(Z_1)_x \cap \operatorname{cl}(Z_2)_x : x \in X \}.$$

Each $(Z_i)_x$ can be viewed as a zero-set of Y so that if Z_1 and Z_2 are disjoint, (5) of Theorem 1.46 shows that $\operatorname{cl}(Z_1)_x$ and $\operatorname{cl}(Z_2)_x$ are disjoint for each x. Hence, Z_1 and Z_2 have disjoint closures in $X \times \beta Y$.

$(3) \Rightarrow (4)$: Given f in $C^*(X \times Y)$, let F be defined as in (4) and let g be the continuous extension of f to $X \times \beta Y$. To show that F is continuous at a point x_0 of X, let $\varepsilon > 0$ be given and for each point p of βY find a basic neighborhood $U_p \times V_p$ of (x_0, p) on which g varies less than ε. The cover $\{V_p\}$ of βY admits a finite subcover $\{V_{p_i} : 1 \leq i \leq n\}$. Then F

varies by less than ε on the neighborhood $\bigcap\{U_{p_i}:1\leqslant i\leqslant n\}$.

$(4)\Rightarrow(5)$: Because γ is a pseudometric, joint continuity of γ will follow from separate continuity. To show that γ is separately continuous, fix a point x_2 in X and note that the function defined by

$$(x_1,y)\mapsto|f(x_1,y)-f(x_2,y)|$$

is continuous on $X\times Y$. Then the associated sup function defined as in (4) is continuous on X and its value at x_1 is just $\gamma(x_1,x_2)$.

$(5)\Rightarrow(6)$: To show that the family $\{f_y:y\in Y\}$ is equicontinuous at a point x_0 of X, let $\varepsilon>0$ be given and use the continuity of the pseudometric γ of (5) to choose a neighborhood U of x_0 such that

$$\sup\{|f(x,y)-f(x_0,y)|:y\in Y\}<\varepsilon$$

whenever x is in U. Then U is the required neighborhood of x_0.

$(6)\Leftrightarrow(7)$: This equivalence is clear because both conditions may be stated as follows: For each point x_0 of X and each $\varepsilon>0$, there is a neighborhood U of x_0 such that

$$|f(x,y)-f(x_0,y)|<\varepsilon$$

for all y in Y and all x in U.

$(6)\Rightarrow(8)$: Given f in $C^*(X\times Y)$, x_0 in X, and $\varepsilon>0$, use the equicontinuity of the family $\{f_y\}$ to obtain a neighborhood U of x_0 such that

$$|_yf(x)-{}_yf(x_0)|<\varepsilon$$

for all y in Y and all x in U. Then the required member g of $C^*(Y)$ is obtained by defining $g(y)=f(x_0,y)$.

$(8)\Rightarrow(9)$: If f in $C^*(X\times Y)$ and $\varepsilon>0$ are given, then for every x_0 in X we must find a neighborhood U of x_0 and a finite open cover $\mathcal{V}$ of Y such that $\operatorname{osc}_{U\times V}(f)<\varepsilon$ whenever V is an element of $\mathcal{V}$. By (8), there exists g in $C^*(Y)$ and a neighborhood U of x_0 such that

$$|f(x,y)-g(y)|<\varepsilon/3$$

whenever x is in U and y is in Y. Because g is bounded, there is a finite open cover $\mathcal{V}$ of Y such that $\operatorname{osc}_V(g)<\varepsilon/3$ for each V in $\mathcal{V}$ and an application of the triangle inequality shows that $\mathcal{V}$ is the required covering.

$(9)\Rightarrow(1)$: Let $Z=\mathbf{Z}(g)$ be a zero-set of $X\times Y$ and suppose that x_0 is not in $\pi_X[Z]$. We will produce a neighborhood of x_0 which misses

$\pi_X[Z]$. Since $\{x_0\} \times Y$ is closed, we have that the restriction $g \mid \{x_0\} \times Y$ is bounded away from zero. Hence, putting

$$f(x, y) = |g(x, y)/g(x_0, y)| \wedge 1$$

yields a member f of $C^*(X \times Y)$ which has $\mathbf{Z}(g)$ as its zero-set and which is identically equal to 1 on $\{x_0\} \times Y$. Choose $\varepsilon = 1/2$ and apply (9) to obtain a neighborhood U of x_0 such that $f(x, y) > 1/2$ for all (x, y) in $U \times Y$. Hence, U is a neighborhood of x_0 missing $\pi_X[Z]$, and therefore $\pi_X[Z]$ is closed in X. $\Box$

8.7. Of the eight conditions in Theorem 8.6 which are equivalent to $X \times Y$ being C^*-embedded in $X \times \beta Y$, condition (1) is one of the easiest to verify. Observe that if both projections π_X and π_Y on $X \times Y$ are z-closed, then $X \times Y$ is C^*-embedded in both $\beta X \times Y$ and $X \times \beta Y$. In their 1966 paper, W. W. Comfort and S. Negrepontis call such a pair of spaces X and Y a C^*-*pair*. If in addition, $X \times Y$ is not C^*-embedded in $\beta X \times \beta Y$, X and Y are called a *proper C^*-pair*. They presented the following example of a proper C^*-pair.

Example. Two copies of $\mathbb{N}$ constitute a proper C^*-pair. It is easy to see that $\mathbb{N} \times \mathbb{N}$ is C^*-embedded in both $\beta \mathbb{N} \times \mathbb{N}$ and $\mathbb{N} \times \beta \mathbb{N}$ since both projections must be z-closed. However, the Kronecker delta function defined by

$$\delta(n, m) = \begin{cases} 1 & \text{if } n = m \\ 0 & \text{otherwise} \end{cases}$$

will not extend to $\beta \mathbb{N} \times \beta \mathbb{N}$ since any neighborhood of a point (p, p) in $\mathbb{N}^* \times \mathbb{N}^*$ must meet infinitely many elements which lie on the diagonal of $\mathbb{N} \times \mathbb{N}$ as well as infinitely many off diagonal elements.

8.8. The effort devoted to Theorem 8.6 now begins to pay dividends in the proof of Tamano's 1960A characterization of pseudocompact products. The proof given here is a Comfort-Hager modification of a Glicksberg-Frolík argument for necessity and of a 1968 Stephenson argument for sufficiency.

Theorem (Tamano). *The product $X \times Y$ is pseudocompact if and only if both X and Y are pseudocompact and the projection map π_X of $X \times Y$ onto X is z-closed.*

Proof. Necessity: X and Y are pseudocompact since both are continuous images of $X \times Y$. If π_X fails to be z-closed, then there is a zero-set $Z = \mathbf{Z}(f)$ in $X \times Y$ and a point x_0 of X belonging to $\operatorname{cl} \pi_X[Z] \setminus \pi_X[Z]$. As in the proof of $(9) \Rightarrow (1)$ in Theorem 8.6, we can assume that $f(x_0, y) = 1$ for all y in Y. The proof will be completed by constructing a sequence of open sets of $X \times Y$ which will cluster at a point at which f will fail to be continuous, yielding a contradiction.

Let (x_1, y_1) in Z be chosen. Since f is constantly equal to 1 on $\{x_0\} \times Y$, there exist open neighborhoods V_1 of y_1, U_1 of x_1, and W_1 of x_0 such that

$$(x_1, y_1) \in U_1 \times V_1 \subset \{(x, y) : f(x, y) < 1/3\}, \quad \text{and}$$

$$(x_0, y_1) \in W_1 \times V_1 \subset \{(x, y) : f(x, y) > 2/3\}.$$

Now suppose that for each $k \leqslant n$, a point (x_k, y_k) of Z and open neighborhoods U_k, V_k, and W_k of x_k, y_k, and x_0, respectively, have been chosen so that for all k,

$$U_k \times V_k \subset \{(x, y) : f(x, y) < 1/3\},$$

$$W_k \times V_k \subset \{(x, y) : f(x, y) > 2/3\}, \quad \text{and}$$

$$W_k \cup U_k \subset W_{k-1}.$$

Then $W_n \cap \pi_X[Z] \neq \emptyset$, so that there exists (x_{n+1}, y_{n+1}) in Z and an open neighborhood $U_{n+1} \times G$ of (x_{n+1}, y_{n+1}) such that $U_{n+1} \subset W_n$ and

$$U_{n+1} \times G \subset \{(x, y) : f(x, y) < 1/3\}.$$

Since f is constantly equal to 1 on $\{x_0\} \times Y$, there also exists an open neighborhood $W_{n+1} \times N$ of (x_0, y_{n+1}) such that $W_{n+1} \subset W_n$, and

$$W_{n+1} \times N \subset \{(x, y) : f(x, y) > 2/3\}.$$

Putting $V_{n+1} = G \cap N$ completes the induction step.

Now because $X \times Y$ is pseudocompact, the sequence $\{U_n \times V_n\}$ has a cluster point $(\bar{x}, \bar{y})$. Thus, $f(\bar{x}, \bar{y}) \leqslant 1/3$ by the continuity of f. Since every neighborhood of $(\bar{x}, \bar{y})$ meets each member of some subsequence $\{U_{n_k} \times V_{n_k}\}$, every neighborhood also meets each member of the subsequence $\{W_{n_k} \times V_{n_k}\}$ because

$$U_{n_k} \subset W_{n_k - 1} \subset \cdots \subset W_{n_{k-1}}.$$

But this implies that $f(\bar{x}, \bar{y}) \geqslant 2/3$, which is a contradiction.

Sufficiency: To show that $X \times Y$ is pseudocompact, it suffices to show that if f is in $C^*(X \times Y)$ and f is everywhere positive, then

$$\inf\{f(x, y) : (x, y) \in X \times Y\} > 0.$$

Set $F(x) = \inf\{f(x,y): y \in Y\}$. By assuming (4) of Theorem 8.6, F belongs to $C^*(X)$. Since X is pseudocompact, Proposition 8.5 implies that

$$\inf\{f(x,y): (x,y) \in X \times Y\} = \inf\{F(x): x \in X\} > 0. \quad \square$$

Observe that in the light of Theorem 8.6, Tamano's Theorem is actually nine characterizations rolled into one, a fact which will frequently be useful.

8.9. As Example 8.7 showed, a projection on a product space may be z-closed without either factor being pseudocompact. However, in such a case the image of the projection must be a P-space. The following result has appeared in a number of contexts. The present proof is from Comfort and Hager's paper and is credited there to N.J. Fine.

Lemma. *If the projection π_X of $X \times Y$ onto X is z-closed, then either X is a P-space or Y is pseudocompact.*

Proof. If X fails to be a P-space, then there exists h in $C^*(X)$ and a point x_0 in $\mathbf{Z}(h)$ such that h is not identically equal to 0 on any neighborhood of x_0. If Y fails to be pseudocompact, then by Proposition 8.5 there is an everywhere positive map g in $C^*(Y)$ such that $\inf\{g(y): y \in Y\} = 0$. Define f in $C^*(X \times Y)$ by $f(x,y) = g(y)^{h(x)}$. Theorem 8.6 demands that $F(x) = \inf\{f(x,y): y \in Y\}$ define a continuous function on X. Since $h(x_0) = 0$, we have $F(x_0) = 1$. However, because g is not bounded away from zero, every neighborhood of x_0 contains a point x such that $F(x) < 1/2$. Hence, (4) of Theorem 8.6 fails, and π_X thus cannot be z-closed. $\quad \square$

8.10. The lemma has a number of corollaries. The first is a modification of Tamano's Theorem which is obtained by requiring that one of the factor spaces be infinite. An infinite pseudocompact space cannot be a P-space by Proposition 1.65. Thus, if the projection map π_X of $X \times Y$ onto X is z-closed, the other factor Y must be pseudocompact. When combined with Theorem 8.8, this verifies the

Corollary. *If X is infinite, then the product space $X \times Y$ is pseudocompact if and only if X is pseudocompact and the projection map π_X of $X \times Y$ onto X is z-closed.*

8.11. From Theorem 8.6, we obtain

Corollary. *If $X \times Y$ is C^*-embedded in $X \times \beta Y$, then either Y is pseudocompact or X is a P-space.*

Other corollaries will follow Glicksberg's Theorem or be contained in Exercises.

8.12. The following proof of Glicksberg's Theorem is taken from Comfort and Hager's 1971 paper.

Theorem (Glicksberg). *If X and Y are infinite, then the product space $X \times Y$ is pseudocompact if and only if $X \times Y$ is C^*-embedded in $\beta X \times \beta Y$, i.e. $\beta(X \times Y) = \beta X \times \beta Y$.*

Proof. Necessity: Because $X \times Y$ is pseudocompact, Theorem 8.8 implies that the projection π_X is z-closed. Theorem 8.6 then requires that $X \times Y$ be C^*-embedded in $X \times \beta Y$. Since the pseudocompact space $X \times Y$ is dense in $X \times \beta Y$, the space $X \times \beta Y$ is also pseudocompact. Hence, with βY playing the role of X and X playing the role of Y, Theorem 8.6 implies that $X \times \beta Y$ is C^*-embedded in $\beta X \times \beta Y$. Therefore, $X \times Y$ is C^*-embedded in $\beta X \times \beta Y$.

Sufficiency: $\beta(X \times Y) = \beta X \times \beta Y$ implies that $X \times Y$ is C^*-embedded in both $\beta X \times Y$ and $X \times \beta Y$. Theorem 8.6 then implies that both projections π_X and π_Y are z-closed. Hence, applying Lemma 8.9, we see that either X is pseudocompact or βX is a P-space. But neither of the infinite compact spaces βX and βY are P-spaces, so that both X and Y are pseudocompact. Then Theorem 8.8 implies that the product $X \times Y$ is pseudocompact. □

Partial results along the lines of the preceding theorem had been obtained prior to Glicksberg's solution. In 1952, T. Isiwata showed that if $\beta(X \times X)$ is $\beta X \times \beta X$, then X is totally bounded in any uniform structure on X, which is equivalent to X being pseudocompact [GJ, ex. 15 Q]. In 1957 B, M. Henrikson and J.R. Isbell showed that if $\beta(X \times Y) = \beta X \times \beta Y$, then $X \times Y$ is pseudocompact and they also obtained partial results in the converse direction.

8.13. If X and Y form a proper C^*-pair (Section 8.7), then both projections must be z-closed. If either space were pseudocompact, then Glicksberg's Theorem and Corollary 8.10 would imply that the pair is not proper. Combining these facts with Corollary 8.11 verifies the

Corollary. *If X and Y are a proper C^*-pair, then both X and Y are P-spaces.*

8.14. In his 1960 B paper, Z. Frolík offered another proof of Glicksberg's Theorem and introduced a condition similar to (9) of Theorem 8.6. A mapping f in $C^*(X \times Y)$ is said to satisfy the *rectangle condition* if given $\varepsilon > 0$, there is a finite covering $\{U_i \times V_i\}$ of $X \times Y$ by open rectangles such that for each i, $\mathrm{osc}_{U_i \times V_i}(f) < \varepsilon$. Frolík showed that if every f in $C^*(X \times Y)$ satisfies the rectangle condition, then $X \times Y$ is pseudocompact, and conversely. In his 1966 paper, A. W. Hager showed that the rectangle condition characterizes those maps in $C^*(X \times Y)$

which will extend continuously to $\beta X \times \beta Y$. Here we prove a version of Hager's result which requires the rectangles to be products of cozero-sets.

Theorem. *A member f of $C^*(X \times Y)$ admits a continuous extension to $\beta X \times \beta Y$ if and only if for any $\varepsilon > 0$, there exists a finite open cover $\{U_i \times V_i : 1 \leqslant i \leqslant n\}$ of $X \times Y$ where each U_i and V_i are cozero-sets and such that $\mathrm{osc}_{U_i \times V_i}(f) < \varepsilon$ for each i.*

Proof. Sufficiency: The equivalence of (3) and (9) of Theorem 8.6 shows that f has a continuous extension g to $X \times \beta Y$. Following [GJ, 16.10], we now extend the cozero-set cover $\{V_i\}$ of Y to an open cover of βY. Put

$$V_i^\beta = \beta Y \setminus \mathrm{cl}_{\beta Y}(Y \setminus V_i).$$

Since $\{V_i\}$ is a cover of Y, $\bigcap (Y \setminus V_i) = \emptyset$. Because each $Y \setminus V_i$ is a zero-set, (5) of Theorem 1.46 implies that

$$\beta Y \setminus \bigcup V_i^\beta = \bigcap \mathrm{cl}_{\beta Y}(Y \setminus V_i) = \emptyset$$

so that $\{V_i^\beta\}$ covers βY. Now each point of $U_i \times V_i^\beta$ is a limit point of $U_i \times V_i$. Hence,

$$\mathrm{osc}_{U_i \times V_i^\beta}(g) = \mathrm{osc}_{U_i \times V_i}(f) < \varepsilon.$$

Hence, another application of the equivalence of (3) and (9) of Theorem 8.6 yields an extension h of g to $\beta X \times \beta Y$.

Necessity: If h is the extension of f to $\beta X \times \beta Y$, then the compactness of $\beta X \times \beta Y$ implies that there is a finite covering $\{U_i \times V_i\}$ of $\beta X \times \beta Y$ consisting of products of cozero-sets which satisfies the required oscillation condition. Intersecting these sets with $X \times Y$ yields the result. $\square$

8.15. Corollary. *The product space $X \times Y$ is pseudocompact if and only if for every f in $C^*(X \times Y)$ and $\varepsilon > 0$, there exists a finite open cover $\{U_i \times V_i : 1 \leqslant i \leqslant n\}$ consisting of products of cozero-sets such that $\mathrm{osc}_{U_i \times V_i}(f) < \varepsilon$ for each i.*

8.16. In addition to characterizing the members of $C^*(X \times Y)$ which are continuously extendable to $\beta X \times \beta Y$ in terms of the rectangle condition, in his 1966 paper Hager also characterized this set of mappings in terms of tensor products of function rings. The *tensor product* of the rings $C^*(X)$ and $C^*(Y)$ is denoted by $C^*(X) \otimes C^*(Y)$ and consists of all members of $C^*(X \times Y)$ of the form

$$h(x, y) = \sum f_i(x) g_i(y)$$

where the sum is finite and each f_i and g_i belongs to $C^*(X)$ and $C^*(Y)$, respectively. In the following theorem, uniform closure refers to closure in the topology on $C^*(X \times Y)$ induced by the sup norm.

Theorem (Hager). *The set of mappings contained in $C^*(X \times Y)$ which extend continuously to $\beta X \times \beta Y$ coincides with the uniform closure of $C^*(X) \otimes C^*(Y)$ in $C^*(X \times Y)$.*

Proof. Let $h = \sum f_i \cdot g_i$ be in $C^*(X) \otimes C^*(Y)$ where f_i and g_i belong to $C^*(X)$ and $C^*(Y)$, respectively. Then each f_i and g_i extend to the appropriate compactification βX or βY and $l = \sum \beta(f_i) \cdot \beta(g_i)$ is the appropriate extension of h. Now let the sequence $\{h_n\}$ contained in $C^*(X) \otimes C^*(Y)$ converge uniformly to b in $C^*(X \times Y)$. Then the corresponding extensions $\{l_n\}$ form a Cauchy sequence in $C^*(\beta X \times \beta Y)$ since $X \times Y$ is dense in $\beta X \times \beta Y$. The limit of the sequence $\{l_n\}$ is the appropriate extension of b to $\beta X \times \beta Y$.

Conversely, the tensor product $C(\beta X) \otimes C(\beta Y)$ is easily seen to separate points of $\beta X \times \beta Y$ and to contain constant mappings. Therefore, by the Stone-Weierstrass Theorem, [GJ, 16.4] or [D, p. 282], the subring $C(\beta X) \otimes C(\beta Y)$ is dense in $C(\beta X \times \beta Y)$ so that every element f in $C(\beta X \times \beta Y)$ is the uniform limit of a sequence $\{f_n\}$ contained in $C(\beta X) \otimes C(\beta Y)$. Hence, $f | X \times Y$ is the uniform limit of $\{f_n | X \times Y\}$ and $f_n | X \times Y$ belongs to $C^*(X) \otimes C^*(Y)$. $\Box$

8.17. In addition to the characterization of pseudocompact products given in Theorem 8.8, in 1960A Tamano also gave a characterization in terms of the tensor product $C^*(X) \otimes C^*(Y)$. His result can now be stated as a corollary of the theorems of Glicksberg and Hager.

Corollary (Tamano). *The product space $X \times Y$ is pseudocompact if and only if $C^*(X) \otimes C^*(Y)$ is dense in $C^*(X \times Y)$.*

8.18. Observe that in the statement of Glicksberg's Theorem, we first asserted that $X \times Y$ is C^*-embedded in $\beta X \times \beta Y$ and then wrote that $\beta(X \times Y)$ is equal to $\beta X \times \beta Y$. It is important to clarify what is intended by equality in this setting. We mean equality in the partially ordered set of compactifications as discussed in Section 1.13. Thus, we mean that there is a homeomorphism of $\beta X \times \beta Y$ with $\beta(X \times Y)$ under which $X \times Y$ is pointwise fixed. The following example shows that it is not enough to merely ask that $\beta X \times \beta Y$ be homeomorphic with $\beta(X \times Y)$. The example is taken from the 1959 paper of Gillman and Jerison.

Example. Let $X \oplus Y$ denote the disjoint topological sum of two spaces X and Y and let 3 denote the discrete space having 3 points. Let N_1 and N_2 denote the odd and even integers, respectively, and put $E = N_1 \oplus \beta N_2$. Then $\beta E = \beta \mathbb{N}$. We will show that $\beta(E \times E)$ and $\beta E \times \beta E$

are homeomorphic. However, since E is not pseudocompact, it is clear from Glicksberg's Theorem that they are not equal. The statement that X is homeomorphic with Y will be written $X \approx Y$.

(a) $\beta(\mathbb{N} \times \beta\mathbb{N}) \approx \beta\mathbb{N}$: $\mathbb{N} \times \mathbb{N}$ is dense and C^*-embedded in $\mathbb{N} \times \beta\mathbb{N}$ and therefore in $\beta(\mathbb{N} \times \beta\mathbb{N})$. Hence, $\beta(\mathbb{N} \times \mathbb{N})$ is equal to $\beta(\mathbb{N} \times \beta\mathbb{N})$. Since $\mathbb{N} \times \mathbb{N} \approx \mathbb{N}$, we have that $\beta(\mathbb{N} \times \beta\mathbb{N}) \approx \beta(\mathbb{N} \times \mathbb{N}) \approx \beta\mathbb{N}$.

(b) $\beta(E \times E) \approx \beta E \times \beta E$: Since $E \approx \mathbb{N} \oplus \beta\mathbb{N}$, we have:

$$E \times E \approx (\beta\mathbb{N} \oplus \mathbb{N}) \times (\beta\mathbb{N} \oplus \mathbb{N})$$

$$\approx \beta\mathbb{N} \times \beta\mathbb{N} \oplus \mathbb{N} \times \beta\mathbb{N} \oplus \beta\mathbb{N} \times \mathbb{N} \oplus \mathbb{N} \times \mathbb{N}$$

$$\approx \beta\mathbb{N} \times \beta\mathbb{N} \oplus \mathbb{N} \times \beta\mathbb{N} \oplus \mathbb{N} \times \beta\mathbb{N} \oplus \mathbb{N} .$$

Since β will distribute over finite disjoint sums, this gives:

$$\beta(E \times E) \approx \beta(\beta\mathbb{N} \times \beta\mathbb{N}) \oplus \beta(\mathbb{N} \times \beta\mathbb{N}) \oplus \beta(\mathbb{N} \times \beta\mathbb{N}) \oplus \beta\mathbb{N} .$$

By (a), this now becomes:

$$\beta(E \times E) \approx \beta\mathbb{N} \times \beta\mathbb{N} \oplus \beta\mathbb{N} \oplus \beta\mathbb{N} \oplus \beta\mathbb{N}$$

$$= \beta\mathbb{N} \times \beta\mathbb{N} \oplus \beta\mathbb{N} \times 3$$

$$= \beta\mathbb{N} \times (\beta\mathbb{N} \oplus 3)$$

$$\approx \beta\mathbb{N} \times \beta\mathbb{N}$$

$$\approx \beta E \times \beta E .$$

Hence, $\beta(E \times E)$ and $\beta E \times \beta E$ are homeomorphic.

8.19. In the paper containing the previous example, Gillman and Jerison showed that if both X and Y are first countable, then it is possible to replace equal by homeomorphic in Glicksberg's Theorem. The key to the proof is that no point of X^* can have a countable neighborhood base in βX for any X. This fact follows from Corollary 3.7.

Proposition. *If X and Y are both first countable spaces, then $\beta(X \times Y)$ and $\beta X \times \beta Y$ are homeomorphic if and only if they are equal.*

Proof. Let h be a homeomorphism from $\beta(X \times Y)$ onto $\beta X \times \beta Y$. Since only the points of $X \times Y$ can have countable neighborhood bases in $\beta(X \times Y)$ by Exercise 2F, we see that h maps the subspace $X \times Y$ of $\beta(X \times Y)$ onto the subspace $X \times Y$ of $\beta X \times \beta Y$. Since $X \times Y$ is C^*-embedded in $\beta(X \times Y)$, its homeomorphic image $X \times Y$ is C^*-embedded in $\beta X \times \beta Y$. Thus, $\beta(X \times Y)$ and $\beta X \times \beta Y$ are equal. $\square$

8.20. Combining the previous result with Glicksberg's Theorem gives an immediate corollary.

Corollary. *If X and Y are infinite first countable spaces, then $\beta(X \times Y)$ and $\beta X \times \beta Y$ are homeomorphic if and only if $X \times Y$ is pseudocompact.*

8.21. Glicksberg's Theorem ignited an interest in determining sufficient conditions for the product of two spaces to be pseudocompact. The following proposition is one of the simpler and most useful of the results to be obtained.

Proposition. *The product of two pseudocompact spaces one of which is also locally compact is pseudocompact.*

Proof. Let X and Y be pseudocompact spaces and let Y be locally compact. For a point y of Y, choose a compact neighborhood V of y. Then the projection $\pi_X : X \times V \to X$ is closed since V is compact [D, p. 227] and hence, π_X is z-closed. Therefore, Theorem 8.8 shows that $X \times V$ is pseudocompact. From Glicksberg's Theorem and (7) of Theorem 8.6, we have that if g belongs to $C^*(X \times Y)$, then the assignment

$$\Phi(g)(y) = g_y$$

defines a continuous mapping from V into $C^*(X)$. But it follows from the fact that Y is locally compact that $\Phi(g)$ is continuous on all of Y since we have shown that it is continuous on a neighborhood of each point. Now the equivalence of (1) and (7) of Theorem 8.6 shows that the projection $\pi_X : X \times Y \to X$ is z-closed and $X \times Y$ is pseudocompact by Theorem 8.8. $\square$ ˙

Frolík, 1960 B, and Glicksberg, 1959, both consider sufficient conditions for a product $X \times Y$ to be pseudocompact as do Stephenson, 1968, and Tamano, 1960 A. In addition, Frolík obtained a characterization for the class of spaces such that the product of any space in the class with any pseudocompact space is pseudocompact. In 1960 A, Frolík considers the analogous class of spaces for countably compact products. The classes of spaces introduced by Frolík in 1960 A and 1960 B are also discussed in Isiwata's 1964 paper and Noble's 1969 C paper.

8.22. Recall that a space X is said to be a *retractive space* if X^* is a retract of βX (Section 6.5). In Theorem 6.6 we saw, modulo the Continuum Hypothesis, that any retractive space is pseudocompact. The preceding result together with Glicksberg's Theorem show that the class of retractive spaces is closed under products with compact spaces.

Corollary [CH]. *The product of a retractive space and a compact space is retractive.*

Proof. Let X be retractive and Y be compact. Then the product $X \times Y$ is pseudocompact and we have that $\beta(X \times Y) = \beta X \times \beta Y = \beta X \times Y$. Then if r is a retraction of βX onto X^*, the required retraction of $\beta(X \times Y)$ onto $(X \times Y)^* = X^* \times Y$ is given by $(p,q) \mapsto (r(p), q)$. $\quad\square$

8.23. However, *the product of two retractive spaces need not be retractive.* The following example appears in Comfort's 1965 paper and is based on an argument used in Exercise 4E of Kelley, 1955. Since the example will be both locally compact and pseudocompact, it also shows that *the converse to Theorem 6.6 is false.*

Example. The space ω_1 is surely retractive since its growth in $\beta \omega_1 = \omega_1 + 1$ is a single point. Glicksberg's Theorem together with the fact that ω_1 is locally compact show that $\beta(\omega_1 \times \omega_1) = (\omega_1 + 1) \times (\omega_1 + 1)$. Denote the sets $\{\omega_1\} \times \omega_1$ and $\omega_1 \times \{\omega_1\}$ by E and F, respectively, so that the growth of $\omega_1 \times \omega_1$ is $E \cup F \cup \{(\omega_1, \omega_1)\}$.

Now assume that r is a retraction of $\beta(\omega_1 \times \omega_1)$ onto $(\omega_1 \times \omega_1)^*$. Choose $x_1 < \omega_1$. Then the point $(\omega_1, x_1) = r(\omega_1, x_1)$ does not belong to the closed set $F \cup \{(\omega_1, \omega_1)\}$. Thus, the image of some neighborhood of (ω_1, x_1) misses $F \cup \{(\omega_1, \omega_1)\}$ and we can therefore find x_2 such that $x_1 < x_2 < \omega_1$ and $r(x_2, x_1)$ belongs to E. Since $r(x_2, \omega_1) = (x_2, \omega_1)$ misses $E \cup \{(\omega_1, \omega_1)\}$, we can find x_3 such that $x_2 < x_3 < \omega_1$ and $r(x_2, x_3)$ belongs to F.

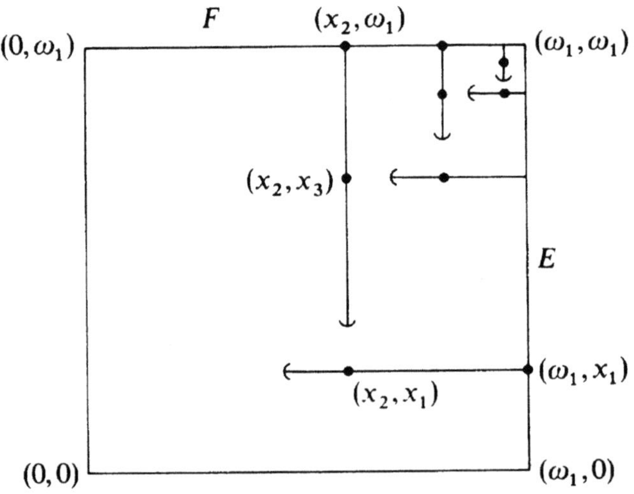

By continuing this process, if we assume that for all $n < 2k$ we have chosen x_n, we can choose x_{2k} such that $x_{2k-1} < x_{2k} < \omega_1$ and $r(x_{2k}, x_{2k-1})$ belongs to E. Then we can choose x_{2k+1} such that $x_{2k} < x_{2k+1} < \omega_1$ and $r(x_{2k}, x_{2k+1})$ belongs to F. By putting $\sigma = \sup\{x_n\}$, we see that

$$\lim(x_{2k}, x_{2k-1}) = \lim(x_{2k}, x_{2k+1}) = (\sigma, \sigma).$$

By continuity, we have that $r(\sigma,\sigma)$ belongs to $\mathrm{cl}\,E \cap \mathrm{cl}\,F$ and therefore that $r(\sigma,\sigma)=(\omega_1,\omega_1)$. Thus the countable sequence $r(x_{2k},x_{2k-1})$ must have the limit (ω_1,ω_1). However, no countable sequence is cofinal in E because E is a copy of ω_1 [GJ, 5.12]. Thus, the assumption that r is a retraction leads to a contradiction.

The Product Theorem for Infinite Products

8.24. Glicksberg's Theorem for finite pseudocompact products can be extended to a pseudocompact product involving an arbitrarily large index set. The extension will be accomplished by applying the theorem for finite products to a finite partial product. The following result is an analogue of (8) in Theorem 8.6 and will be used to single out an appropriate finite partial product. The proof of both the lemma and the necessity portion of the theorem are as in Glicksberg's 1959 paper.

Lemma. *If the product space* $\times\{X_\alpha : \alpha \in \mathscr{A}\}$ *is pseudocompact,* f *is in* $C(\times X_\alpha)$, *and* $\varepsilon > 0$, *then there exists a finite set* F *of indices such that whenever the coordinates of two points* x *and* y *agree for each element of* F, *then* $|f(x) - f(y)| < \varepsilon$.

Proof. Suppose that no such finite set exists. Then if F_0 is any finite set of indices, there are points x_1 and y_1 whose coordinates agree for the indices of F_0 and such that $|f(x_1) - f(y_1)| > \varepsilon$. Choose disjoint basic neighborhoods of x_1 and y_1 which are contained in

$$f^{\leftarrow}(f(x_1) - \varepsilon/4, f(x_1) + \varepsilon/4)$$

and

$$f^{\leftarrow}(f(y_1) - \varepsilon/4, f(y_1) + \varepsilon/4),$$

respectively. Since basic neighborhoods in a product space restrict only finitely many coordinates, we can assume that x_1 and y_1 differ in only finitely many coordinates, say those belonging to F_1, by asking only that $|f(x_1) - f(y_1)| > \varepsilon/2$.

Now by considering the finite set $F_0 \cup F_1$, we can in the same way obtain two points x_2 and y_2 such that their coordinates agree in $F_0 \cup F_1$, differ in only a finite set F_2, and such that $|f(x_2) - f(y_2)| > \varepsilon/2$. Continue this process to obtain sequences $\{x_i\}$, $\{y_i\}$, and $\{F_i\}$ such that $|f(x_i) - f(y_i)| > \varepsilon/2$, x_i and y_i differ only in those coordinates belonging to F_i, and the F_i are pairwise disjoint finite sets of indices.

By continuity, we can obtain basic open neighborhoods U_i and V_i of x_i and y_i, respectively, such that $|f(x) - f(y)| > \varepsilon/4$ whenever x belongs to U_i and y belongs to V_i, and such that U_i and V_i restrict each factor space X_α equally except for those whose index belongs to F_i. Since the product is pseudocompact, $\{U_i\}$ has a cluster point x_0 (Proposition 8.5). If W is any basic neighborhood of x_0, then the set E of coordinates which are restricted by W is finite, and so for some $n \geqslant 1$,

$$E \cap \left(\bigcup F_i \right) \subseteq \bigcup \{ F_i : 1 \leqslant i \leqslant n \} .$$

Thus, W meets both U_i and V_i for all $i > n$ so that W contains points x and y such that $|f(x) - f(y)| > \varepsilon/4$, which contradicts the continuity of f. $\quad\square$

8.25. In the proof of Glicksberg's Theorem, we will use the lemma to write $\bigtimes X_\alpha = X \times Y$ where Y is a partial product over a finite index set. We can then apply the theorem for the finite case to Y. The proof is then completed by using the equivalence of (1), (3), (6), and (7) in Theorem 8.6.

Theorem (Glicksberg). *If $\{X_\alpha : \alpha \in \mathscr{A}\}$ is any family of spaces such that $\bigtimes \{X_\alpha : \alpha \in \mathscr{A}, \alpha \neq \alpha_0\}$ is infinite for each α_0, then $\bigtimes X_\alpha$ is pseudocompact if and only if $\bigtimes X_\alpha$ is C^*-embedded in $\bigtimes(\beta X_\alpha)$, i.e. $\beta(\bigtimes X_\alpha) = \bigtimes(\beta X_\alpha)$.*

Proof. Necessity: The proof will be by contradiction. We will show that any member of $C^*(\bigtimes X_\alpha)$ which fails to extend continuously to $\bigtimes(\beta X_\alpha)$ cannot itself be continuous. If f in $C^*(\bigtimes X_\alpha)$ does not admit a continuous extension to $\bigtimes(\beta X_\alpha)$, there exists a net $\{x_\delta\}$ in $\bigtimes X_\alpha$ converging to a point p in $\bigtimes(\beta X_\alpha)$ such that

$$\lim \sup \{ f(x_\delta) \} - \lim \inf \{ f(x_\delta) \} = a > 0 .$$

Let $0 < 3\varepsilon < a$ and let $\{\alpha_1, \ldots, \alpha_n\}$ be the finite set of indices obtained for this f by the lemma. Put $X = \bigtimes\{X_\alpha : \alpha \neq \alpha_i, 1 \leqslant i \leqslant n\}$ and $Y = \bigtimes\{X_{\alpha_i} : 1 \leqslant i \leqslant n\}$. Theorem 8.8 implies that the projection $\pi_X : X \times Y \to X$ is z-closed. Therefore, (3) of Theorem 8.6 shows that f extends to a mapping g in $C^*(X \times \beta Y)$. The Glicksberg Theorem for finite products, (8.12), shows that

$$X \times \beta Y = X \times \beta(\bigtimes X_{\alpha_i}) = X \times (\bigtimes(\beta X_{\alpha_i})) .$$

The product $X \times \beta Y$ is also pseudocompact and we can again apply Theorem 8.6, this time with βY in the role of X and X in the role of Y. By (6) of Theorem 8.6, the family $\{{}_x g : x \in X\}$ is an equicontinuous family on βY. By (7) of Theorem 8.6, if h belongs to $C^*(X \times \beta Y)$ and p

is a point of βY, then the assignment $\Phi(h)(p) = h_p$ defines a mapping of βY into $C^*(X)$. Thus, there exists a neighborhood U of the point $\{p_{\alpha_1}, \ldots, p_{\alpha_n}\}$ in βY such that both $\mathrm{osc}_U({}_xg) < \varepsilon$ and $\mathrm{osc}_{X \times U}(g) < \varepsilon$. Further, the neighborhood $X \times U$ of p contains points $(x_{\delta_1}, y_{\delta_1})$ and $(x_{\delta_2}, y_{\delta_2})$ of the net such that

$$a - \varepsilon < |f(x_{\delta_1}, y_{\delta_1}) - f(x_{\delta_2}, y_{\delta_2})|$$
$$\leqslant |f(x_{\delta_1}, y_{\delta_1}) - f(x_{\delta_1}, y_{\delta_2})| + |f(x_{\delta_1}, y_{\delta_2}) - f(x_{\delta_2}, y_{\delta_2})|$$
$$= |_{x_{\delta_1}}g(y_{\delta_1}) - {}_{x_{\delta_1}}g(y_{\delta_2})| + |g(x_{\delta_1}, y_{\delta_2}) - g(x_{\delta_2}, y_{\delta_2})|$$
$$< 2\varepsilon.$$

Hence, we obtain $3\varepsilon < a < 3\varepsilon$, which contradicts the assumption that f had no extension to $\bigtimes(\beta X_\alpha)$.

Sufficiency: It is easy to see that the property of the theorem is inherited by partial products, i.e. that if $\mathcal{B}$ is any subset of $\mathcal{A}$, then $\beta(\bigtimes\{X_\alpha : \alpha \in \mathcal{B}\}) = \bigtimes\{\beta X_\alpha : \alpha \in \mathcal{B}\}$. Further, if we put $X = \bigtimes\{X_\alpha : \alpha \in \mathcal{B}\}$ and $Y = \bigtimes\{X_\alpha : \alpha \in \mathcal{A} \setminus \mathcal{B}\}$, we have that

$$\beta(X \times Y) = \bigtimes\{\beta X_\alpha : \alpha \in \mathcal{A}\} = \beta X \times \beta Y.$$

Thus, we can show that $\bigtimes\{X_\alpha : \alpha \in \mathcal{A}\}$ is pseudocompact by showing that $X \times Y$ is pseudocompact. This can now be accomplished in the same manner as in Theorem 8.12 since $\beta(X \times Y) = \beta X \times \beta Y$. □

8.26. As did his theorem for finite products, Glicksberg's Theorem for infinite products stimulated interest in finding conditions under which products would be pseudocompact. Glicksberg conducted such an investigation and one of his results is the following which gives a criterion for the pseudocompactness of a product in terms of countable partial products.

Proposition. *The pseudocompactness of the product is equivalent to the pseudocompactness of every countable partial product.*

Proof. If the product is pseudocompact, then every partial product is the continuous image of the product and therefore must be pseudocompact.

Now assume that every countable partial product of $\bigtimes X_\alpha$ is pseudocompact. We will show that any sequence of non-void open subsets of $\bigtimes X_\alpha$ has a cluster point and that therefore Proposition 8.5 establishes that the product is pseudocompact. We can assume that the members of the sequence of open sets are basic open sets. Therefore, each set restricts only finitely many factor spaces. Let $\{\alpha_i\}$ be the countable set of indices corresponding to the factor spaces restricted by one or more of the open sets. Then the projection $\pi: \bigtimes X_\alpha \to \bigtimes X_{\alpha_i}$ sends the open sets to open sets in the countable partial product and the sequence of images therefore has a cluster point $\{x_{\alpha_i}\}$. An arbitrary choice of co-ordinates in the remaining factors yields a cluster point of the original sequence. $\square$

8.27. Example 8.4 showed that a subspace of $\beta\mathbb{N}$ consisting of $\mathbb{N}$ and the orbit Σ_p of a point p of $\mathbb{N}^*$ is pseudocompact. By choosing two points of different types, two pseudocompact spaces were obtained whose product failed to be pseudocompact. That argument can be extended to show that the preceding criterion for pseudocompactness of a product cannot be improved, i.e. we will construct a non-pseudo-compact product such that every finite partial product is pseudocompact. The first step is to demonstrate that any finite power of the space $\mathbb{N} \cup \Sigma_p$ is pseudocompact. The proposition and subsequent example both appear in W. W. Comfort's 1967 A paper.

Proposition. *If p belongs to $\mathbb{N}^*$ and m is a positive integer, then $(\mathbb{N} \cup \Sigma_p)^m$ is pseudocompact.*

Proof. The proof will be by induction on m. The first step of the argument has already been accomplished in Example 8.4. Now assume that the theorem has been demonstrated for all integers less than m and that it fails for m. Then there is an unbounded real-valued mapping f on $(\mathbb{N} \cup \Sigma_p)^m$ and a sequence $\{x_n\}$ can be chosen from $\mathbb{N}^m$ such that $f(x_n) > n$. The set $\{x_n^1\}$ of first coordinates of the sequence is infinite. If not, infinitely many of the points have the same first coordinate so that $(\mathbb{N} \cup \Sigma_p)^m$ contains a copy of $(\mathbb{N} \cup \Sigma_p)^{m-1}$ on which f is unbounded. By considering only a subsequence if necessary, we may assume that $x_i^1 \neq x_j^1$ if $i \neq j$. Repeating this argument $(m-1)$ times, we can assume that for all $1 \leqslant k \leqslant m$, we have $x_i^k \neq x_j^k$ whenever $i \neq j$. Finally, by discarding infinitely many points of the sequence, we may also assume that for each k, the set $\{x_n^k\}$ has infinite complement in $\mathbb{N}$.

We will now use the properties of our adjusted sequence to locate a point q of $(\mathbb{N} \cup \Sigma_p)^m$ such that f is unbounded on every neighborhood of q, thus contradicting the assumption that f is continuous. Since the sets $\{x_n^1\}$ and $\{x_n^k\}$ are both infinite and both have infinite complements in $\mathbb{N}$, we can find a permutation σ_k of $\mathbb{N}$ such that $\sigma_k(x_n^1) = x_n^k$. Because Σ_p is dense in $\mathbb{N}^*$, there is a point q^1 of Σ_p in the closure of $\{x_n^1\}$. Put $q^k = \beta(\sigma_k)(q^1)$ for each k, $1 \leqslant k \leqslant m$, and let q be the point whose k-th coordinate is q^k. If $U = U_1 \times \cdots \times U_m$ is any basic neighborhood of q, put $V_k = \beta(\sigma_k)^\leftarrow(U_k)$ so that $V = \bigcap\{V_k : 1 \leqslant k \leqslant m\}$ is a neighborhood of q^1.

Then there are infinitely many integers n for which x_n^1 belongs to V, and for every such n, we have

$$x_n^k = \sigma_k(x_n^1) \in \beta(\sigma_k)[V_k] = U_k$$

so that the point x_n belongs to U. Hence, f is unbounded on U, so that f cannot be continuous at q.   □

8.28. Example. Choose a sequence $\{t_n\}$ of distinct types of ultrafilters and for each n, choose p_n in $\mathbb{N}^*$ such that $\tau(p_n) = t_n$. Then the definition of type shows that $\Sigma_{p_i} \cap \Sigma_{p_j} = \emptyset$ whenever $i \neq j$. For each n, put $X_n = \mathbb{N} \cup (\bigcup\{\Sigma_{p_i} : i \neq n\})$. We will show that $X = \underset{}{\times} X_n$ is not pseudocompact but that every finite partial product is pseudocompact.
(a) X is not pseudocompact: Let $A_k = \{x \in X : x^n = k, 1 \leqslant n \leqslant k\}$. The sequence $\{A_k\}$ consists of disjoint sets so that the following function on X is well-defined:

$$f(x) = \begin{cases} k & \text{if } x \in A_k \\ 0 & \text{if } x \notin \bigcup A_k . \end{cases}$$

The function f is unbounded and is continuous at each point of $\bigcup A_k$ since f is constant on each open set A_k. To show that f is continuous, it remains to show that $f^{\leftarrow}(0)$ is open. Choose x in $f^{\leftarrow}(0)$ and let m be the smallest integer such that $x^m \neq x^1$. Such an integer must exist since x cannot belong to A_k and because $\bigcap \Sigma_{p_i} = \emptyset$, no point of X can have every coordinate equal to the same point of $\mathbb{N}^*$. Choose disjoint open neighborhoods U and V of x^1 and x^m, respectively. Then the set

$$W = (\pi_1^{\leftarrow}(U) \cap \pi_m^{\leftarrow}(V)) \setminus \bigcup\{A_k : 1 \leqslant k \leqslant m\}$$

is open because each A_k is closed. Further, it is clear that x belongs to W and that f is constantly equal to 0 on W. Hence, f is continuous and X fails to be pseudocompact.

To show that each of the finite partial products is pseudocompact, it suffices to show that for each m
(b) $\underset{}{\times}\{X_n : 1 \leqslant n \leqslant m\}$ is pseudocompact: We need only to produce a dense pseudocompact subspace. For any integer $k > m$, $(\mathbb{N} \cup \Sigma_{p_k})^m$ is such a subspace as was shown in the preceding proposition.

8.29. If $\{X_\alpha\}$ is an uncountable family of spaces and x is a point of $\underset{}{\times} X_\alpha$, then the Σ-*product* of $\{X_\alpha\}$ with *base point* x is the subspace of $\underset{}{\times} X_\alpha$ consisting of those points which differ from x at only countably many coordinates. Observe that the restriction to an uncountable family is necessary only to assure that a Σ-product will be a proper subspace of the product. This terminology was introduced in 1959 by H.H. Corson, who investigated normality of Σ-products and gave an example of a normal space X whose Hewitt-Nachbin realcompactification υX (Section 1.53) fails to be normal.

By combining the methods which he used to prove Lemma 8.24 and Theorem 8.25, Glicksberg showed that the Stone-Čech compactification of a Σ-product of a family of non-trivial compact spaces is the product of the family. A Σ-product of compact spaces proves to be countably compact which assures that each sequence must cluster. Thus, in the proof we will see that the technique of obtaining a family of open sets which must have a cluster point in a pseudocompact space as was done in Lemma 8.24 can be replaced by sequences in Σ-products.

Theorem (Glicksberg). *If S is a Σ-product of an uncountable family $\{X_\alpha\}$ of compact spaces each of which has at least two points, then $\beta S = \underset{}{\times} X_\alpha$.*

Proof. S is clearly dense in the compact space $\times X_\alpha$. S is countably compact since any sequence in S has a cluster point in $\times X_\alpha$, and it is clear that the cluster point can differ from the base point in only countably many coordinates and therefore belongs to S.

As in Lemma 8.24, for any member f of $C(S)$ and $\varepsilon > 0$, there is a finite set F of indices such that if points x and y agree on the coordinate set F, then $|f(x) - f(y)| < \varepsilon$. If not, since we may alter countably many coordinates of a point in S and remain in S, we can obtain sequences $\{x_i\}$ and $\{y_i\}$ of points of S and a sequence $\{F_i\}$ of disjoint finite sets of coordinates such that $|f(x_i) - f(y_i)| > \varepsilon/2$ and x_i and y_i differ only in those coordinates which belong to F_i as in the proof of Lemma 8.24. Because S is countably compact, the sequence $\{x_i\}$ must cluster to a point x_0 of S. Because any basic neighborhood W of x_0 is the restriction to S of a basic neighborhood in the product topology, W restricts only a finite set E of coordinates. Hence, for some $n \geqslant 1$,

$$E \cap (\bigcup F_i) \subseteq \bigcup \{F_i : 1 \leqslant i \leqslant n\}.$$

Hence, if $j > n$ and W contains x_j, W must also contain y_j so that x_0 is also a cluster point of the sequence $\{y_i\}$. Thus, W contains points x and y such that $|f(x) - f(y)| > \varepsilon/4$, which contradicts the continuity of f.

We use the result of the previous paragraph to show that S is C^*-embedded in $\times X_\alpha$ as was done in the proof of Theorem 8.25. If a member f of $C^*(S)$ fails to admit a continuous extension to $\times X_\alpha$, then there exists a point x_1 in $\times X_\alpha$ and a net $\{x_\delta\}$ in S converging to x_1 such that

$$\limsup \{f(x_\delta)\} - \liminf \{f(x_\delta)\} = a > 0.$$

We choose $\varepsilon > 0$ such that $0 < 3\varepsilon < a$ and obtain the finite set $\{\alpha_i\}$

of indices as outlined above. Put $Y = \times \{X_{\alpha_i} : 1 \leqslant i \leqslant n\}$. Let X be the Σ-product of the remaining spaces X_α where $\alpha \neq \alpha_i$ for any i and the base point is the restriction of the base point of S. Then we can write $S = X \times Y$. Since Y is compact, $\beta Y = Y = \times \{X_{\alpha_i} : 1 \leqslant i \leqslant n\}$. Because $S = X \times Y$ is countably compact and therefore pseudocompact, the proof can now be completed exactly as in Theorem 8.25. $\quad \square$

8.30. Some of the more interesting applications of the previous result occur when each X_α is a compact topological group. Then the usual topological product is also a topological group when given the algebraic structure of the group product. If the base point is obtained by choosing the identity element in each coordinate, then the Σ-product is easily seen to also be a topological group whose Stone-Čech compactification is the whole product. One interesting modification of such an example enabled Glicksberg to describe a space which is homeomorphic to its growth.

Example. Let G be an uncountable product of the two element discrete group $\{0, 1\}$ and let S be the Σ-product with base point the identity element of the product group. Then $\beta S = G$. Let the element of G which has every coordinate equal to 1 be denoted by u and let H be a maximal subgroup of G containing S but missing u. Then for any element x not in H, $\{x\} \cup H$ generates G so that $u = x + h$ for some element h of H. Thus, $G \backslash H$ is the coset $x + H$ and is also the growth H^* of H. Because translation by any element of the group is a homeomorphism in a topological group, H^* is homeomorphic to H since H^* is just the translation of H by x. A further consequence of the fact that translation is a homeomorphism is that any topological group is homogeneous. Thus, G, S, H, and H^* are all homogeneous.

8.31. The preceding example also yields information about density. Recall from Chapter 5 that the density of a space Y is the smallest cardinal number which can be the cardinality of some dense subspace of Y. If X is dense in Y, then it is apparent that the density of X is at least as great as that of Y. The following adaptation of the preceding example is due to Comfort in 1963 and shows that the density of X may be strictly greater than that of βX.

Example. Let G and S be as in the preceding example and assume in addition that the cardinality of the indexing set of the product is $\mathfrak{c}$. Then we have shown that $\beta S = G$. Since G is the product of $\mathfrak{c}$ separable spaces, G is separable [D, p. 175]. It remains to show that S is not separable so that the density of S is uncountable. Let $\{s_i\}$ be any countable subset of S. We will show that $\{s_i\}$ cannot be dense in S. There is some index α for which $s_i^\alpha = 0$ for all i. Let $U = \{g \in G : g^\alpha = 1\}$. Then U is an open set of G which meets S but which fails to contain any member of the set $\{s_i\}$. Hence, S is not separable.

8.32. Before leaving the subject of pseudocompact products, we should remark that there is one class of pseudocompact spaces which is closed under products. In their 1966 paper, W. W. Comfort and K. Ross showed that any product of pseudocompact topological groups is pseudocompact. By applying Glicksberg's Theorem for finite products to the square of a pseudocompact group and using the continuity of the group operations, one can show that *the Stone-Čech compactification of a pseudocompact group is a topological group*. Thus, Example 8.30 is one instance of a more general theorem since both S and G are topological groups and $\beta S = G$.

Assorted Product Theorems

8.33. We have seen that many spaces contain closed copies of $\mathbb{N}$. In particular, the spaces which fail to be countably compact can be characterized as those spaces which contain closed copies of $\mathbb{N}$. Since the product of a family of such spaces will contain as a closed subspace a product of copies of $\mathbb{N}$, we can expect that the following result will be useful in the investigation of such products. The result appears in A. H. Stone's 1948 paper.

Theorem (A. H. Stone). *An uncountable product of copies of* $\mathbb{N}$ *fails to be normal.*

Proof. We can think of the product of copies of $\mathbb{N}$ indexed by an uncountable index set $\mathscr{A}$ as the set $\mathbb{N}^{\mathscr{A}}$, i.e. the functions on $\mathscr{A}$ taking their values in $\mathbb{N}$. We will assume that a basic neighborhood U of a point $x = \{x^\alpha\}$ in $\mathbb{N}^{\mathscr{A}}$ is obtained by restricting a finite set $F(U) = \{\alpha_1, \ldots, \alpha_j\}$ of coordinates to a single point x^{α_i}.

For each positive integer k, let B_k be the set of all points $x = \{x^\alpha\}$ such that for each integer n other than k, $x^\alpha = n$ for at most one α in $\mathscr{A}$. The sets B_k are closed and are pairwise disjoint since $\mathscr{A}$ is uncountable. Thus, if $\mathbb{N}^{\mathscr{A}}$ were normal, there would exist disjoint open sets V and W containing B_1 and B_2, respectively. We will show that this assumption leads to a contradiction.

We will define inductively three sequences: a sequence $\{x_n\}$ of points of B_1, a strictly increasing sequence $\{m_n\}$ of positive integers, and a sequence $\{\alpha_n\}$ of elements of $\mathscr{A}$. Define x_1 to be the point which is constantly equal to 1. Then x_1 belongs to B_1 and therefore there is a basic neighborhood U_1 of x_1 contained in V. Let $F(U_1) = \{\alpha_1, \ldots, \alpha_{m_1}\}$ be the set of coordinates restricted by U_1 and m_1 be the number of coordinates restricted. Now suppose that x_n and U_n have been obtained so that x_n belongs to B_1, U_n is a basic neighborhood of x_n contained in V, and $F(U_n) = \{\alpha_1, \ldots, \alpha_{m_1}, \ldots, \alpha_{m_n}\}$ is the set of coordinates restricted by U_n. Define x_{n+1} as follows:

$$x_{n+1}^\alpha = \begin{cases} k & \text{if } \alpha = \alpha_k \text{ for some } k, \quad 1 \leqslant k \leqslant m_n \\ 1 & \text{otherwise.} \end{cases}$$

Then x_{n+1} belongs to B_1 and there is a basic neighborhood U_{n+1} of x_{n+1} contained in V. We can also assume that $F(U_{n+1})$ contains $F(U_n)$ as a proper subset so that $F(U_{n+1})$ has m_{n+1} elements and $m_{n+1} > m_n$. Write

$$F(U_{n+1}) \backslash F(U_n) = \{\alpha_{m_n+1}, \ldots, \alpha_{m_{n+1}}\}.$$

The induction is now complete.

Define a point y in $\mathbb{N}^\mathscr{A}$ by:

$$y^\alpha = \begin{cases} k & \text{if } \alpha = \alpha_k \text{ for some } k \geqslant 1 \\ 2 & \text{otherwise.} \end{cases}$$

Thus, y belongs to B_2 and hence there is a basic neighborhood G of y contained in W. Since $F(G)$ is finite, there is some integer n such that α_k is not in $F(G)$ for all $k > m_n$. Finally, define z in $\mathbb{N}^\mathscr{A}$ by

$$z^\alpha = \begin{cases} k & \text{if } \alpha = \alpha_k \text{ for some } k, \ k \leqslant m_n \\ 1 & \text{if } \alpha = \alpha_k \text{ for some } k, \ m_n < k \leqslant m_{n+1} \\ 2 & \text{otherwise.} \end{cases}$$

Then because z agrees with both y and x_{n+1} on the coordinates $\alpha_1, \ldots, \alpha_{m_{n+1}}$, we have that

$$z \in U_{n+1} \cap G \subset W \cap V,$$

which is a contradiction. $\quad\square$

8.34. Because an uncountable product of non-countably compact spaces will contain $\mathbb{N}^\mathscr{A}$ as a closed, non-normal subspace, we have verified the

Corollary (A. H. Stone). *If a product of non-empty T_1-spaces is normal, all but at most a countable number of the factor spaces is countably compact.*

8.35. The previous result together with Glicksberg's Theorem for infinite products enables us to establish the following result which appeared in N. Noble's 1971 paper and shows that any T_1-space such that every power of the space is normal must be compact. The proof appears in a paper of S. P. Franklin and the present author.

Theorem (N. Noble). *If each power of a T_1-space is normal, then the space is compact.*

Proof. Let X be such a space and let X^m be an uncountable power of X. Since $(X^m)^m = X^m$, the preceding corollary shows that X^m is countably compact and therefore is pseudocompact. Because X is a T_1-space and is normal, X is completely regular and we can consider βX. Glicksberg's Theorem shows that $\beta(X^m) = (\beta X)^m$. The proof will be completed by choosing a particular uncountable index set.

If X fails to be compact, there is a point p in X^* and we can identify p with the unique free z-ultrafilter on X which converges to p. Consider the product set X^p which consists of the functions defined on the collection of zero-sets belonging to p and taking their values in X. The proof is completed by showing that the assumption that p is free will provide disjoint closed subsets of X^p which cannot be completely separated. Write Δ for the diagonal, i.e. the set of all constant functions. Write C for the set of choice functions, i.e.

$$C = \{x \in X^p : x_Z \in Z \text{ for } Z \in p\}\,.$$

The sets Δ and C are closed and are disjoint since p is free. Therefore they are completely separated by a mapping f in $C(X^p)$. But f cannot extend continuously to $\beta(X^p)$ since the function in $(\beta X)^p = \beta(X^p)$ which is constantly equal to p belongs to the closure in $(\beta X)^p$ of both Δ and C. Thus, it must be that $\Delta \cap C \neq \emptyset$, i.e. that p is fixed and X is compact. $\quad\Box$

8.36. Recall from Chapter 6 that a paracompact space is one in which every open cover has a locally finite open refinement. Since a finite subcover is also a locally finite refinement, the class of paracompact spaces is another class containing the compact spaces. The equivalence of the first two conditions in the following theorem was demonstrated by H. Tamano in 1962. The methods used in the proof will be reminescent of those used earlier in the chapter.

Theorem (Tamano). *The following are equivalent for any space X:*
  (1) $X \times \beta X$ *is normal.*
  (2) X *is paracompact.*
  (3) *If K is a compact space, $X \times K$ is paracompact.*

Proof. The following technical condition is also equivalent to the three stated in the theorem. It is included to smooth the proof of $(1) \Rightarrow (2)$.

(4) *For each compact subspace F of X^*, there is a locally finite open cover $\{U_\alpha:\alpha\in\mathscr{A}\}$ of X such that $(\mathrm{cl}_{\beta X}\,U_\alpha)\cap F=\emptyset$ for each α in $\mathscr{A}$.*

(2)$\Rightarrow$(3): Assume that X is paracompact and let $\mathscr{U}$ be an open cover of $X\times K$. For each point x of X, there is a finite subfamily $\{U_i^x:1\leqslant i\leqslant n_x\}$ which covers the compact space $\{x\}\times K$. Since K is compact, we can choose an open neighborhood V_x of x such that $V_x\times K$ is contained in $\bigcup\{U_i^x:1\leqslant i\leqslant n_x\}$ [D, p. 228]. Repeating this for each point of X yields an open cover $\{V_x:x\in X\}$ of X. Since X is paracompact, the cover has a locally finite open refinement $\{W_\beta:\beta\in\mathscr{B}\}$. For each W_β choose x in X such that W_β is contained in V_x and consider the open sets

$$(W_\beta\times K)\cap U_i^x,\quad 1\leqslant i\leqslant n_x.$$

Running through the set $\mathscr{B}$ of indices yields an open cover $\mathscr{V}$ of $X\times K$. For each (x,k) in $X\times K$, there is a neighborhood U of x which meets only finitely many elements of $\{W_\beta\}$. Hence, $U\times K$ is a neighborhood of (x,k) which meets only finitely many sets of $\mathscr{V}$. Thus, $\mathscr{V}$ is a locally finite open refinement of $\mathscr{U}$ and $X\times K$ is paracompact.

(3)$\Rightarrow$(1): This is immediate from the observation that a paracompact space is normal.

(1)$\Rightarrow$(4): Assume that $X\times\beta X$ is normal and that F is a compact subset of X^*. Then the diagonal $\Delta=\{(x,x)\in X\times\beta X:x\in X\}$ and $X\times F$ are disjoint closed subsets of $X\times\beta X$. Thus there is a map $f:X\times\beta X\to I$ such that f is constantly 0 on Δ and 1 on $X\times F$. Placing βX in the role of Y, the equivalence of (3) and (5) of Theorem 8.6 implies that

$$\gamma(x_1,x_2)\equiv\sup\{|f(x_1,p)-f(x_2,p)|:p\in\beta X\}$$

defines a continuous pseudometric on X. Thus γ induces a pseudometric topology on X which is weaker than the original topology. The open cover of X consisting of the spheres

$$B(x,1/2)=\{y\in X:\gamma(x,y)<1/2\}$$

has a locally finite open refinement $\{U_\alpha:\alpha\in A\}$ in the pseudometric topology which is also open and locally finite in the original topology. If y belongs to $B(x,1/2)$, then $\gamma(x,y)<1/2$ so that

$$_xf(y)=|f(x,y)-f(y,y)|<1/2.$$

Thus, $_xf(p)\leqslant1/2$ for each p in $\mathrm{cl}_{\beta X}B(x,1/2)$ and since $_xf(p)=1$ for each point p of F, $\mathrm{cl}_{\beta X}B(x,1/2)\cap F=\emptyset$. Hence, each member of the locally finite open cover $\{U_\alpha\}$ satisfies $\mathrm{cl}_{\beta X}U_\alpha\cap F=\emptyset$.

(4)$\Rightarrow$(2): Let $\mathcal{V}=\{V_\alpha:\alpha\in\mathcal{A}\}$ be an open cover of X. For every V_α, choose an open set W_α of βX such that $V_\alpha=X\cap W_\alpha$. Put $F=\bigcap\{\beta X\setminus W_\alpha:\alpha\in\mathcal{A}\}$. Then F is a compact subspace of X^* so that by (4) there exists a locally finite open cover $\{U_\beta:\beta\in\mathcal{B}\}$ of X such that $\mathrm{cl}_{\beta X}\,U_\beta\cap F=\emptyset$ for each U_β. Thus, $\mathrm{cl}_{\beta X}\,U_\beta$ is contained in $\bigcup\{W_\alpha:\alpha\in\mathcal{A}\}$ and since $\mathrm{cl}_{\beta X}\,U_\beta$ is compact, it can be covered by a finite subfamily $\{W_{\alpha_1},\dots,W_{\alpha_{n_\beta}}\}$. Define $\mathcal{U}$ to be the cover $\{U_\beta\cap V_{\alpha_i}:\beta\in\mathcal{B},\,\alpha_1,\dots,\alpha_{1_{n_\beta}}\}$. $\mathcal{U}$ is then easily seen to be the required locally finite refinement of $\mathcal{V}$. $\square$

8.37. Example. In Section 6.17, we noted that the ordinal space ω_1 is normal but not paracompact. Since $\beta\omega_1=\omega_1+1$ [GJ, 5.12], Theorem 8.36 implies that $\omega_1\times(\omega_1+1)$ is not normal. Hence, we also see that the product of a normal space with a compact space need not be normal.

8.38. The rather intriguing condition (1) of Theorem 8.36 is made even more so by observing that there is an analogous characterization of P-spaces:

Theorem (Negrepontis). *X is a P-space if and only if $X\times\beta X$ is an F-space.*

The theorem appears in the 1969 B paper of S. Negrepontis. Since the infinite compact space βX cannot be a P-space, one direction of the proof is immediate from Proposition 1.65. The other direction of the proof requires machinery which we have not developed and will not be given.

The υ-Analogue: An Open Question

8.39. Once Glicksberg's Theorem has been proven, it is natural to seek a characterization of pairs of spaces X and Y such that $\upsilon(X\times Y)=\upsilon X\times\upsilon Y$. Because a space S is pseudocompact if and only if $\upsilon S=\beta S$ (Corollary 1.58), we can interpret Glicksberg's Theorem as providing one sufficient condition. However, the pseudocompactness of the product is hardly necessary since $\mathbb{R}\times\mathbb{R}$ is not pseudocompact and yet

$$\upsilon\mathbb{R}\times\upsilon\mathbb{R}=\mathbb{R}\times\mathbb{R}=\upsilon(\mathbb{R}\times\mathbb{R})\,.$$

The problem appears to be substantially more difficult than the corresponding one for the Stone-Čech compactification. One complication is the relationship between realcompactness and measurable cardinals which is described in [GJ, Chapters 12 and 15]. A discrete space is realcompact if and only if its cardinality is non-measurable and this fact causes cardinality restrictions to enter into the problem. One approach to the problem based on the assumption that a measurable cardinal exists will be discussed briefly in Example 10.24(g).

The problem is approachable from the point of view of uniform spaces as is shown in the 1960 paper of N. Onuchic and the 1969A paper of A.W. Hager. In his 1964 text, *Uniform Spaces*, J.R. Isbell proves Glicksberg's Theorem from the point of view of uniformities. A uniform space is said to be *fine* if the uniformity is the finest one which is compatible with the topology of the space. Isbell discusses the question of when the product of two fine spaces will be fine, which is a related question, and proves a product theorem in the context of uniformities which includes Glicksberg's Theorem as a special case. In his 1972 paper, A.W. Hager proves a still more general result concerning the uniform completion of products.

In his 1968B paper, W.W. Comfort shows that if X is a k-space (see Exercise 8H), υY is locally compact, and neither X nor υY contains a compact subspace of measurable cardinal, then $\upsilon X \times \upsilon Y = \upsilon(X \times Y)$. Comfort's results together with other related results are also presented in the thesis of M. Weir. In his 1972A paper, Hušek approaches the problem from the point of view of function spaces.

In 1969, T. Isiwata showed that υ will distribute over $X \times Y$ if X is a continuous image of a first countable space, Y is pseudocompact, and Y has non-measurable cardinal. In his 1970 paper, W.G. McArthur characterizes the pairs X and Y for which $\upsilon X \times \upsilon Y = \upsilon(X \times Y)$ in terms of a condition similar to that of Theorem 8.14. Another condition characterizing pairs of spaces for which this equality holds is given in the 1971 paper of H. Buckwalter.

The case where one of the factors is a linearly ordered topological space is considered by McArthur in his 1973 paper. There he shows that if X is a separable realcompact space and Y is linearly ordered, then $\upsilon(X \times Y) = X \times \upsilon Y$. Further, he shows that if both X and Y are well-ordered spaces, then the equality will always hold.

Exercises

8A. *Pseudocompact subspaces of βX*

For H contained in $\beta X \setminus \upsilon X$, define $Y = X \cup H$.
1. If Y is pseudocompact, then H is dense in $\beta X \setminus \upsilon X$.
2. If H is dense in $\beta X \setminus \upsilon X$ and if υX is locally compact, then Y is pseudo-compact.

Reference: In his 1967B paper, W. W. Comfort characterizes those spaces X for which υX is locally compact.

8B. *Closed projections*

Let X be a P-space and Y be Lindelöf.
1. The projection $\pi_X : X \times Y \to X$ is a closed map. [If $U_i \times V_i$ is a countable cover of $\{x\} \times Y$, $\bigcap U_i$ is a neighborhood of x.]
2. $X \times Y$ is C^*-embedded in $X \times \beta Y$.

Reference: Comfort and Negrepontis, 1966. The thesis and paper of A. K. Misra contain more results relating P-spaces and Lindelöf spaces.

8C. *Proper C^*-pairs*

1. Let $S \subset X \subset \beta S$ and suppose that X and Y are a proper C^*-pair for some space Y. Then $X \subset \upsilon S$. [Theorem 1.53, Corollary 8.13.]
2. Let S be realcompact and $S \subset X \subset \beta S$. Then if X and Y are a proper C^*-pair for some space Y, X is equal to S.
3. Let X be σ-compact and $S \subset X \subset \beta S$. Then if X and Y are a proper C^*-pair for some space Y, $X = S = \mathbb{N}$. [Proposition 1.65.]

Reference: Comfort and Negrepontis, 1966.

8D. *Subspace of $\beta(X \times Y)$*

Suppose that $\beta(X \times Y)$ has a subspace S such that $\beta(X \times Y)\backslash S$ is an F-space and $|S| < |\beta Y|$.
1. If e is an embedding of $\beta X \times \beta Y$ into $\beta(X \times Y)$, then there exists a point q of βY such that $\beta X \times \{q\}$ misses $e^{\leftarrow}(S)$.
2. If X is not an F-space, then $\beta(X \times Y)$ contains no copy of $\beta X \times \beta Y$. $[e[\beta X \times \{q\}] \subset \beta(X \times Y)\backslash S.]$
3. $\beta(\mathbb{R} \times \mathbb{R})$ contains no copy of $\beta\mathbb{R} \times \beta\mathbb{R}$.

Reference: Gillman and Jerison, 1959, contains this result and several similar results.

8E. *$\beta\mathbb{N}$ contains 2^c disjoint copies of $\beta\mathbb{N}$*

1. If p belongs to $\beta\mathbb{N}$, then $\mathbb{N} \times \{p\}$ is C^*-embedded in $\mathbb{N} \times \beta\mathbb{N}$.
2. If p and q are distinct, then $\mathbb{N} \times \{p\}$ and $\mathbb{N} \times \{q\}$ are completely separated in $\mathbb{N} \times \beta\mathbb{N}$. [Theorem 8.36.]
3. $\beta\mathbb{N}$ contains 2^c mutually disjoint copies of itself. [8.18(a).]

Reference: Gillman and Jerison, 1959. Note that in the proof of Theorem 3.45, we showed that $\mathbb{N}^*$ contains 2^c disjoint copies of $\beta\mathbb{N}$ by a different method.

8 F. *C-embedding of $X \times Y$ in $\upsilon X \times \upsilon Y$*

A subspace X of S is said to be G_δ-*dense* in S if every non-empty G_δ of S meets X.

1. If X is G_δ-dense in S and Y is G_δ-dense in T, then $X \times Y$ is G_δ-dense in $S \times T$.
2. If $X \times Y$ is C^*-embedded in $\upsilon X \times \upsilon Y$, then $X \times Y$ is C-embedded in $\upsilon X \times \upsilon Y$. [Theorem 1.3.]

Reference: Comfort and Negrepontis, 1966.

8 G. *Pseudocompact products and k-spaces*

A space X is a *k-space* if each subset F of X such that $F \cap K$ is closed in K for every compact subspace K is closed in X.

1. A function is continuous on a k-space if and only if its restriction to every compact subspace is continuous.
2. If X and Y are pseudocompact and Y is a k-space, then $X \times Y$ is pseudocompact. [Modify Proposition 8.21.]
3. Any first countable space is a k-space. [A convergent sequence together with its limit point is compact.]
4. Any locally compact space is a k-space.
5. Let p belong to $\mathbb{N}^*$. The subspace $\mathbb{N} \cup \{p\}$ of $\beta\mathbb{N}$ is not a k-space. [$\mathbb{N}$ is not closed.]

Reference: (2) appears in Tamano, 1960A. k-spaces are discussed in [D, pp. 247—249].

8 H. *Products of metric spaces*

1. A normal space is pseudocompact if and only if it is countably compact.
2. A metric space is pseudocompact if and only if it is compact.
3. The following are equivalent for a product of non-empty metric spaces:
    (a) The product is normal.
    (b) The product is paracompact.
    (c) At most countably many of the factors fail to be compact.

Reference: (3) appears in A. H. Stone, 1948.

8 I. *Pseudocompact subspaces of $\beta\mathbb{N}$*

Every pseudocompact subspace of $\beta\mathbb{N}$ which contains $\mathbb{N}$ has cardinality at least $\mathfrak{c}$. [Lemma 8.3 and Theorem 5.12.]
Reference: S. Mrówka, 1959.

Chapter 9. Local Connectedness, Continua, and X^*

9.1. A space is *connected* if it is not the union of two disjoint non-empty closed sets. An equivalent condition is that there does not exist a continuous mapping of the space onto the discrete space having two points. From the latter condition it is easy to see that a space *X is connected if and only if βX is connected.*

A space is *locally connected* if every point has a base of open connected neighborhoods. The relationship between the local connectedness of a space and the local connectedness of its Stone-Čech compactification is not as clear-cut as for connectedness. For example, the space $\mathbb{N}$ is locally connected, but it is clear that $\beta\mathbb{N}$ is not locally connected. In this chapter, we will use the lattice structure of $C^*(X)$ to characterize the class of spaces which have locally connected Stone-Čech compactifications as the class of locally connected pseudocompact spaces.

The growth of the ray $A = [1, \infty)$ will be shown to be an indecomposable continuum, i.e. a compact connected space which is not the union of two closed connected proper subspaces. Among the interesting properties of this space which we will explore is the fact that it can be mapped onto any compact connected metric space.

We will then use the fact that A^* is an indecomposable continuum to characterize a class of spaces whose growths are also indecomposable continua as well as a larger class whose growths are continua, but not necessarily indecomposable. As a consequence of these characterizations, we will see that for $n \geq 2$, the growth of $\mathbb{R}^n$ is a continuum which is decomposable.

Compactifications of Locally Connected Spaces

9.2. Suppose that a space X is the disjoint union of n non-empty closed subsets F_i, $1 \leq i \leq n$. Then each F_i is also open and the characteristic function χ_i of each F_i belongs to $C^*(X)$. Thus, if f is in $C^*(X)$, we can write $f_i = \chi_i f$ so that $f = f_1 + \cdots + f_n$. Further, no two of the mappings $\{f_i\}$ are non-zero at the same point, and hence we have $|f_i| \wedge |f_j| = 0$ whenever $i \neq j$. When f can be expressed in this form, the set $\{f_i : 1 \leqslant i \leqslant n\}$ is called a decomposition of f. Our characterization of the class of spaces having locally connected Stone-Čech compactifications will be in terms of a certain type of decomposition.

Let E be a normed linear lattice. A subset $\{p_i : 1 \leqslant i \leqslant n\}$ of E is said to be a *decomposition* of a point x of E if $x = p_1 + \cdots + p_n$ and $|p_i| \wedge |p_j| = 0$ whenever $i \neq j$. A member x of E is said to be *well behaved* if for every $\varepsilon > 0$, x admits a decomposition $\{p_i : 1 \leqslant i \leqslant n\}$ such that for $i = 1, \ldots, n-1$, p_i admits no proper decomposition and $\|p_n\| < \varepsilon$. The lattice E is called *well behaved* if every point of E is well behaved.

If $f = p_1 + \cdots + p_n$ is a decomposition of a member f of $C^*(X)$, then the cozero-set of f is the disjoint union of the cozero-sets of the p_i, and therefore is not connected. We will ultimately use the existence of mappings which do not admit proper decompositions to produce a base of connected cozero-sets.

If S is C^*-embedded in a space X, then it is easy to see that a mapping in $C^*(S)$ can be extended to one in $C^*(X)$ is such a way that the norm of the mapping is preserved. Hence, $C^*(X)$ contains a lattice isomorphic copy of $C^*(S)$. Since we frequently deal with C^*-embedded copies of $\mathbb{N}$, the following example of a normed linear lattice which is isomorphic to $C^*(\mathbb{N})$ will be useful.

Example. The normed linear lattice l_∞ consisting of all bounded sequences taken with the sup norm is not well behaved. This is clear since any sequence containing more than two non-zero terms will admit a proper decomposition. Note also that because l_∞ is isomorphic to $C^*(\mathbb{N})$, it is also isomorphic to $C(\beta\mathbb{N})$.

9.3. We now show that βX is locally connected exactly when X is both locally connected and pseudocompact. In his 1956B paper, B. Banaschewski showed that X must be locally connected and pseudocompact whenever βX is locally connected. The converse was established by M. Henriksen and J.R. Isbell in their 1957A paper. The characterization of pseudocompact locally connected spaces in terms of well behaved linear lattices was introduced by D.E. Wulbert in his 1969 papers.

Theorem. *The following are equivalent:*
   (1) *X is locally connected and pseudocompact.*
   (2) *$C^*(X)$ is a well behaved linear lattice.*
   (3) *βX is locally connected.*

The proof will use the characterization of locally connected spaces as those spaces in which the components of each open subspace are open [D, p. 113].

Proof. (1)$\Rightarrow$(2): Let X be locally connected and pseudocompact and let f belong to $C^*(X)$. Let $\mathscr{G}$ be the family of components of the cozero-set of f. Then $\mathscr{G}$ is a family of open sets because X is locally connected. For $\varepsilon > 0$, consider the family $\mathscr{F} = \{G \cap \{x \in X : |f(x)| > \varepsilon\} : G \in \mathscr{G}\}$. The family $\mathscr{F}$ is locally finite: If $f(x) = 0$, then $\{x \in X : |f(x)| < \varepsilon\}$ is a neighborhood of x which misses each member of $\mathscr{F}$. If $|f(x)| > 0$, then x has a neighborhood which is contained in some member of $\mathscr{G}$ and which therefore meets at most one member of $\mathscr{F}$.

Now because X is pseudocompact, Proposition 5.5 implies that the locally finite family $\mathscr{F}$ can contain only finitely many non-empty sets. Suppose that f will admit no well behaved decomposition. Then we can generate an infinite sequence $\{f_i\}$ of members of $C^*(X)$ such that each f_i is a restriction of f, each has norm greater than or equal to ε, and the cozero-sets $\{\mathbf{Cz}(f_i)\}$ form a pairwise disjoint family. But then the non-empty members of the family

$$\{\mathbf{Cz}(f_i) \cap F : F \in \mathscr{F}, i = 1, 2, \ldots\}$$

form an infinite, locally finite family of non-empty open sets, which cannot exist in a pseudocompact space. Hence, $C^*(X)$ is well behaved.

(2)$\Rightarrow$(1): Assume that $C^*(X)$ is a well behaved linear lattice. We will first show that X is pseudocompact. Any non-pseudocompact space contains a closed, C^*-embedded copy of $\mathbb{N}$ (Lemma 4.5). Hence, $C^*(X)$ contains a lattice isomorphic copy of l_∞, the space of all bounded sequences. Since we have seen that l_∞ is not well behaved, $C^*(X)$ cannot be well behaved. Hence, X is pseudocompact.

Now we establish that X is locally connected. Let x belong to X and let U be a neighborhood of x. We will exhibit a connected neighborhood of x which is contained in U. There exists f in $C^*(X)$ such that $f(x) = \|f\| = 1$ and such that f vanishes on the complement of U. Further, since f is well behaved, we can assume that f admits no proper decomposition. Hence, the cozero-set of f is a connected neighborhood of x contained in U, and X is locally connected.

(2)$\Leftrightarrow$(3): This is immediate since $C^*(X)$ and $C^*(\beta X)$ are lattice isomorphic. $\square$

9.4. Examples. *$\beta\mathbb{R}$ is connected, but not locally connected.*

$\beta\mathbb{Q}$ is neither connected nor locally connected.

9.5. The class of locally connected pseudocompact spaces has several interesting properties which are not shared by the class of locally connected spaces. Note that the continuous image of a locally connected space need not be locally connected. For instance, any discrete space is locally connected, but the discrete space of cardinality 2^c will map onto $\beta\mathbb{N}$ and $\beta\mathbb{N}$ fails to be locally connected.

However, a quotient of a locally connected space is locally connected [D, p. 125]. The following result appears in the 1967 paper of J. de Groot and R. H. McDowell and shows that the quotient hypothesis is not needed in the class of pseudocompact locally connected spaces.

Proposition. *A continuous image of a pseudocompact, locally connected space is locally connected.*

Proof. Let X be pseudocompact and locally connected and let f map X onto Y. Then βX is locally connected and βY is a continuous image of βX under the extension $\beta(f)$. Because βX is compact, $\beta(f)$ is a closed map and is therefore a quotient mapping. But a quotient of a locally connected space is locally connected, hence βY is locally connected. But then we have seen that Y must be locally connected. $\Box$

9.6. Corollary. *If K_1 and K_2 are compactifications of X and K_2 is greater than K_1, then K_1 is locally connected if K_2 is.*

9.7. The next result is also due to de Groot and McDowell.

Theorem. *The following are equivalent:*
   (1) *X is locally connected and pseudocompact.*
   (2) *Every space in which X is dense is locally connected.*
   (3) *Every compactification of X is locally connected.*

Proof. (1)$\Rightarrow$(2): Let X be dense in Y. Then X is also dense in βY and therefore βY is a continuous image of βX. But βX is locally connected, so that Proposition 9.5 implies that βY is locally connected. Hence, Y is also locally connected. (2)$\Rightarrow$(3) is immediate and Theorem 9.3 shows that (1) follows from (3). $\Box$

9.8. Certain results concerning the local connectedness of X and βX can be achieved without assuming that either space is locally connected. The next three results appear in Henriksen and Isbell's 1957A paper. The first one was established for normal spaces by A. D. Wallace in 1951.

Lemma. *An open subset U of βX is connected if and only if $X \cap U$ is connected.*

Proof. The proof will be accomplished by proving both contrapositives.

   If U is not connected, then $U = V \cup W$ where V and W are disjoint non-empty open sets. Then $U \cap X = (V \cap X) \cup (W \cap X)$, so that $U \cap X$ is also disconnected.

Now suppose that $U \cap X$ is disconnected, i.e. that $U \cap X = V \cup W$ where V and W are disjoint, non-empty open subsets of X. Since X is dense in βX, U is contained in $\mathrm{cl}_{\beta X} V \cup \mathrm{cl}_{\beta X} W$. If no point of U belongs to both of these closures, then we have

$$U = ((\mathrm{cl}\, V) \cap U) \cup ((\mathrm{cl}\, W) \cap U)$$

and U is not connected. Now suppose that a point p of U belongs to both closures. Let g in $C(\beta X)$ be such that $g(p) = 0$ and $g[\beta X \setminus U] = \{1\}$. Define a function f on X by:

$$f(x) = \begin{cases} 1/2 & \text{if } x \in W \quad \text{and} \quad g(x) < 1/2 \\ g(x) & \text{otherwise}. \end{cases}$$

Because V and W are complementary clopen subsets of $U \cap X$, it is easy to verify that f is continuous. Now the extension $\beta(f)$ of f must coincide with g on $\mathrm{cl}_{\beta X} V$ and be greater than or equal to $1/2$ on $\mathrm{cl}_{\beta X} W$. But this contradicts the assumption that p belongs to both closures. Hence, U is not connected. $\quad \Box$

9.9. The following corollary is immediate and corresponds to Exercise 2F.2 which states an analogous result for points of first countability.

Corollary. *βX is locally connected at a point of X if and only if X is locally connected at the point.*

9.10. In Corollary 1.58, we saw that $\upsilon X = \beta X$ if and only if X is pseudocompact. Thus, the following result shows that βX cannot be locally connected whenever υX is not equal to βX and indicates the importance of pseudocompactness in (1) of Theorem 9.3. Notice that the preceding Corollary combined with the following proposition yield another proof that X must be locally connected and pseudocompact whenever βX is locally connected.

Proposition. *βX is not locally connected at any point of $\beta X \setminus \upsilon X$.*

Proof. If p belongs to $\beta X \setminus \upsilon X$, then there is an f in $C(X)$ such that $f^{\alpha}(p) = \infty$ (Section 1.53). For $i = 0, 1, 2, 3$, let Z_i be the set of all x in X such that $n \leq f(x) \leq n+1$ for some integer n such that $n \equiv i \pmod 4$. The sets Z_i cover X so that p belongs to $\mathrm{cl}_{\beta X} Z_j$ for some j. Then let $k = 0, 1, 2$, or 3 be congruent to $j+2$ modulo 4. Then Z_k and Z_j are disjoint zero-sets and therefore p cannot belong to $\mathrm{cl}_{\beta X} Z_k$. Thus, there is a neighborhood U of p which misses Z_k. If βX is locally connected at p, then there is a connected neighborhood V of p such that V is contained in U. By Lemma 9.8, $V \cap X$ is connected. However, f is necessarily unbounded on $V \cap X$ and $V \cap X$ misses Z_k. Thus, $f|V \cap X$ cannot take any values between $4n+k$ and $4n+k+1$ for any n, so that the image of $V \cap X$ is not connected. This contradicts the assumption that $V \cap X$ is connected and therefore that βX is locally connected at p. $\square$

A space is *connected im kleinen at p* if every neighborhood of p contains a connected neighborhood of p. This property is strictly weaker than local connectedness, but if a space is connected im kleinen at every point, then it is locally connected. In his 1972 B paper, R. G. Woods has used an approach similar to that of the proof of Proposition 9.10 to prove the analogous result for connectedness im kleinen.

9.11. Examples. *$\beta \mathbb{N}$, $\beta \mathbb{Q}$, and $\beta \mathbb{R}$ are not locally connected at any point in their respective growths.*

A Non-Metric Indecomposable Continuum

9.12. A *continuum* is a compact, connected set. Common examples of continua are the closed unit interval I and the circle S^1. In the remaining sections of the chapter, we will consider several recent results concerning Stone-Čech compactifications whose growths are continua. A continuum is said to be *decomposable* if it can be written as the union of two proper subcontinua. Otherwise, a continuum is said to be *indecomposable*. Both of the examples given above are decomposable metric continua. We will show that the growth of the interval $[1, \infty)$ is a non-metric indecomposable continuum. We will first need a preliminary result on the intersection of continua. The proof is taken from the text of S. Willard.

Proposition. *The intersection of a decreasing family of continua is a continuum.*

Proof. Let $\{K_\alpha : \alpha < \beta\}$ be a family of continua in a space X such that $K_{\alpha+1}$ is contained in K_α for each α. Then $\bigcap K_\alpha$ is a closed subset of each K_α and is therefore compact. Now suppose that the intersection fails to be connected, i.e. that $\bigcap K_\alpha = F \cup H$ where F and H are disjoint, non-empty closed sets. F and H are contained in K_1 so that we can replace X by K_1 and assume that X is compact. Thus, there exist disjoint open sets U and V of X containing F and H, respectively. For each α, K_α cannot be contained in $U \cup V$ since $U \cap K_\alpha$ and $V \cap K_\alpha$ would disconnect K_α. Hence, we can choose a point x_α in $K_\alpha \setminus (U \cup V)$ for each α and the resulting net $\{x_\alpha : \alpha < \beta\}$ must cluster at some point z of X.

If W is any neighborhood of z, then W meets each K_α so that z belongs to $\operatorname{cl} K_\alpha = K_\alpha$ for each α. But then we have

$$z \in \bigcap K_\alpha \subset U \cup V,$$

and this is a contradiction since the net $\{x_\alpha\}$ is never in the neighborhood $U \cup V$ of z. $\quad\square$

9.13. The growth of $\beta\mathbb{R}$ consists of two pieces, one at each end of $\mathbb{R}$. It is easy to see that the two pieces are the growths of the rays $A = [1, \infty)$ and $B = (-\infty, -1]$. The next result shows that each piece is a non-metric indecomposable continuum. The theorem appears in the dissertations of both D. P. Bellamy and R. G. Woods and also in the 1971 paper of Bellamy from which the present proof is taken.

Theorem (Bellamy and Woods). $A^* = \beta A \setminus A$ *is a non-metric indecomposable continuum.*

Proof. A^* is a continuum: Put $A_n = [n, \infty)$. Then $\operatorname{cl}_{\beta\mathbb{R}} A_n$ is a continuum since it is compact and is the closure of a connected set. Then we have

$$A^* = \bigcap \operatorname{cl}_{\beta\mathbb{R}} A_n$$

and the previous proposition implies that A^* is a continuum.

A^* is non-metric: Because A is σ-compact and locally compact, Proposition 1.62 implies that A^* is a compact F-space. A^* is clearly infinite, and so contains a copy of $\beta\mathbb{N}$ (Proposition 1.64). No point of $\mathbb{N}^*$ is a G_δ in $\beta\mathbb{N}$ (Corollary 3.7) so that no point of the growth $\mathbb{N}^*$ is a G_δ in A^*. Hence, A^* fails to be first countable and therefore cannot be metric.

A^* is indecomposable: Suppose that F and H are proper closed subsets of A^* such that $A^* = F \cup H$. To show that A^* is not the union of two proper subcontinua, it will be sufficient to show that F is not connected. We will construct a mapping of F onto the discrete space $\{0, 1\}$ to establish that F fails to be connected.

Choose x in $A^* \setminus H$ and y in $A^* \setminus F$. Then there exist open βA-neighborhoods U and V of x and y, respectively, such that

$$\operatorname{cl}_{\beta A} U \cap \operatorname{cl}_{\beta A} V = (\operatorname{cl}_{\beta A} U) \cap H = (\operatorname{cl}_{\beta A} V) \cap F = \emptyset.$$

Note also that $U \cap A$ and $V \cap A$ are both unbounded. We now choose by induction three infinite sequences of distinct points of A to enable us to construct the previously mentioned mapping. Choose p_1 in $U \cap A$ and $q_1 > p_1$ with q_1 in $V \cap A$. Because V is open, it is possible to choose

$s_1 > q_1$ so that the interval (q_1, s_1) is contained in V. Now assume that $p_k, q_k,$ and s_k have been choosen for all $k < n$ such that for each k.

    (1) $p_k \in U$,

    (2) The interval (q_k, s_k) is contained in V,

    (3) $p_k < q_k < s_k$, and if $k < n-1$, then $s_k < p_{k+1}$.

Since $U \cap A$ is unbounded, we can choose p_n in $U \cap A$ such that $p_n > s_{n-1}$. Because $V \cap A$ is unbounded, we can choose q_n in $V \cap A$ such that $q_n > p_n$. Finally, because $V \cap A$ is open, we can choose $s_n > q_n$ such that (q_n, s_n) is contained in V, and this completes the induction.

Each of the sequences is unbounded, since otherwise they have a common supremum which must lie in $\mathrm{cl}_{\beta A} U \cap \mathrm{cl}_{\beta A} V$, which is impossible. Now define $f: A \to I$ as follows:

$$(1) \qquad f(r) = \begin{cases} 0 & \text{if } r \leqslant q_1 \\ 0 & \text{if } r \in [s_{k-1}, q_k] \quad \text{for } k \text{ odd} \\ 1 & \text{if } r \in [s_{k-1}, q_k] \quad \text{for } k \text{ even} \end{cases}$$

    (2) f is linear on the intervals $[q_k, s_k]$.

A portion of the mapping f is illustrated below:

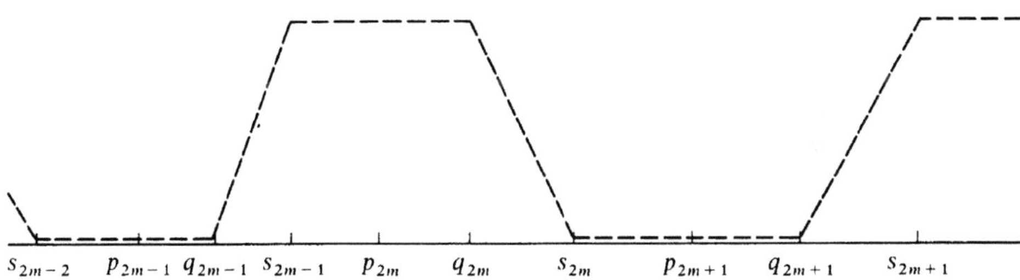

Consider the extension $\beta(f)$ of f. We will show that $\beta(f)$ takes only the values 0 and 1 on F. The zero-set $\mathbf{Z}(\beta(f))$ is a closed set containing the subsequence $\{p_{2k+1}\}$. Thus, $\mathbf{Z}(\beta(f))$ meets A^*. Further, any limit point of $\{p_{2k+1}\}$ is in $\mathrm{cl}_{\beta A} U$ and therefore is not in H. Hence, $\mathbf{Z}(\beta(f))$ meets F. By a similar argument, $\beta(f)^{\leftarrow}(1)$ also meets F. Now suppose that a is a point of A^* such that $\beta(f)(a)$ belongs to $(0, 1)$. Then a is a limit point of $f^{\leftarrow}(0, 1)$ and we also have that

$$f^{\leftarrow}(0, 1) = \bigcup \{(q_k, s_k) : k = 1, 2, \ldots\} \subset V$$

so that a belongs to $\mathrm{cl}_{\beta A} V$. But then a is not in F so that $\beta(f)[F] = \{0, 1\}$, and F is not connected. $\quad \square$

9.14. We will consider several properties of A^* and its subcontinua. A continuum K is said to be *irreducible* about a subset S of K if no proper subcontinuum of K contains S. In the event that $S = \{x, y\}$, the continuum is said to be *irreducible between x and y*. We see easily that the only pair of points between which the continuum I is irreducible is the pair $\{0, 1\}$. The circle S^1 is not irreducible between any pair of points. In fact, S^1 fails to be irreducible about any subset which is not dense.

To obtain a less trivial example of irreducibility, we return to βA.

Proposition. *The continuum βA is irreducible between 1 and any point of A^*.*

Proof. Let p be a point of A^*. We will show that any proper closed subset F of βA which includes both 1 and p fails to be connected and that therefore the only continuum containing both points is βA itself. Since A is dense in βA, there exists a point a in A such that a lies outside of F. Then the set

$$U = [1, a) \cap F = [1, a] \cap F$$

is clopen in F and non-empty so that F fails to be connected.   □

9.15. If a continuum C maps into a continuum K in such a way that the image of C contains a set about which K is irreducible, then the map must be onto K since a continuous image of a continuum is a continuum. Bellamy used this fact and the irreducibility of βA between 1 and a point of A^* to establish the following

Theorem. *Any non-degenerate subcontinuum of A^* can be mapped onto βA and consequently has cardinality 2^c.*

Proof. The second statement follows immediately from the first and the observation that the cardinality of βA is 2^c.

Let K be any non-degenerate subcontinuum of A^* and choose distinct points x and y of K. Let U and V be open neighborhoods in βA of x and y, respectively, such that $\operatorname{cl} U \cap \operatorname{cl} V = \emptyset$ and $\inf(U \cap A) < \inf(V \cap A)$. By modifying the procedure used in the proof of Theorem 9.13, obtain sequences $\{p_i\}$, $\{q_i\}$, $\{r_i\}$, and $\{s_i\}$ in A such that

$$p_i < q_i < r_i < s_i < p_{i+1}$$

for each i and

$$U \cap A \subset \bigcup \{[p_i, q_i] : i < \omega_0\}$$
$$V \cap A \subset \bigcup \{[r_i, s_i] : i < \omega_0\} \, .$$

Let η be the inclusion map of A into βA and define f from A into A by:

$$f(r)=\begin{cases} 1 & \text{if } r\leqslant q_1 \\ 1 & \text{if } r\in[p_i,q_i] \quad \text{for some } i \\ i & \text{if } r\in[r_i,s_i] \quad \text{for some } i, \end{cases}$$

and extend f linearly to the intervals $[q_i,r_i]$ and $[s_i,p_{i+1}]$ for each i. Now consider the map $g=\beta(\eta\circ f):\beta A\to\beta A$. The proof is completed by showing that g maps x to 1 and y into A^*. Then the image of K is a continuum containing 1 and meeting A^* and must therefore be all of βA since βA is irreducible between 1 and any point of A^*. Since x belongs to $\mathrm{cl}_{\beta A}(U\cap A)$, $g(x)$ is in $g[\mathrm{cl}_{\beta A}(U\cap A)]$. However,

$$g[\mathrm{cl}_{\beta A}(U\cap A)]=\mathrm{cl}_{\beta A}(g[U\cap A])$$
$$=\mathrm{cl}_{\beta A}((\eta\circ f)[U\cap A])$$
$$=\{1\}$$

so that $g(x)=1$. To show that $g(y)$ is in A^*, note that y is in $\mathrm{cl}_{\beta A}(V\cap[r_i,\infty))$ for all i so that $g(y)$ belongs to $g[\mathrm{cl}_{\beta A}(V\cap[r_i,\infty))]$ for all i. Further,

$$g[\mathrm{cl}_{\beta A}(V\cap[r_i,\infty))]=\mathrm{cl}_{\beta A}(g[V\cap[r_i,\infty)])$$
$$\subseteq\mathrm{cl}(g[\bigcup\{[r_j,s_j]:i\leqslant j<\omega_0\}])$$
$$=\mathrm{cl}(\{n\in\mathbb{N}:n\geqslant i\}).$$

Thus, $g(y)$ is in A^*, and $g[K]$ is all of βA. $\square$

9.16. In addition to the preceding result, Bellamy also shows in his 1971 paper that any non-degenerate subcontinuum will map onto any metric continuum. This fact is an immediate consequence of the following four lemmas. The first of these assures the existence of a subcontinuum which is irreducible about a given proper subset.

Lemma. *Any proper subset of a continuum is contained in a subcontinuum which is irreducible about the subset.*

Proof. Let S be a proper subset of a continuum K. Let $\mathcal{K}$ be the set of all subcontinua containing S and define a partial order on $\mathcal{K}$ by $K_1\geqslant K_2$ if K_1 is contained in K_2. It follows from Proposition 9.12 that each chain in $\mathcal{K}$ has the intersection of its members as an upper bound. Thus, $\mathcal{K}$ contains a maximal element and any maximal element of $\mathcal{K}$ must be irreducible about S. $\square$

9.17. The next three lemmas are due to Bellamy and describe a mapping of any non-degenerate subcontinuum of A^* onto any metric continuum as the composition of two mappings. The first step is to obtain a metric compactification of A having as its growth the given metric continuum. In order to obtain this compactification we will use the fact that any metric continuum is separable and therefore can be embedded into the Hilbert cube $I^{\aleph_0}$ (Exercise 6 H). In addition, we will use the fact that $I^{\aleph_0}$ is *arcwise connected*, i. e. that for any two points of $I^{\aleph_0}$ there is a homeomorphism of I into $I^{\aleph_0}$ that sends 0 to one point and 1 to the other. From this fact together with the observation that every point of $I^{\aleph_0}$ has a neighborhood base consisting of copies of $I^{\aleph_0}$, we see that $I^{\aleph_0}$ is locally arcwise connected.

Lemma. *If K is a metric continuum, then there exists a metric compactification S of A such that $S \backslash A$ is a copy of K.*

Proof. As was noted above, K can be embedded into the Hilbert cube so we will assume that K is a subspace of $I^{\aleph_0}$. We will embed A into $I^{\aleph_0} \times I$ by means of a mapping h in such a way that K is the boundary of $h[A]$. Let d be any metric on $I^{\aleph_0}$. For each point x belonging to K and each positive integer n, let $U(x,n)$ be an open, connected neighborhood of x contained in the ball $\{y \in I^{\aleph_0} : d(y,x) < 1/n\}$ and having compact closure. Put $U_n = \bigcup \{U(x,n) : x \in K\}$. Because K is a continuum, any two points of U_n can be connected by traversing only finitely many arcs. Thus, each U_n is arcwise connected. Let $\{a_n : n \geqslant 1\}$ be a countable dense subset of K and let $f : A \to I^{\aleph_0}$ be a mapping such that $f|[n, n+1]$ is an arc in U_n connecting a_n and a_{n+1}.

Now define $h : A \to I^{\aleph_0} \times I$ by $h(t) = (f(t), 1/t)$. Thus, h is one-to-one and $h[A] = \bigcup \{h[n, n+1] : n \geqslant 1\}$ is a union of arcs so that $h[A]$ is homeomorphic to A. Because $f[[n, n+1]]$ is contained in U_n for each n, every unbounded set of $f[A]$ has its limit points contained in K. Further, because $f[A]$ contains a dense subset of K, $\mathrm{cl}(f[A]) = f[A] \cup K$. Thus, the subset $h[A] \cup (K \times \{0\})$ is the required compactification of A. $\quad\square$

9.18. If a continuum K is irreducible between two points a and b, then we noted earlier that a mapping of another continuum C into K will be onto if the image of C contains a and b. The next lemma indicates that we can use this argument again to map a subcontinuum of A^* onto any metric continuum.

Lemma. *Any metric continuum is a retract of a metric continuum which is irreducible between two of its points.*

Proof. Let K be a metric continuum. Since a compact metric space is separable [D, p. 187], there exists a countable dense subset $\{a_i\}$ of K.

Using Lemma 9.16, let L_i be a subcontinuum of K which is irreducible between a_i and a_{i+1}. Consider the subspace C of the product $K \times I$ defined as follows:

$$C = (K \times \{0\}) \cup (\bigcup\{L_i \times \{1/(i+1)\} : i \geqslant 1\}) \cup (\bigcup\{\{a_i\} \times [1/(i+1), 1/i] : i \geqslant 1\})$$

C is a continuum since $\{(x,t) \in C : t > 0\}$ is clearly connected and has C as·its closure. C is illustrated below for the case in which K is an interval:

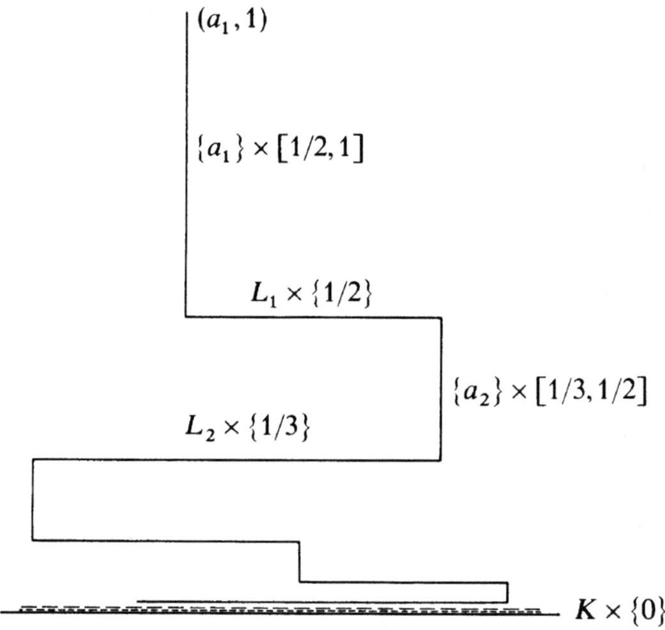

Let x be any point in K. We will show that C is irreducible between the points $(a_1, 1)$ and $(x, 0)$. If P is any subcontinuum containing these two points, then the projection of P onto I contains $\{0, 1\}$ and therefore is all of I. Thus, P must contain $\{a_i\} \times (1/(i+1), 1/i)$ for all i. Then because P is closed, it must therefore contain the closed interval $\{a_i\} \times [1/(i+1), 1/i]$ for all i. The set $P_i = \{(x, 1/i) : (x, 1/i) \in P\}$ is connected for each i: If not, we can write $P_i = U \cup V$ with both U and V open in P_i and $(a_{i-1}, 1/i)$ in U and $(a_i, 1/i)$ in V. Then the sets $U \bigcup \{(x, t) \in P : t > 1/i\}$ and $V \bigcup \{(x, t) \in P : t < 1/i\}$ are disjoint open sets in P and their union is all of P, which is impossible because P is connected.

Thus, P_i is a subcontinuum of $L_{i-1} \times \{1/i\}$ and therefore is all of $L_{i-1} \times \{1/i\}$ since L_{i-1} is irreducible between a_i and a_{i-1}. Hence, P contains all points of C whose second coordinate is positive and therefore is a closed dense set which must be all of C.

Finally, if we identify K with $K \times \{0\}$ and consider the projection of C onto K, K is a retract of C.   ☐

9.19. From the previous lemma, in order to show that a non-degenerate subcontinuum of A^* maps onto any metric continuum, we need only show that it maps onto a metric continuum which is irreducible between two points. Our proof of this fact will be an adaptation of the proof of Theorem 9.15.

Lemma. *Any non-degenerate subcontinuum of A^* maps onto any metric continuum which is irreducible between two points.*

Proof. Let K be a non-degenerate subcontinuum of A^*, let C be any metric continuum which is irreducible between two points a and b, and let S be a metric compactification of A such that $S \setminus A$ is a copy of C. Let x and y be distinct points of K. We will use a procedure similar to that of Theorem 9.15 to construct a mapping g of βA onto C such that $g(x) = a$ and $g(y) = b$.

Let $\{a_i\}$ and $\{b_i\}$ be sequences in A such that when A is considered as a subspace of S,

$$\lim a_i = a \, ,$$

$$\lim b_i = b \, , \quad \text{and}$$

$$a_i < b_i < a_{i+1} \quad \text{for all } i < \omega_0 .$$

Let U and V be neighborhoods of x and y, respectively, such that $\mathrm{cl}_{\beta A} U \cap \mathrm{cl}_{\beta A} V = \emptyset$. Choose sequences $\{p_i\}$, $\{q_i\}$, $\{r_i\}$, and $\{s_i\}$ as in the proof of Theorem 9.15 and define a mapping $f : A \to A$ so that

$$f(t) = \begin{cases} a_i & \text{for } t \in [p_i, q_i] \\ b_i & \text{for } t \in [r_i, s_i] \end{cases}$$

and extend f linearly elsewhere. Let g be the extension $g = \beta(e \circ f) : \beta A \to S$ where e is the inclusion $e : A \to S$. By arguments similar to those of Theorem 9.15, $g(x) = a$ and $g(y) = b$. Therefore, since x and y belong to K and C is irreducible between a and b, $g[K] = C$.   ☐

9.20. The theorem is now obvious by composing the two mappings just constructed.

Theorem (Bellamy). *Any non-degenerate subcontinuum of A^* maps onto any metric continuum.*

9.21. A subset C of a continuum K is called a *composant* if there is a point p in K for which C is the set of points x in K such that p and x are contained in some proper subcontinuum of K. Stated another way,

C is the set of points x of K such that K is not irreducible between p and x. The point p is called the *defining point* of C and if we wish to single out a particular defining point, we will let C_p denote the composant with defining point p.

Because any two points of the circle S^1 belong to a proper subcontinuum, S^1 itself is a composant and any point can be chosen as the defining point. In the case of the interval I, the composant determined by an endpoint is the half-open interval obtained by removing the other endpoint. However, the composant defined by an interior point is the whole interval. The example of I shows that distinct composants need not be disjoint. However, the following result, which is adapted from the Hocking and Young text, *Topology*, shows that this cannot be the case in an indecomposable continuum.

Proposition. *Distinct composants of an indecomposable continuum are disjoint.*

Proof. Let C_1 and C_2 be distinct composants of an indecomposable continuum K with defining points p_1 and p_2, respectively. If x is a point belonging to both C_1 and C_2, then there is a proper subcontinuum K_1 containing x and p_1 and another K_2 containing x and p_2. Because K is indecomposable, $K_1 \cup K_2$ is a proper subset of K and is a continuum since K_1 and K_2 have the point x in common. Similarly, if y is any point of C_2, there is a proper subcontinuum K_3 containing p_2 and y so that $(K_1 \cup K_2) \cup K_3$ a proper subcontinuum containing both p_1 and y. Thus, y belongs to C_1 and C_2 is therefore a subset of C_1. By a symmetric argument, C_1 is a subset of C_2. Thus, if two composants of K are distinct, they are disjoint. □

One immediate consequence of the proposition is that any point of a composant of an indecomposable continuum can be chosen as the defining point of the composant.

9.22. The composants of the indecomposable continuum A^* are considered in M. E. Rudin's 1970 paper from which the next two results are taken. In the first of these she uses the fact that $\mathbb{N}^*$ contains 2^c types of points to describe a subset G of 2^c points of $\mathbb{N}^*$ such that when $\beta\mathbb{N}$ is considered as a subspace of βA, each point of G belongs to a different composant of A^*. The Continuum Hypothesis is assumed in both results.

Lemma [CH]. *There is a subset G of $\mathbb{N}^*$ having cardinality 2^c such that if p and q are distinct points of G, f maps $\mathbb{N}$ onto itself, and $\beta(f)(p)$ is in $\mathbb{N}^*$, then $\beta(f)(p) \neq \beta(f)(q)$.*

Proof. Using the Continuum Hypothesis, let $\{f_\alpha : \alpha < \omega_1\}$ be the set of all mappings of $\mathbb{N}$ onto itself. Assume that in the indexing we have chosen f_0 to be the identity. For each ordinal α, let $C(\alpha, 2)$ denote the mappings of α, taken with the interval topology, into the two point discrete space $2 = \{0, 1\}$. For each α, $0 < \alpha < \omega_1$, and each g in $C(\alpha, 2)$ we will choose an infinite subset M_g of $\mathbb{N}$ in the following way:

Suppose that $C(1, 2) = \{h, k\}$ where h and k are defined by $h(0) = 0$ and $k(0) = 1$. Choose M_h to be the even integers and M_k to be the odd integers. Now suppose that the construction has been completed for all $\beta < \alpha$ for some $\alpha < \omega_1$. If $\alpha = \gamma + 1$ is a non-limit ordinal and g belongs to $C(\alpha, 2)$, then the set $M_{g|\gamma}$ has been chosen. There is an infinite subset M of $M_{g|\gamma}$ such that either $f_\alpha[M]$ is finite or $f_\alpha | M$ is one-to-one. Let M_g consist of alternate terms of M, starting with the first if $g(\gamma) = 1$ and the second if $g(\gamma) = 0$. In the limit ordinal case, assume that for all $\beta < \alpha$, the sets $M_{g|\beta}$ have been chosen so that for $\gamma < \beta$, $M_{g|\beta} \setminus M_{g|\gamma}$ is finite. This is clearly the case for $\alpha = \omega_0$ since in that case, $M_{g|\beta}$ is a subset of $M_{g|\gamma}$ if $\gamma < \beta$. Now let $\{\alpha_n : n < \omega_0\}$ be a strictly increasing sequence such that $\sup \alpha_n = \alpha$. Choose x_0 in $M_{g|\alpha_0}$ and in general, choose

$$ x_n \in M_{g|\alpha_0} \cap \cdots \cap M_{g|\alpha_n} \setminus \{x_0, \ldots, x_{n-1}\} . $$

Let M_g be an infinite subset of $\{x_n : n < \omega_0\}$ such that either $f_\alpha[M_g]$ is finite or $f_\alpha | M_g$ is one-to-one. Now if $\beta < \alpha$, there exists n such that $\beta \leqslant \alpha_n$. Since $M_{g|\alpha_n} \setminus M_{g|\beta}$ is finite and

$$ M_g \setminus M_{g|\beta} \subset (M_{g|\alpha_n} \setminus M_{g|\beta}) \cup \{x_0, \ldots, x_{n-1}\} , $$

$M_g \setminus M_{g|\beta}$ is finite. This completes the induction.

Note that for each g in $C(\omega_1, 2)$, the sets $\{M_{g|\alpha} : 0 < \alpha < \omega_1\}$ have been chosen in such a way that they have the finite intersection property. Therefore, we can choose a point p_g in $\mathbb{N}^*$ such that the corresponding ultrafilter on $\mathbb{N}$ contains $\{M_{g|\alpha} : 0 < \alpha < \omega_1\}$. The points $\{p_g : g \in C(\omega_1, 2)\}$ are distinct since for distinct g and h in $C(\omega_1, 2)$, there is a least α such that $g(\alpha) \neq h(\alpha)$, and since such an α must be non-limit, $M_{g|\alpha+1} \cap M_{h|\alpha+1} = \emptyset$. Now suppose that g is in $C(\omega_1, 2)$ and that f_α is a mapping of $\mathbb{N}$ onto itself. Then either $f_\alpha[M_{g|\alpha+1}]$ is finite or $f_\alpha | M_{g|\alpha+1}$ is one-to-one. In the first case, $\beta(f_\alpha)(p_g)$ is in $\mathbb{N}$, not $\mathbb{N}^*$, and in the second case, there is a permutation σ of $\mathbb{N}$ such that $\beta(\sigma)(p_g) = \beta(f_\alpha)(p_g)$. In this latter case, we can use the fact that there are only $\mathfrak{c}$ permutations of $\mathbb{N}$ or in the terminology of Chapter 3, only $\mathfrak{c}$ ultrafilters of a given type, to choose the required subset G from the set $\{p_g : g \in C(\omega_1, 2)\}$. Using the Continuum Hypothesis, $C(\omega_1, 2)$ has cardinality $2^{\mathfrak{c}}$ and we can partition $\{p_g : g \in C(\omega_1, 2)\}$ into $2^{\mathfrak{c}}$ subsets consisting of points having the same type. Obtain G by choosing one point from each element of the partition. Thus, if p_g and p_h are in G and f_α maps $\mathbb{N}$ onto $\mathbb{N}$, then $\beta(f_\alpha)(p_g) \neq \beta(f_\alpha)(p_h)$ even if both are in $\mathbb{N}^*$, since there are permutations σ and γ of $\mathbb{N}$ such that $\beta(f_\alpha)(p_g) = \beta(\sigma)(p_g)$ and $\beta(f_\alpha)(p_h) = \beta(\gamma)(p_h)$ and $\beta(\sigma)(p_g)$ cannot equal $\beta(\gamma)(p_h)$ since p_g and p_h are of different types. $\square$

9.23. Theorem [CH] (M. E. Rudin). *A^* has exactly 2^c composants.*

Proof. Because distinct composants of A^* are disjoint and the cardinality of A^* is 2^c, A^* can have no more than 2^c composants. To obtain a lower bound for the number of composants, we will show that when $\mathbb{N}$ and $\beta\mathbb{N}$ are considered as subspaces of βA, then any distinct pair of points belonging to the set G described in the previous lemma must belong to different composants. For any point p of $\mathbb{N}^*$, let A^p denote the corresponding free ultrafilter on $\mathbb{N}$.

Suppose that K is a proper subcontinuum of A^* containing distinct points p and q of G. Then there is a point r belonging to $A^* \setminus K$. Choose open subsets U and V of βA such that

$$r \in U \subset \mathrm{cl}\, U \subset V \subset \mathrm{cl}\, V \subset \beta A \setminus K .$$

Since p and q both belong to $\beta A \setminus \mathrm{cl}\, V$, there exist M in A^p and L in A^q such that $(M \cup L) \cap V = \emptyset$. Let $I_1, I_2, \ldots, I_n, \ldots$ denote the components of $A \setminus \mathrm{cl}\, U$ which meet $\mathbb{N} \setminus V$. Since A is locally connected, each I_n is open. Evidently, both $A \cap U$ and $A \setminus U$ are unbounded in A so that there must be infinitely many such components. Choose an f mapping $\mathbb{N}$ onto itself such that $f(n) = i$ if n belongs to I_i. Now by the properties of G, there exist M' in A^p and L' in A^q such that $f[M'] \cap f[L'] = \emptyset$. Let $X = \bigcup\{I_i : i \in f[M \cap M']\}$ and $Y = \bigcup\{I_i : i \in f[L \cap L']\}$. The sets $X \setminus V$ and $A \setminus (X \cup V)$ are closed and disjoint in A so that their closures in βA are also disjoint. Since the subcontinuum K misses V, we have that

$$K \subset \mathrm{cl}(X \setminus V) \cup \mathrm{cl}(A \setminus (X \cup V)) .$$

If we can demonstrate that p belongs to $\mathrm{cl}(X \setminus V)$ while q belongs to $\mathrm{cl}(A \setminus (X \cup V))$, we will have shown that K is not connected, and the resulting contradiction establishes that p and q cannot lie in the same composant of A^*. From the definition of f,

$$p \in \mathrm{cl}(M \cap M') \subset \mathrm{cl}(X \setminus V), \quad \text{and}$$

$$q \in \mathrm{cl}(L \cap L') \subset \mathrm{cl}(Y \setminus V) \subset \mathrm{cl}(A \setminus (X \cup V)) .$$

Thus, the proof is complete and A^* has exactly 2^c composants. $\square$

Continua as Growths

9.24. Theorem 9.13 has provided us with one example, the ray $[1, \infty)$, of a space whose growth is an indecomposable continuum. In his 1972 paper, R. F. Dickman describes a class of spaces which share this property. The remainder of the chapter will be devoted to Dickman's results and will conclude with a proof that the growth of $\mathbb{R}^n$ is a decomposable continuum for $n \geqslant 2$. The term *generalized continuum* will refer to a locally compact connected metric space. A subset of a space will be said to be *relatively compact* if its closure is compact. A connected space X will be said to have the *strong complementation property* if for every non-relatively compact connected open subset U of X, $X \setminus U$ is compact. Dickman showed that among the locally connected generalized continua, the spaces having indecomposable continua as growths are precisely those having the strong complementation property.

9.25. The first of the preliminary results which we will require to prove the theorem stated above is the following lemma which will be used to recognize decomposable continua, i.e. those which are the union of two proper subcontinua.

Lemma. *A continuum K is decomposable if it contains a proper subcontinuum C which satisfies either of the following conditions:*
   (a) *C has non-empty interior and $K \setminus C$ is connected,*
   (b) *$K \setminus C$ is not connected.*

Proof. In case (a), $\mathrm{cl}(K \setminus C)$ is connected and is a proper subcontinuum of K because it excludes the interior of C. Hence, $K = C \cup \mathrm{cl}(K \setminus C)$ is a decomposition of K.

We will first consider (b) in the case where C is a singleton, i.e. $C = \{p\}$. Since $K \setminus \{p\}$ is not connected, it can be expressed as the union $K \setminus \{p\} = U \cup V$ where U and V are disjoint, non-empty open sets. We will show that $U \cup \{p\}$ and $V \cup \{p\}$ are both subcontinuum so that K is decomposable. By defining a function f by

$$f(x) = \begin{cases} x & \text{for } x \in U \cup \{p\} \\ p & \text{for } x \in V \end{cases}$$

we obtain a mapping of the connected compact space K onto $U \cup \{p\}$ so that $U \cup \{p\}$ is a subcontinuum. Similarly, $V \cup \{p\}$ is a subcontinuum.

If C is not a singleton, then we can still express $K \setminus C$ as the disjoint union of open sets U and V. Let π be the quotient mapping on K that shrinks C to a point $\{p\}$. Then $\pi[K] = U \cup \{p\} \cup V$ is a continuum, and from above both $U \cup \{p\}$ and $V \cup \{p\}$ are continua. Since V is open, all accumulation points of U must belong to $U \cup C$ so that $U \cup C$ is compact. If $U \cup C$ is not connected, then it contains a clopen subset W. Since C is connected, W must miss C. But then $\pi[W]$ is a clopen subset of $U \cup \{p\}$, which is impossible. Hence, $U \cup C$ and $V \cup C$ are both subcontinua of K, and K is decomposable. $\square$

9.26. A *ray* in a space X is a copy of $[1, \infty)$. The key property of a locally connected generalized continuum which we will require is that such a space must contain a closed ray. The next three results will be devoted to showing this fact. The first of these is a modification of a lemma which appeared in the 1963 paper of E. Duda.

Lemma. *A compact subset K of a locally connected generalized continuum X is contained in a compact subset C such that $X \backslash C$ has only finitely many components.*

Proof. For each point x in K, let V_x be an open neighborhood of x such that cl V_x is compact. Choose a finite subcover of the resulting cover of K and let V be the union of the members of this subcover. Then V is relatively compact. Since X is metric and therefore normal, we can choose W to be an open neighborhood of K such that

$$K \subset W \subset \operatorname{cl} W \subset V \subset \operatorname{cl} V.$$

Put $\partial W = \operatorname{cl} W \backslash W$ and consider the components of $X \backslash \partial W$. Each component is open in X since X is locally connected. Each component is closed in $X \backslash \partial W$ because it is a component of this set. However, because X is connected, no component is closed in X. Thus, each component has an accumulation point in ∂W and no component can be contained in $X \backslash \operatorname{cl} V$. Hence, every component is either contained in V or meets $\partial V = \operatorname{cl} V \backslash V$. Let C be the union of the components which miss ∂V. Since no component can meet both W and $X \backslash \partial W$, W is contained in C. Since ∂V is compact, only finitely many of the components, say $\{U_1, \ldots, U_n\}$, can meet ∂V. Thus, the set $C = X \backslash \bigcup \{U_i : 1 \leqslant i \leqslant n\}$ is closed and contained in $\operatorname{cl} V$. Hence, C is compact. Since the U_i's are the only components of $X \backslash C$, C is the required compact set. $\square$

9.27. We now apply this result to obtain the following proposition which appears in Dickman's 1967 paper.

Proposition. *The one point compactification of a non-compact locally connected generalized continuum is locally connected.*

Proof. Let X be such a generalized continuum and let ∞ denote the added point in the one-point compactification αX of X. Since X is locally connected, to show that αX is locally connected it is sufficient to show that any neighborhood of ∞ contains a connected neighborhood of ∞. Thus, for K any compact subset of X, we must find a connected neighborhood of ∞ contained in $\alpha X \backslash K$. Using the lemma, let C be a compact subspace of X containing K and such that $X \backslash C$ has only finitely many components, $\{U_i : 1 \leqslant i \leqslant n\}$. Without loss of generality, we can assume that there exists an integer p, $1 \leqslant p < n$ such that each U_i is relatively compact for $1 \leqslant i \leqslant p$ and fails to be relatively compact for $p+1 \leqslant i \leqslant n$. As in the proof of the lemma, all of the accumulation points in X of each U_i lie in C. Therefore, $C \cup U_1 \cup \cdots \cup U_p$ is compact. Since each U_i is not relatively compact for $i > p$, ∞ belongs to $\mathrm{cl}_{\alpha X} U_i$ so that each $U_i \cup \{\infty\}$ is connected. Thus, $\bigcup \{U_i : p < i \leqslant n\} \cup \{\infty\}$ is a connected set whose complement is compact, and thus is a connected neighborhood of ∞.　◻

9.28. Lemma. *Any non-compact locally connected generalized continuum contains a closed ray.*

Proof. Let X be such a space and let ∞ denote the added point in the one point compactification αX of X. If we can show that αX is arcwise connected, then if x is any point of X, there is an arc I connecting x to ∞ and $I \backslash \{\infty\}$ is a ray in X. To show that αX is arcwise connected, we will use the theorem that a locally connected, connected, compact metric space is arcwise connected. For a proof of this result, see p. 219 of Willard's text, *General Topology*. Thus, to complete the proof we need to show that αX satisfies the hypotheses of this theorem. From the previous lemma, αX is locally connected. The compactification is clearly compact and is connected because X is connected. From [D, p. 241], the paracompact locally compact space X is the topological sum of a family of σ-compact spaces. However, because X is connected, there can be only one summand and X is therefore σ-compact. Thus, X is Lindelöf, and since it is also a metric space, it is second countable. From Exercise 6H, to show that αX is metrizable we need only show that it is second countable. Because X is σ-compact, it follows that αX has a countable neighborhood base at ∞, and αX is therefore second countable since we have already seen that X is second countable. Hence, all of the required conditions are satisfied and the proof is complete.　◻

9.29. The following lemma rephrases the strong complementation property in terms of closed rays. It says essentially that the portion of a locally connected generalized continuum which determines its growth in βX is a ray.

Lemma. *If X is a non-compact, locally connected generalized continuum,* *then X has the strong complementation property if and only if the complement of every closed ray contained in X is relatively compact.*

Proof. Assume that X has the strong complementation property and that R is a closed ray in X. Because X is a metric space, to show that $X \backslash R$ is relatively compact it is sufficient to show that every closed subset of $X \backslash R$ is compact. Let F be a closed subset of $X \backslash R$. Since X is locally connected, we can choose a family $\{U_n : n < \omega_0\}$ of open connected sets such that for all n, $U_n \cap F = \emptyset$, $U_n \cap U_{n+1} \neq \emptyset$, and $U_n \cap R \neq \emptyset$, and R is contained in $U = \bigcup U_n$. Thus, U is a connected open subset containing R and missing F. Since U contains the closed ray R, U is not relatively compact. Therefore, the strong complementation property implies that $X \backslash U$ is compact and thus that F is compact.

Now assume that the complement of any closed ray in X is relatively compact. Let U be any non-relatively compact connected open subset of X. To demonstrate that X has the strong complementation property, we must show that $X \backslash U$ is compact. By Lemma 9.28, X contains a closed ray, so that there is a homeomorphism h of $[1, \infty)$ onto a closed subset S of X. By assumption, $\mathrm{cl}(X \backslash S)$ is compact so that the restriction $h^{\leftarrow} | S \cap \mathrm{cl}(X \backslash S)$ attains its supremum. Thus, there exists a point t in $[0, \infty)$ such that $h(t)$ is the last point on S that lies in $\mathrm{cl}(X \backslash S)$. Thus, $h(t, \infty)$ is an open subset of X. Observe that for any s in $[t, \infty)$, $X \backslash h[s, \infty)$ is relatively compact. Thus, the proof would be complete if we could show an s in $[t, \infty)$ exists such that $X \backslash U$ is contained in $X \backslash h[s, \infty)$. Hence, we will show that

(*) There exists an s in $[t, \infty)$ such that $h[s, \infty)$ is a subset of U: Begin by noting that for every s in $[t, \infty)$, $h[s, \infty) \cap U \neq \emptyset$, since otherwise U would be contained in $X \backslash h[s, \infty)$ and thus would be relatively compact. If for every s in $[t, \infty)$, $h[s, \infty)$ is not a subset of U, then there exists r in $[t, \infty)$ such that $h(r)$ is not in U and both of the sets $h[t, r)$ and $h(r, \infty)$ meet U. Put $U_1 = U \cap (X \backslash h[r, \infty))$ and $U_2 = U \cap (h(r, \infty))$. Both sets are open in X since $r > t$, and are non-empty. But then we have $U = U_1 \cup U_2$ which is impossible since U is connected and U_1 and U_2 are disjoint. Hence, there is some s such that $h[s, \infty)$ lies within U. $\square$

9.30. The previous lemma is now employed to prove our main theorem. Note that the first part of the proof shows that the class of spaces having indecomposable continua as growths are those which consist of a single ray "sticking out" of a compact space. Thus, only the ray affects the growth.

Theorem (Dickman). *If X is a non-compact locally connected generalized continuum, then X^* is an indecomposable continuum if and only if X has the strong complementation property.*

Proof. Suppose that X has the strong complementation property. By Lemma 9.28, there exists a closed ray R in X. Since R is closed and therefore C^*-embedded in the normal space X, $\mathrm{cl}_{\beta X} R$ is a copy of βR. Thus, from Theorem 9.13, $R^* = \mathrm{cl}_{\beta X} R \setminus R$ is an indecomposable continuum contained in X^*. By the previous lemma, $\mathrm{cl}_X(X \setminus R)$ is compact and hence $\mathrm{cl}_{\beta X}(X \setminus R)$ misses X^*. Thus, $X^* = R^*$ and is therefore an indecomposable continuum.

Now assume that X^* is an indecomposable continuum but that X does not have the strong complementation property. By Lemmas 9.28 and 9.29, there exists a closed ray in X such that $\mathrm{cl}_X(X \setminus R)$ is not compact. Then there exists a non-compact closed subset F contained in $X \setminus R$. As in the proof of Lemma 9.29, we can obtain an open connected subset U of X whose closure misses F. As in Lemma 9.28, X is σ-compact, and therefore can be written as a union $X = \bigcup \{K_i : i < \omega_0\}$ where each K_i is compact and K_i is a subset of int K_{i+1} for all i. Let $\{R_i : i < \omega_0\}$ be a sequence of subrays of R such that R_i is contained in $X \setminus K_i$. For each i, let G_i be the component of $U \cap (X \setminus K_i)$ that contains R_i and put $H_i = (X \setminus G_i) \cap (X \setminus K_i)$. Set $G = \bigcap \{\mathrm{cl}_{\beta X} G_i : i < \omega_0\}$ and $H = \bigcap \{\mathrm{cl}_{\beta X} H_i : i < \omega_0\}$. Observe that $X^* = H \cup G$. Further, G is a non-empty continuum by Proposition 9.12. We will show that G is not all of X^* and that G has an interior point so that we can apply Lemma 9.25 to show that X^* cannot be an indecomposable continuum.

Since F is not compact, there exists a point p in $\mathrm{cl}_{\beta X} F \cap X^*$. Since X is normal, it follows from Corollary 1.15 that $\mathrm{cl}_{\beta X} F \cap \mathrm{cl}_{\beta X}(\mathrm{cl}_X U) = \emptyset$. Since p is not in $\mathrm{cl}_{\beta X} U$ which contains G, G is a proper subcontinuum of X^*. Similarly, there exists a point r in $\mathrm{cl}_{\beta X} R_1 \cap X^*$ and r cannot belong to $\mathrm{cl}_{\beta X} H_1$. Since H is a subset of $\mathrm{cl}_{\beta X} H_1$, $X^* \setminus \mathrm{cl}_{\beta X} H_1$ is an X^*-neighborhood of r contained within G. Thus, G is a proper subcontinuum of X^* and has non-empty interior. By Lemma 9.25, X^* cannot be an indecomposable continuum.. $\Box$

9.31. In addition to characterizing the non-compact locally connected generalized continua which have indecomposable continua as growths, the proof of the theorem also shows that all such growths are homeomorphic.

Corollary. *If X is a non-compact locally connected generalized continuum and X^* is an indecomposable continuum, then X^* is homeomorphic to A^*, where $A = [1, \infty)$.*

9.32. Theorem 9.30 describes only those locally connected generalized continua which give rise to indecomposable continua as growths. Dickman also provides a condition to characterize those whose growths are merely continua. A connected space is said to have the *complementa-*

tion property if the complement of every compact subset has at most one non-relatively compact component. Then we have the following

Theorem. *If X is a locally connected non-compact generalized continuum, then X^* is a continuum if and only if X has the complementation property.*

Proof. Assume that X has the complementation property. Let K be any compact subspace of X and using Lemma 9.26, let C be a compact subspace containing K such that $X\backslash C$ has only finitely many components. The complementation property thus implies that $X\backslash C$ has exactly one non-relatively compact component, V. Thus, $\mathrm{cl}_{\beta X} V$ is connected and contains X^*.

Now as in the proof of Lemma 9.28, X is σ-compact and so is the union of a sequence $\{K_i : i < \omega_0\}$ of compact subspaces such that K_i is contained in int K_{i+1}. For each K_i, obtain a V_i as we did the V for K above. The V_i are decreasing and must have empty intersection so that $X^* = \bigcap \{\mathrm{cl}_{\beta X} V_i : i < \omega_0\}$ and is therefore connected by Proposition 9.12.

Now assume that X^* is connected, but that X fails to have the complementation property. Then there exists a compact subspace K of X such that $X\backslash K$ has more than one non-relatively compact component. Again using Lemma 9.26, let C be a compact subspace containing K such that $X\backslash C$ has only finitely many components. Then every component of $X\backslash C$ is contained in a component of $X\backslash K$. For each non-relatively compact component U of $X\backslash K$, $U\backslash C$ is the union of finitely many components of $X\backslash C$, so that $X\backslash C$ also has more than one non-conditionally compact component. Let $\{U_1, \ldots, U_n\}$ be these components. Then $X\backslash \bigcup \{U_i : 1 \leqslant i \leqslant n\}$ is compact and X^* is contained in $\bigcup \{X^* \cap \mathrm{cl}_{\beta X} U_i : 1 \leqslant i \leqslant n\}$. But this shows that X^* is the union of a finite pairwise disjoint family of at least two non-empty closed subsets, contradicting the assumption that X^* is connected. $\square$

9.33. It follows immediately from the two preceding theorems that in locally connected non-compact generalized continua, the strong complementation property implies the complementation property. However, it is easy to see that the converse fails.

Examples. The growth $\mathbb{R}^*$ of $\mathbb{R}$ is easily seen from Theorem 9.13 to be the disjoint union of two indecomposable continua. Thus $\mathbb{R}$ has neither of the complementation properties. Now consider $\mathbb{R}^n$ for $n \geqslant 2$. Since any compact subspace of $\mathbb{R}^n$ is contained in a ball of finite radius and the complement of the ball is connected, $\mathbb{R}^n$ is easily seen to satisfy the complementation property. However, $\mathbb{R}^n$ contains two disjoint closed rays so that it cannot satisfy the strong complementation property. Thus, $(\mathbb{R}^n)^*$ *is a decomposable continuum for each* $n \geqslant 2$. This result was obtained by R. G. Woods in his thesis.

Exercises

9A. *Locally connected continua are decomposable*

1. No locally connected continuum can be indecomposable. [Lemma 9.25.]
2. The growth of $[1, \infty)$ is not locally connected.

9B. *Locally connected one-point compactifications*

If X is normal, locally compact, connected, and locally connected but not compact, then αX is locally connected.
[Modify Lemma 9.26 and Proposition 9.27.]

9C. *Non-comparable types with respect to $\leqslant$*

1. If p and q belong to $\mathbb{N}^*$ and there is a function g of $\mathbb{N}$ onto itself such that $\beta(g)(p) = q$, then there is a function f of $\mathbb{N}$ onto itself such that $\beta(f)(p) = \beta(f)(q)$.
2. Assume the Continuum Hypothesis. Then there exist 2^c types which are non-comparable under the partial order $\leqslant$ of Exercise 7C. [Lemma 9.22.]

9D. *Spaces having the complementation property*

Let X be a separable non-compact locally connected generalized continuum. Let S^1 be the unit circle in the complex plane and let $f : X \to S^1$ be a mapping whose fibers have compact boundaries.

1. αX is first countable. [X is σ-compact.]
2. If $\{x_i\}$ is any sequence in X which has no convergent subsequences, then $\{f(x_i)\}$ has only one limit point in S^1 if X has the complementation property. [Distinct limit points lie in different components of $S^1 \setminus \{a, b\}$ for some a and b. Consider the components of $X \setminus \partial(f^{\leftarrow}(\{a, b\}))$.]
3. The images under f of any two sequences having no convergent subsequences have the same limit point. Thus, f extends continuously to αX.
4. If R is a relatively compact open subset of X such that $X \setminus \mathrm{cl}\, R$ has only finitely many components, then any two of the components are completely separated. [First use Tietze's Theorem to separate their boundaries as subspaces of $\mathrm{cl}\, R$.]
5. If g completely separates two non-relatively compact components of $X \setminus \mathrm{cl}\, R$, then $h(x) = e^{ig(x)}$ maps X into S^1 and has no continuous extension to αX.
6. X has the complementation property if and only if every mapping of X into S^1 whose fibers have compact boundaries extends continuously to αX.

Reference: Dickman, 1967.

Chapter 10. βX in Categorical Perspective

10.1. Extensions of topological structures abound. In Chapter 1 we investigated the extensions βX and υX. Metric and uniform completions are also well known constructions, which, although not strictly topological, are similar. All satisfy a factorization property similar to the requirement that a mapping f of X to a compact space K extend to βX:

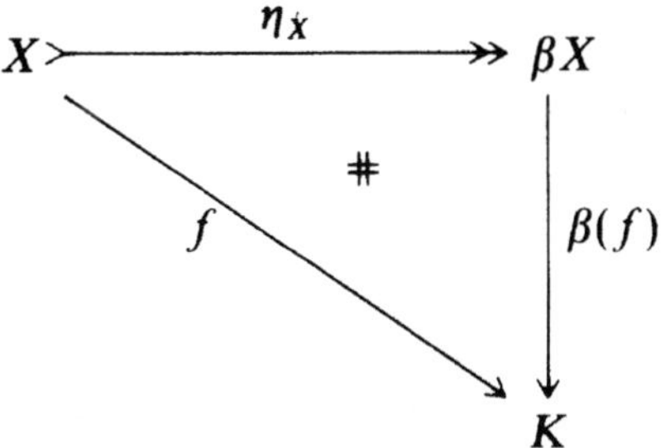

In the first portion of this chapter, we will relate the existence of such a diagram to properties of the appropriate classes of spaces. A more complete treatment of this area is given in H. Herrlich's 1968 Lecture Notes.

In the second part of the chapter, we will place the familar construction of the adjunction space in a categorical context. We will apply our findings to the construction of certain Stone-Čech compactifications.

In the third part of the chapter, we will examine the class of perfect mappings from a categorical standpoint and consider classes of spaces which are inversely preserved under perfect mappings.

In the final portion of the chapter, we will consider projective spaces and projective covers. This final topic will serve to relate the Stone-Čech compactification to Boolean algebras and the class of extremally disconnected spaces.

Though much of the language of the chapter will be categorical in nature, most of the proofs will be topological. Much of the chapter is based on the 1970 notes of S. P. Franklin. A comprehensive survey of the area is also provided in the 1971 paper of H. Herrlich.

For much of the chapter, we will be dealing with Hausdorff spaces and *it will be necessary to suspend the presumption that all spaces mentioned are completely regular.*

Categories and Functors

10.2. All the categories to be considered here will be made up of a class of spaces together with the mappings between spaces belonging to the class. For example, we will be speaking of the categories of completely regular spaces, of compact spaces, of realcompact spaces, et cetera. It is helpful when considering the following definition to keep such examples in mind.

A *category* $\mathscr{C}$ consists of two classes, a class $o\mathscr{C}$ called the *objects* of $\mathscr{C}$ and a class called the *morphisms* of $\mathscr{C}$, together with the following axioms which link the two classes:

    (1) The composition $h \circ g \circ f$ of three morphisms is defined whenever the compositions $h \circ g$ and $g \circ f$ are defined.

    (2) Composition of morphisms is associative, i.e. $(h \circ g) \circ f = h \circ (g \circ f)$ and both compositions are defined if either is defined.

    (3) There is a bijection which assigns to each object X an identity morphism 1_X, and for each morphism f there are two identity morphisms 1_X and 1_Y such that $f = f \circ 1_X$ and $f = 1_Y \circ f$.

10.3. The objects in many common categories are sets together with an algebraic or topological structure. The appropriate morphisms are usually just the structure preserving functions between the sets. Our examples will be of this type.

Examples. (a) Any class of topological spaces together with the continuous functions between them form a category. Most of our discussion will center around the category where the spaces are Hausdorff.

    (b) The classes of all sets and functions form a category.

    (c) Any of the classes of algebraic objects, such as groups, rings, or vector spaces will form a category when the class of morphisms is taken to be the appropriate class of homomorphisms.

(d) Topological structure can be combined with an algebraic structure to yield such examples as the category of topological groups or topological vector spaces. For instance, a *topological group* is a group $(G, +)$ together with a topology τ on G such that the group operation $+$ from $G \times G$ to G and the formation of inverses are both continuous. The appropriate morphisms are the continuous group homomorphisms.

(e) The class of Boolean algebras together with Boolean algebra homomorphisms forms a category.

10.4. A category $\mathscr{B}$ is a *subcategory* of a category $\mathscr{C}$ if every object and morphism of $\mathscr{B}$ is also an object or morphism of $\mathscr{C}$. The subcategory $\mathscr{B}$ is said to be a *full subcategory* if every morphism in $\mathscr{C}$ between two objects of $\mathscr{B}$ is also a morphism in $\mathscr{B}$. We will consider two examples of non-full subcategories below, otherwise, *all of the subcategories which we will discuss will be presumed to be full.* Because our examples consist mainly of subclasses of topological spaces together with *all* the mappings between them, this will not be a serious restriction.

Examples. (a) Throughout most of the preceding chapters, we have been restricting our attention to the full subcategory of the category of all topological spaces which consists of all completely regular spaces and all mappings between them. In the present chapter we will be concerned mainly with subcategories of the category of Hausdorff spaces.

(b) The category of all complete Boolean algebras together with all complete Boolean algebra homomorphisms is a subcategory of the category of all Boolean algebras and all homomorphisms. Note that one must verify that the composition of complete homomorphisms is complete in order to establish the previous statement. We will show that this provides an example of a non-full subcategory. Let L be the Boolean algebra of all subsets of $\beta\mathbb{N}$ and consider the inclusion of $CO(\beta\mathbb{N})$ into L. The algebra L is clearly complete and $CO(\beta\mathbb{N})$ is complete because $\beta\mathbb{N}$ is both zero-dimensional and extremally disconnected (Proposition 2.5.) However, in L the supremum is merely the union of the sets, so that the inclusion is clearly not a complete homomorphism.

(c) The category of all Abelian topological groups and all continuous homomorphisms is a subcategory of the category of topological groups.

10.5. Let $f: A \to B$ be a morphism in a category $\mathscr{C}$. A morphism $r: B \to A$ such that $f \circ r = 1_B$ is called a *right inverse* of f. A *left inverse* of f is a morphism $l: B \to A$ such that $l \circ f = 1_A$. An *isomorphism* is a morphism which has both a left and a right inverse. If f is an isomorphism, then the following string of equalities shows that the inverses r and l of f must be equal:

$$l = l \circ 1_B = l \circ (f \circ r) = (l \circ f) \circ r = 1_A \circ r = r \,.$$

Thus, if f is an isomorphism, we will write $f^{\leftarrow} = l = r$ and refer to $f^{\leftarrow}$ as the *inverse* of f. There are many familiar examples of isomorphisms.

Examples. (a) We will mainly be considering the four categories of topological spaces obtained by specifying that the objects be Hausdorff, completely regular, compact, or realcompact. Since each of these properties is a topological invariant, the isomorphisms in each category are the homeomorphisms.

(b) In the category of metric spaces and uniformly continuous maps, the isomorphisms are the *isometries*, i.e. the distance preserving homeomorphisms. Note therefore that this is a non-full subcategory of the category of metric spaces and all mappings between metric spaces. This situation arises because the property of being a metric space is not a topological invariant. However, the property of being metrizable is a topological invariant and in the category of metrizable spaces, the isomorphisms are the homeomorphisms.

(c) In the category of groups, the isomorphisms are the bijective homomorphisms.

10.6. A subcategory $\mathscr{B}$ of $\mathscr{C}$ is a *replete subcategory* of $\mathscr{C}$ if every object A of $\mathscr{C}$ which is isomorphic to an object B of $\mathscr{B}$ must belong to $\mathscr{B}$ as does the isomorphism. Thus, *any subcategory of the category of topological spaces which is obtained by specifying that the objects possess a topological invariant is replete.* For this reason, *all of the subcategories which we will discuss will be presumed to be replete.*

However, just to note that there are subcategories which fail to be replete, we observe that two of our previous examples are not replete.

Examples. (a) The subcategory of metric spaces and uniformly continuous maps is not replete because two metric spaces can be topologically isomorphic without the homeomorphism being an isometry.

(b) The subcategory of complete Boolean algebras and complete homomorphisms is not replete. Observe that if the injection discussed in Example 10.4(b) is thought of as a homomorphism onto its image, then neither it nor its image is complete.

10.7. A *functor* from a category $\mathscr{A}$ to a category $\mathscr{C}$ is a rule F which assigns to each object A and morphism f of $\mathscr{A}$ an object FA and morphism $F(f)$ of $\mathscr{C}$ such that:

    (1) F preserves identities, i.e. $F(1_A) = 1_{FA}$,

    (2) F preserves composition, i.e. if $f \circ g$ is defined in $\mathscr{A}$, then $F(f) \circ F(g)$ is defined in $\mathscr{C}$ and is equal to $F(f \circ g)$.

There are many examples of functors, several of which we are already acquainted with.

Examples. (a) For every set S, let DS be the topological space obtained by putting the discrete topology on S. If $f: S \to T$ is a function between sets and $D(f)$ is the same function regarded as a mapping of DS to DT, then D is a functor from the category of sets to the category of topological spaces.

(b) For any topological space X, let UX be the points or underlying set of X and for a mapping, let U "forget" the continuity of the map leaving only a function. Then U is a functor from the category of topological spaces to that of sets. U is frequently called a *forgetful functor* or *underlying set functor*.

(c) In Sections 1.52 and 1.55 we saw that the Stone-Čech compactification and the Hewitt-Nachbin realcompactification are both functorial, i.e. both can be used to define a functor. It is also easy to see that the completely regular space ρX described in Theorem 1.6 can be used to define a functor from the category of all topological spaces to the category of completely regular spaces. We will see that these three functors have much in common. These and similar functors will be the principal objects of investigation in this chapter.

10.8. The definition of functor requires that a functor preserves compositions, i.e. that $F(f \circ g) = F(f) \circ F(g)$. If the definition is altered by requiring that compositions be reversed, i.e. that $F(f \circ g) = F(g) \circ F(f)$, then F is called a *contravariant functor*.

Examples. (a) Our discussion in Section 2.10 shows that the assignment of the Stone space to a Boolean algebra and the mapping of the Stone spaces induced by a Boolean algebra homomorphism defines a contravariant functor from the category of Boolean algebras and homomorphisms to the category of compact totally disconnected spaces and the mappings between them.

(b) If X is any set, let $\mathscr{P}X$ denote the power set of X. Then for any function $f: X \to Y$ between two sets, define $\mathscr{P}(f)$ to be the function $f^{\leftarrow}: \mathscr{P}Y \to \mathscr{P}X$. Then $\mathscr{P}$ is a contravariant functor from the category of sets to itself and is called the *power set functor*. Observe that if X

and Y are topological spaces, then the statement that f is continuous is a statement about the image of f under the power set functor.

10.9. Let $f: A \to B$ and $g: B \to A$ be morphisms in a category $\mathscr{A}$. If $F: \mathscr{A} \to \mathscr{C}$ is a functor and $f \circ g = 1_B$ and $g \circ f = 1_A$, we must have that $F(f) \circ F(g) = 1_{FB}$ and $F(g) \circ F(f) = 1_{FA}$. A similar statement can be made for a contravariant functor, so that we have verified the

Proposition. *A functor or contravariant functor preserves isomorphisms.*

Reflective Subcategories of the Category of Hausdorff Spaces

10.10. We now consider a particular type of functor which includes the Stone-Čech compactification and the Hewitt-Nachbin realcompactification as examples. A functor r from a category $\mathscr{C}$ to a subcategory $\mathscr{R}$ of $\mathscr{C}$ is a *reflective functor* if there is a morphism $\eta_C: C \to rC$ and every morphism from C to an object R of $\mathscr{R}$ factors uniquely through rC via η_C so that the following diagram commutes:

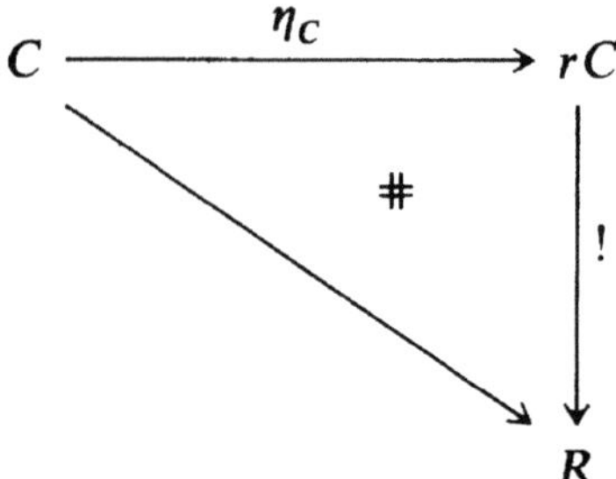

If $r: \mathscr{C} \to \mathscr{R}$ is a reflective functor, the subcategory $\mathscr{R}$ is called a *reflective subcategory*. The object rC is called the *reflection* of C in $\mathscr{R}$. The symbol "!" indicates that the morphism is required to be unique.

We have verified in Chapter 1 that βX and υX can be viewed as the images of objects under reflective functors. In this chapter, we will place these two reflective functors in a common categorical context and describe the topological characteristics of the categories of compact and realcompact spaces which give rise to their similarity. We will also reexamine the role of the category of completely regular spaces from a categorical standpoint.

10.11. We will find that many of the properties of reflective functors and subcategories have already been verified in the case of β and the category of compact spaces. The following result is such a property.

Proposition. *The reflection of an object is unique up to isomorphism.*

Proof. Let $\mathscr{R}$ be a reflective subcategory of $\mathscr{C}$. Let rX be the reflection of an object X. Let Y be any object of $\mathscr{R}$ and let $f: X \to Y$ be a morphism such that any morphism from X to an object of $\mathscr{R}$ factors uniquely through Y via f. Then we have the following commutative diagram in which the morphisms h and g exist and are unique by the reflective properties of Y and rX, respectively. Because rX is an object of $\mathscr{R}$, the morphism η_X must factor uniquely through η_X. One such factorization is 1_{rX} and the diagram shows that $h \circ g$ is another factorization. Thus, we must have $h \circ g = 1_{rX}$ since there can be only one factorization.

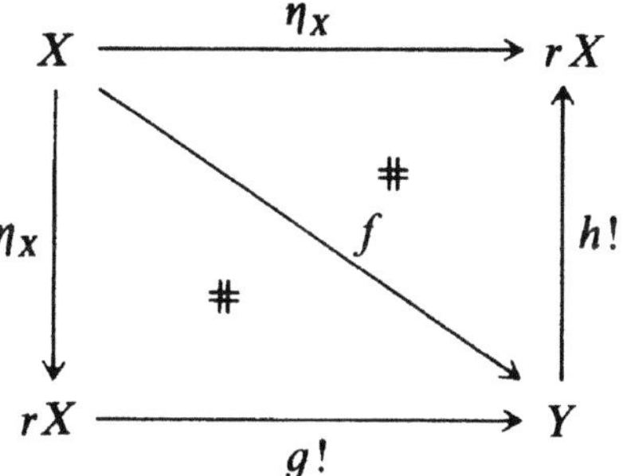

Repeating the argument with rX and Y interchanged will yield $g \circ h = 1_Y$. Hence, rX and Y are isomorphic. ☐

10.12. If R is an object of $\mathscr{R}$, any morphism will factor through the identity 1_R. Hence, the following corollary is immediate:

Corollary. *Any object in a reflective subcategory is isomorphic to its reflection.*

10.13. In Chapter 1, we saw that the factorization of a mapping of X through the mapping $\eta_X : X \to \beta X$ is unique because $\eta_X[X]$ is dense in βX and two maps which agree on a dense subspace of a Hausdorff space are equal. We will see that η_X is a specific instance of a particular type of morphism. A morphism $e: A \twoheadrightarrow B$ is an *epimorphism* if for every pair of morphisms

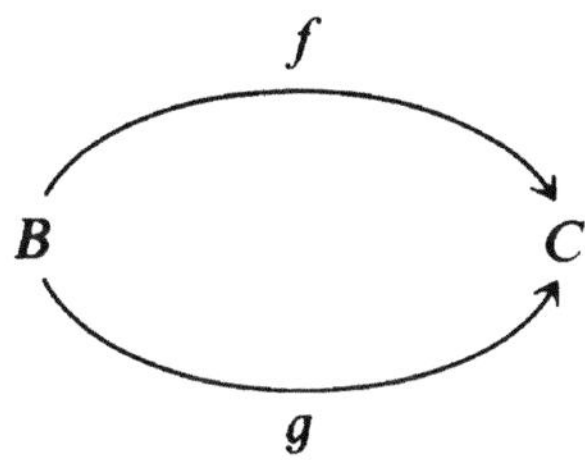

the equality $f \circ e = g \circ e$ implies that $f = g$. The doubleheaded arrow will be used to indicate an epimorphism. A reflective functor r is said to be *epi-reflective* if the morphism $\eta_X : X \to rX$ is an epimorphism. Thus, the preceding discussion shows that the mappings having dense images are epimorphisms in the category of Hausdorff spaces. Our previous considerations also show that β is an epi-reflective functor and that the category of compact spaces is an epi-reflective subcategory of the category of completely regular spaces. However, a closer analysis of the category of Hausdorff spaces will show that the compact spaces actually form an epi-reflective subcategory of the category of Hausdorff spaces. Of course, if X fails to be a completely regular space, the mapping η_X must fail to be an embedding since a subspace of a compact space must be completely regular.

10.14. The investigation of epi-reflective subcategories of the category of Hausdorff spaces will require an examination of properties of epimorphisms. The following result shows that one need only show that an epimorphism has a left-inverse in order to show that it is an isomorphism.

Proposition. *An epimorphism with a left inverse is an isomorphism.*

Proof. Let $e : X \twoheadrightarrow Y$ be an epimorphism in a category $\mathscr{C}$ and let $l : Y \to X$ be a left inverse for e. Then we have that:

$$1_Y \circ e = 1_Y \circ e \circ 1_X = e \circ (l \circ e) = (e \circ l) \circ e .$$

Now since e is an epimorphism, the equality of $1_Y \circ e$ and $(e \circ l) \circ e$ implies that $1_Y = e \circ l$ so that l is also a right inverse of e. Hence, e is an isomorphism. $\square$

Every concept in category theory has a dual concept which is obtained by reversing the arrows in the appropriate diagram. The dual of epimorphism is monomorphism. A morphism $m : C \rightarrowtail B$ is a *monomorphism* if for every pair of morphisms

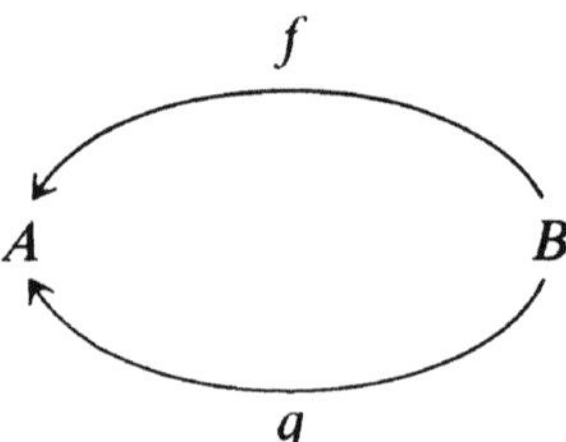

the equality $m \circ f = m \circ g$ implies that $f = g$. Thus a monomorphism is a morphism that is left-cancellable while an epimorphism is one that is right-cancellable. A double-headed arrow indicates an epimorphism while a tail on the arrow signals a monomorphism.

The relationship between epimorphisms and isomorphisms has its dual for monomorphisms.

Dual Proposition. *A monomorphism with a right inverse is an isomorphism.*

The preceding results and definitions can be recalled and motivated from the familiar situation in the category of sets and functions. A function is an epimorphism (resp. monomorphism) if and only if it is onto (resp. one-to-one). Further, a function is onto (resp. one-to-one) exactly when it has a right (resp. left) inverse. Thus, the concepts of epimorphism and monomorphism are generalizations of the familiar onto and one-to-one functions, respectively.

We will frequently not mention the dual of various concepts which we will consider, although a further discussion of duals will be carried out in the exercises.

10.15. We will, however, need to discuss both the concept of the product and its dual, the coproduct. Let $\{A_\alpha \colon \alpha \in \mathscr{A}\}$ be any family of objects of a category $\mathscr{C}$. Then an object A of $\mathscr{C}$ together with a family of morphisms $\{\pi_\alpha \colon A \to A_\alpha\}$ is called the *product* of the family $\{A_\alpha\}$ if for every object B of $\mathscr{C}$ and family of morphisms $\{f_\alpha \colon B \to A_\alpha\}$, there is a unique morphism $b \colon B \to A$ such that the diagram

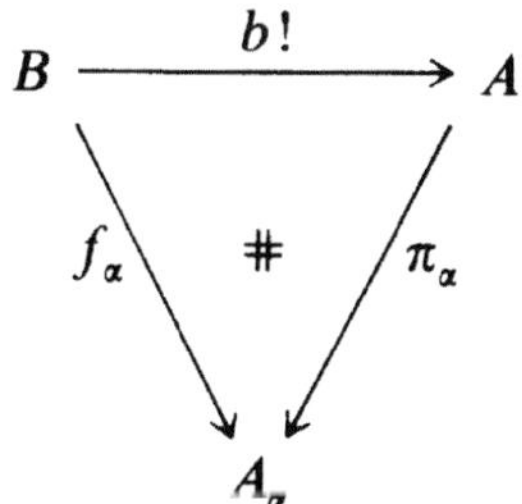

commutes for each α. The product of the family $\{A_\alpha\}$ will be denoted by $\underset{\cdot}{\times} A_\alpha$. In the category of topological spaces, the product is the usual topological product.

Dually, an object A of $\mathscr{C}$ together with a family of morphisms $\{i_\alpha \colon A_\alpha \to A\}$ is called the *coproduct* of the family $\{A_\alpha\}$ if for every object B and family of morphisms $\{g_\alpha \colon A_\alpha \to B\}$ there exists a unique morphism $l \colon A \to B$ such that the diagram

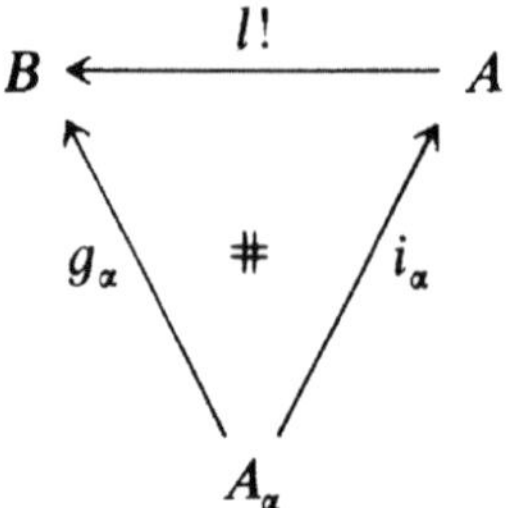

commutes for each α. The coproduct of the family $\{A_\alpha\}$ will be denoted by $\oplus A_\alpha$. In the category of topological spaces, the coproduct is the disjoint topological sum.

It may happen that products or coproducts exist in both a category $\mathscr{C}$ and a subcategory $\mathscr{A}$ of $\mathscr{C}$, but that the constructions of the product or coproduct in the two categories are different. For example, we have seen that the compact spaces form a reflective subcategory of the category of completely regular spaces. It is easy to see that the disjoint topological sum will be the coproduct for completely regular spaces, but not for compact spaces. However, the following result shows that coproducts do exist in the category of compact spaces.

Proposition. *If* $r:\mathscr{C}\to\mathscr{R}$ *is a reflective functor, then the coproduct in* $\mathscr{R}$ *of a family* $\{R_\alpha\}$ *of* $\mathscr{R}$ *is* $r(\oplus R_\alpha)$ *where* $\oplus R_\alpha$ *is the coproduct of the family in* $\mathscr{C}$.

Proof. To show that $r(\oplus R_\alpha)$ is the coproduct in $\mathscr{R}$ of the family $\{R_\alpha\}$, we must exhibit a family of morphisms $\{j_\alpha:R_\alpha\to r(\oplus R_\alpha)\}$ such that if R is any object of $\mathscr{R}$ and $\{g_\alpha:R_\alpha\to R\}$ is any family of morphisms, then there exists a unique morphism $h:r(\oplus R_\alpha)\to R$ such that $g_\alpha=h\circ j_\alpha$ for each α.

Let $\{i_\alpha:R_\alpha\to\oplus R_\alpha\}$ be the coproduct morphisms in $\mathscr{C}$. Then there exists a unique morphism l such that $g_\alpha=l\circ i_\alpha$ for each α. Since R is an object of $\mathscr{R}$ and $r(\oplus R_\alpha)$ is the reflection of $\oplus R_\alpha$, we have the following diagram:

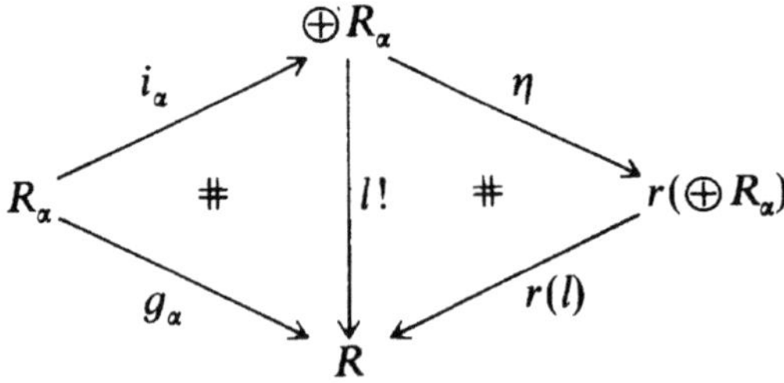

Thus, we have $g_\alpha=r(l)\circ\eta\circ i_\alpha$ for each α. Hence, the family $\{\eta\circ i_\alpha\}$ is the required family of morphisms and $r(l)$ is the unique morphism which must exist to show that $r(\oplus R_\alpha)$ is the coproduct in $\mathscr{R}$. $\square$

The proposition shows that the coproduct of a family of compact spaces in the category of compact spaces is the Stone-Čech compactification of their disjoint topological sum. Hence, $\beta\mathbb{N}$ can be interpreted as the coproduct of a family of countably many singletons.

10.16. However, we now show that the product in $\mathscr{C}$ of a family in $\mathscr{R}$ is also the product in $\mathscr{R}$ if $\mathscr{R}$ is epi-reflective.

Proposition. *If $\mathscr{R}$ is an epi-reflective subcategory of $\mathscr{C}$, then the product in $\mathscr{C}$ of objects of $\mathscr{R}$ belongs to $\mathscr{R}$.*

Proof. Let $\{R_\alpha\}$ be a family of objects of $\mathscr{R}$. We show that $\bigtimes R_\alpha$ is in $\mathscr{R}$ by showing that it is isomorphic to its reflection $r(\bigtimes R_\alpha)$. We have the following diagram:

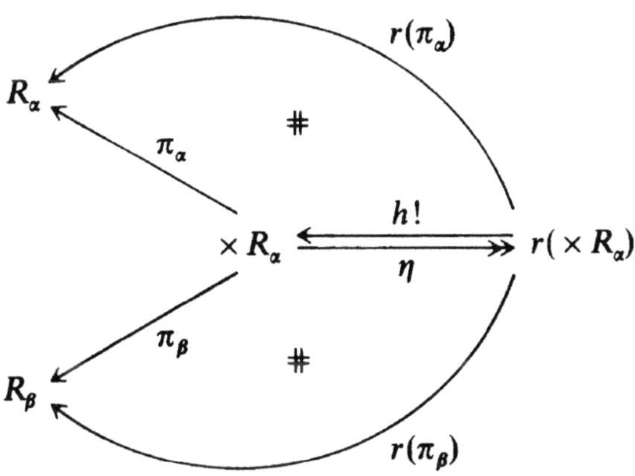

The morphism $r(\pi_\alpha)$ exists and satisfies $\pi_\alpha = r(\pi_\alpha)\circ\eta$ for each α because $r(\bigtimes R_\alpha)$ is the reflection of $\bigtimes R_\alpha$. Then the definition of product implies that there exists a unique morphism h such that $r(\pi_\alpha) = \pi_\alpha \circ h$. We will show that h and η are inverses of each other. For every α, $r(\pi_\alpha)\circ\eta$ is a morphism from $\bigtimes R_\alpha$ to R_α. Thus, from the defining property of the product, there exists a unique morphism $b : \bigtimes R_\alpha \to \bigtimes R_\alpha$ such that $r(\pi_\alpha)\circ\eta = \pi_\alpha \circ b$ for every α. However, we have that

$$r(\pi_\alpha)\circ\eta = \pi_\alpha\circ(h\circ\eta)$$

and also that

$$r(\pi_\alpha)\circ\eta = \pi_\alpha\circ 1_{\times R_\alpha}.$$

Hence, we must have $h\circ\eta = 1_{\times R_\alpha}$. Thus, η is an epimorphism with a left inverse and is therefore an isomorphism (Proposition 10.14). $\square$

10.17. Further description of epi-reflective subcategories of the category of Hausdorff spaces will rely on the categorical properties of three types of mappings: one-to-one maps, maps with dense range, and closed embeddings. We first show that the one-to-one maps are exactly the monomorphisms.

Proposition. *The monomorphisms in the category of Hausdorff spaces are the one-to-one mappings.*

Proof. We first show that a one-to-one mapping is a monomorphism. In the following diagram, let m be a one-to-one map and assume that $m \circ g = m \circ f$.

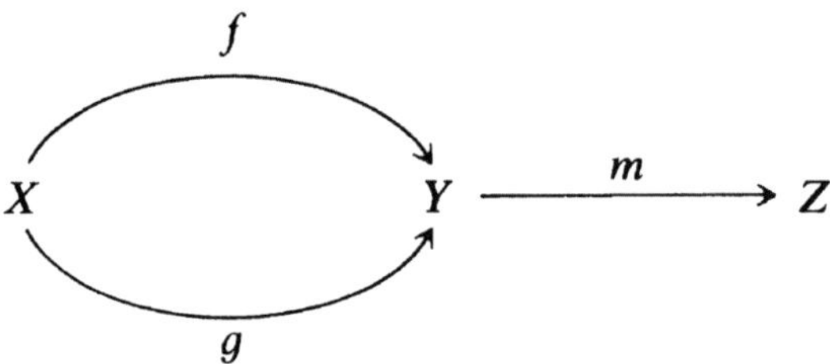

Then for every x in X, we have $m(f(x)) = m(g(x))$. Since m is one-to-one, this implies that $f(x) = g(x)$. Hence, $f = g$ and m is a monomorphism.

Conversely, assume in the diagram that m is a monomorphism. Let y_1 and y_2 be distinct points of Y and let X be the singleton space. Let f send the point of X to y_1 and g send the point of X to y_2. Because m is a monomorphism, $m \circ f \neq m \circ g$. Hence, $m(y_1) \neq m(y_2)$ and m is one-to-one. $\square$

Note that the preceding proof does not use the Hausdorff hypothesis and so we have actually shown that the one-to-one maps are also the monomorphisms in the larger category of all topological spaces.

10.18. Any map f between two spaces can be factored through the closure of its image as is illustrated by the diagram:

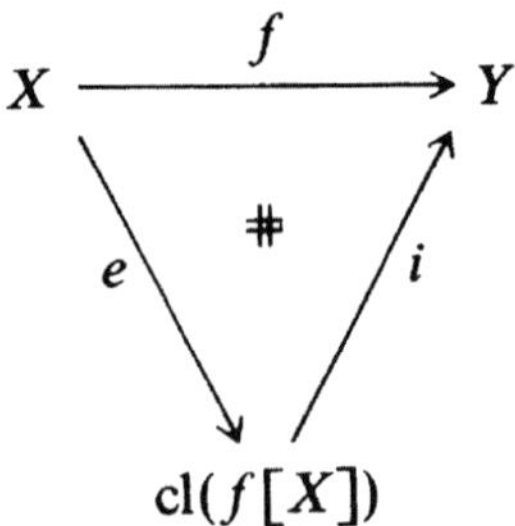

In Theorem 10.21 we will see that this factorization will be important in the characterization of epi-reflective subcategories. We have already seen that if Y is Hausdorff, then e is an epimorphism. We now see that every epimorphism in the category of Hausdorff spaces is of this form, i.e. is a map with dense range.

Proposition. *The epimorphisms in the category of Hausdorff spaces are the mappings with dense range.*

Proof. We have already seen that the maps with dense range are epimorphisms because two maps which agree on a dense subspace of a Hausdorff space are equal.

Conversely, suppose that $f: X \to Y$ is a mapping such that $f[X]$ is not dense in Y, i.e. such that $\mathrm{cl}(f[X])$ is a proper subspace of Y. We will show that f cannot be an epimorphism by constructing a space Z and two maps l_1 and l_2 of Y into Z which agree on $f[X]$ but which are not equal. To construct Z, we first make two disjoint copies of Y by writing $Y_1 = Y \times \{1\}$ and $Y_2 = Y \times \{2\}$. Both Y_1 and Y_2 taken with the product topology are homeomorphic to Y. Let $h_i: Y \to Y_i$ be defined by $h_i(y) = (y, i)$ for each $i = 1, 2$. Let $Y_1 \oplus Y_2$ be the disjoint sum or co-product of Y_1 and Y_2 and for $i = 1, 2$ let $i_k: Y_k \to Y_1 \oplus Y_2$ be the inclusion map. In each of the spaces Y_i, $h_i[\mathrm{cl}(f[X])]$ is a copy of $\mathrm{cl}(f[X])$. Thus, $i_1 \circ h_1[\mathrm{cl}(f[X])] \cup i_2 \circ h_2[\mathrm{cl}(f[X])]$ is the union of two copies of $\mathrm{cl}(f[X])$ which are contained in $Y_1 \oplus Y_2$. We will take Z to be the space obtained by joining Y_1 and Y_2 along the corresponding copies of $\mathrm{cl}(f[X])$. More precisely, let Z be the image of the quotient map q obtained by identifying $i_1 \circ h_1(y) = i_1(y, 1)$ and $i_2 \circ h_2(y) = i_2(y, 2)$ if y belongs to $\mathrm{cl}(f[X])$.

Then we have the following diagram:

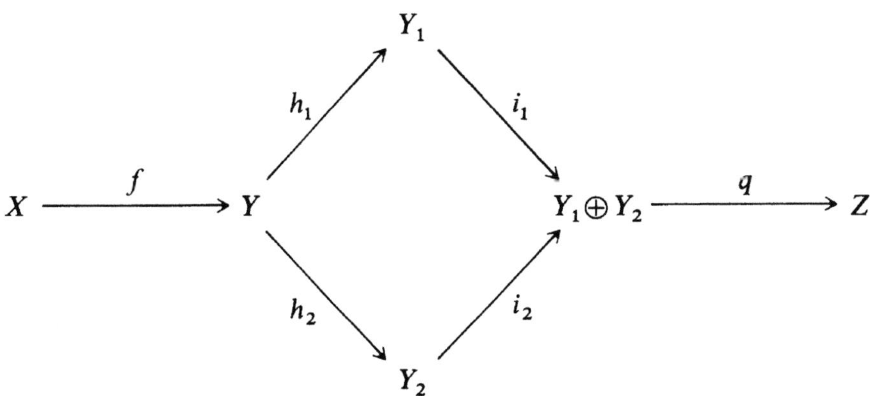

Now if x is a point of X, the two maps $i_1 \circ h_1$ and $i_2 \circ h_2$ split the point $f(x)$ in two and the map q joins the halves together again. Thus, we see that

$((q \circ i_1 \circ h_1) \circ f)(x) = ((q \circ i_2 \circ h_2) \circ f)(x)$. Hence, $(q \circ i_1 \circ h_1) \circ f = (q \circ i_2 \circ h_2) \circ f$. However, any point lying outside of $\mathrm{cl}(f[X])$ in Y is split by $i_1 \circ h_1$ and $i_2 \circ h_2$ but is not joined again by q. Hence, $q \circ i_1 \circ h_1 \neq q \circ i_2 \circ h_2$.

This would show that f cannot be an epimorphism and would complete the proof except that we have not shown that the construction of Z has kept us within the category of Hausdorff spaces. Hence it remains to show that the quotient space Z is Hausdorff.

Let p and r be distinct points of Z. We will find disjoint neighborhoods of p and r in each of six cases. However, only four of the cases will require distinct arguments.

Case 1: p and r both belong to $(q \circ i_1 \circ h_1)[Y \backslash \mathrm{cl}(f[X])]$: Since $\mathrm{cl}(f[X])$ is closed, there exist open sets U_p and U_r of Y containing $(q \circ i_1 \circ h_1)^{\leftarrow}(p)$ and $(q \circ i_1 \circ h_1)^{\leftarrow}(r)$, respectively, and missing $\mathrm{cl}(f[X])$. Since Y is Hausdorff, there exist disjoint neighborhoods V_p and V_r of $(q \circ i_1 \circ h_1)^{\leftarrow}(p)$ and $(q \circ i_1 \circ h_1)^{\leftarrow}(r)$, respectively. Then the required disjoint neighborhoods of p and r are $(q \circ i_1 \circ h_1)[V_p \cap U_p]$ and $(q \circ i_1 \circ h_1)[V_r \cap U_r]$.

Case 2: p and r both belong to $(q \circ i_2 \circ h_2)[Y \backslash \mathrm{cl}(f[X])]$: This is the same as Case 1 with only the index changed.

Case 3: p belongs to $(q \circ i_1 \circ h_1)[Y \backslash \mathrm{cl}(f[X])]$ and r belongs to $(q \circ i_2 \circ h_2)[Y \backslash \mathrm{cl}(f[X])]$: This case is easy since the two given sets containing the points are already disjoint open sets.

Case 4: $p \in (q \circ i_1 \circ h_1)[Y \backslash \mathrm{cl}(f[X])]$ and $r = q \circ i_1 \circ h_1(y)$ for some $y \in \mathrm{cl}(f[X])$: Note that the subscript does not matter for r because the two maps must agree on $\mathrm{cl}(f[X])$. There exist disjoint open subsets U and V in Y with $(q \circ i_1 \circ h_1)^{\leftarrow}(p)$ in U and y in V with U missing $\mathrm{cl}(f[X])$. Now p is in $(q \circ i_1 \circ h_1)[U]$ and r belongs to $q(i_1 \circ h_1[V] \cup i_2 \circ h_2[V])$ and these sets are disjoint. But they are also both open since $i_1 \circ h_1[U]$ and $i_1 \circ h_1(V) \cup i_2 \circ h_2(V)$ are both open and saturated in $Y_1 \oplus Y_2$.

Case 5: $p \in (q \circ i_2 \circ h_2)[Y \backslash \mathrm{cl}(f[X])]$ and $r = q \circ i_1 \circ h_1(y)$ for some $y \in \mathrm{cl}(f[X])$: This is the same as Case 4 with an index changed.

Case 6: $p = q \circ i_1 \circ h_1(x)$ and $r = q \circ i_1 \circ h_1(y)$ for distinct points x and y of $\mathrm{cl}(f[X])$: In this case, x and y have disjoint neighborhoods U and V, respectively, in Y. Then the disjoint neighborhoods of p and r in Z are given by

$$q[(i_1 \circ h_1[U]) \cup (i_2 \circ h_2[U])]$$

and

$$q[(i_1 \circ h_1[V]) \cup (i_2 \circ h_2[V])].$$

Hence, Z is Hausdorff. $\quad\square$

10.19. When a mapping is factored through the closure of its image, the second factor is a closed embedding. These maps also have a categorical characterization in the category of Hausdorff spaces. A monomorphism $m^{\#}$ is an *extremal monomorphism* if whenever $m^{\#}$ can be factored as illustrated

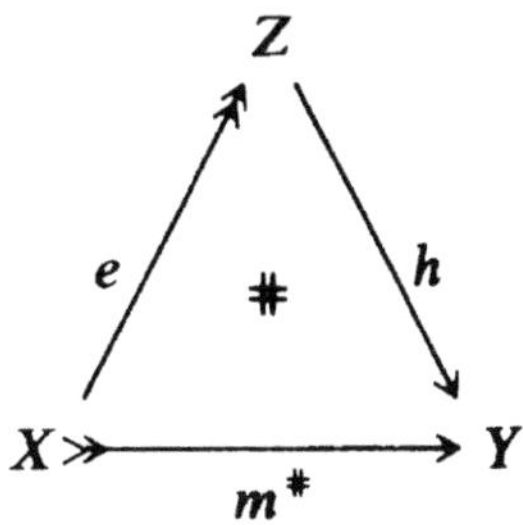

so that e is an epimorphism, then e is an isomorphism. An extremal monomorphism will be indicated by the "double tail" and a "$\#$" on the name of the morphism. In the diagram, the object X is said to be an *extremal subobject* of Y. We will show that the extremal monomorphisms in the category of Hausdorff spaces are the closed embeddings. Thus, the extremal subobjects in the category of Hausdorff spaces are the closed subspaces.

Proposition. *The extremal monomorphisms in the category of Hausdorff spaces are the closed embeddings.*

Proof. We first show that an extremal monomorphism $m^{\#} : X \to Y$ is a closed embedding. We can factor $m^{\#}$ through the closure of its range:

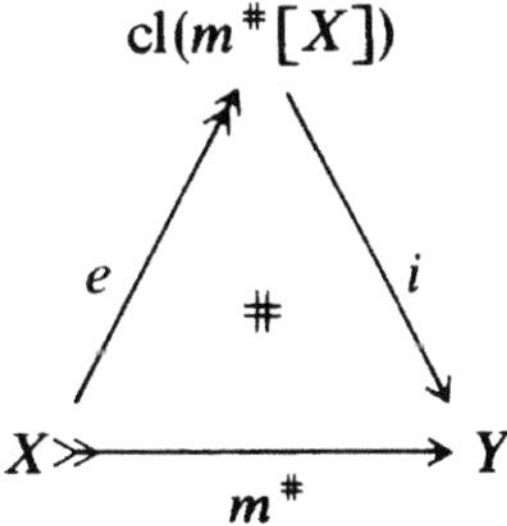

Since the range of e is dense, e is an epimorphism. But then e must be an isomorphism since $m^{\#}$ is an extremal monomorphism. Thus, $m^{\#}$ is a closed embedding.

Now let $m : X \to Y$ be a closed embedding. Assume that $m = h \circ e$ is a factorization of m where e is an epimorphism. The map m is a monomorphism because it is one-to-one. Thus, it remains to show that e is an isomorphism. We can also factor m through its image, thus obtaining the diagram:

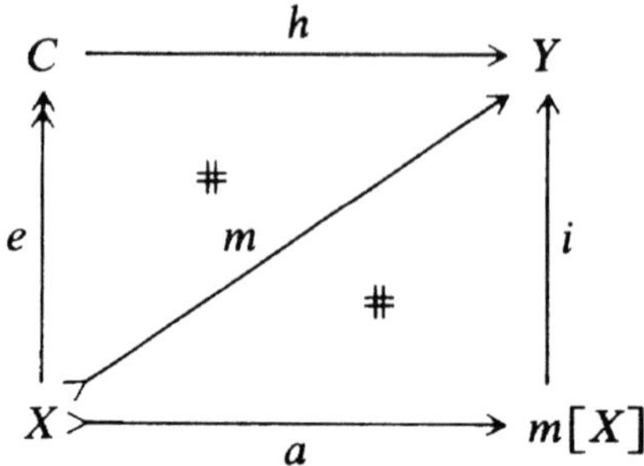

We will show that the epimorphism e is an isomorphism by obtaining a left-inverse for e. Because e has dense range and $m[X]$ is closed in Y, $h[C]$ is contained in $m[X]$. Thus, if we define $h':C\to m[X]$ by $h'(x)=h(x)$, we have that $h=i\circ h'$. But then we also have

$$i\circ h'\circ e=h\circ e=i\circ a$$

where i is a monomorphism. Therefore, $h'\circ e=a$. Since a is an isomorphism, we can write

$$1_X=(a^\leftarrow\circ h')\circ e.$$

Thus, e is an epimorphism with a left-inverse and is therefore an isomorphism (Proposition 10.14).   □

10.20. In Proposition 10.16, we saw that an epi-reflective subcategory is closed under the formation of products. The next result shows that an epi-reflective subcategory is also closed under extremal subobjects.

Proposition. *Epi-reflective subcategories are closed under extremal subobjects. Hence, epi-reflective subcategories of the category of Hausdorff spaces are closed hereditary.*

Proof. Let $\mathscr{R}$ be an epi-reflective subcategory of $\mathscr{C}$. Let Y belong to $\mathscr{R}$ and let $m^{\#}:X\to Y$ be an extremal monomorphism. Then we have the following commutative diagram:

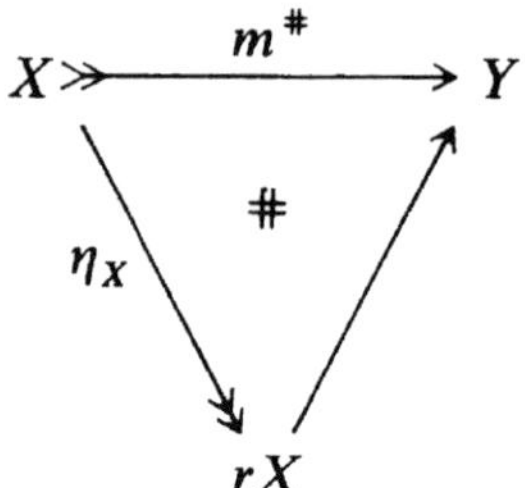

But η_X is therefore an isomorphism because $m^{\#}$ is an extremal mono-morphism. $\square$

10.21. The following theorem characterizes the epi-reflective sub-categories of the category of Hausdorff spaces as the productive and closed hereditary subcategories. The present form of the theorem appears in the 1965 paper of J.F. Kennison. More general categorical formulations of the theorem appear in the following papers: J.R. Isbell, 1964 B; Kennison, 1965; and S. Baron, 1969. Reflections and coreflections in topology are discussed in detail in the 1968 Lecture Notes of H. Herrlich.

Theorem. *The epi-reflective subcategories of the category of Hausdorff spaces are those subcategories which are closed under products and closed subspaces.*

Proof. The necessity of the conditions follows from Propositions 10.16 and 10.20.

For sufficiency, assume that $\mathscr{A}$ is a productive and closed hereditary subcategory of the category of Hausdorff spaces and let X be a Hausdorff space. We will obtain a space rX in $\mathscr{A}$ which will be the reflection of X. Because X is Hausdorff, there is only a set of pairwise non-homeomorphic Hausdorff spaces which can contain a dense image of X. This follows from the facts that if $e: X \to Y$ has dense range, then $|Y| \leqslant 2^{2^{|X|}}$ and there are only a set of topologies on a set of given cardinality. Thus, there exists a set of epimorphisms $\{f_\alpha : X \to A_\alpha\}$ with A_α in $\mathscr{A}$ such that if $f: X \to A$ is an epimorphism with A in $\mathscr{A}$, then there is a homeomorphism i and an index α such that the following diagram commutes:

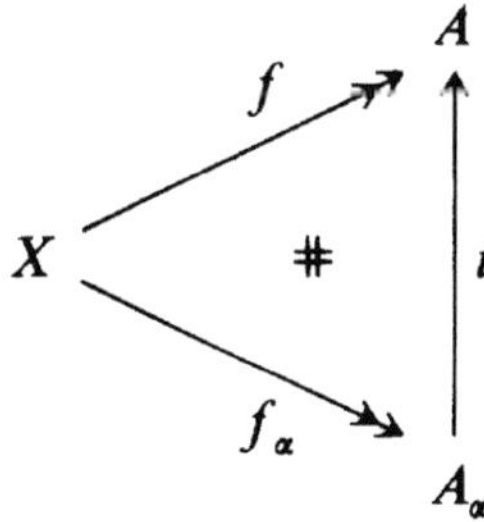

Such a family is called a *skeleton* of epimorphisms.

Consider the following diagram:

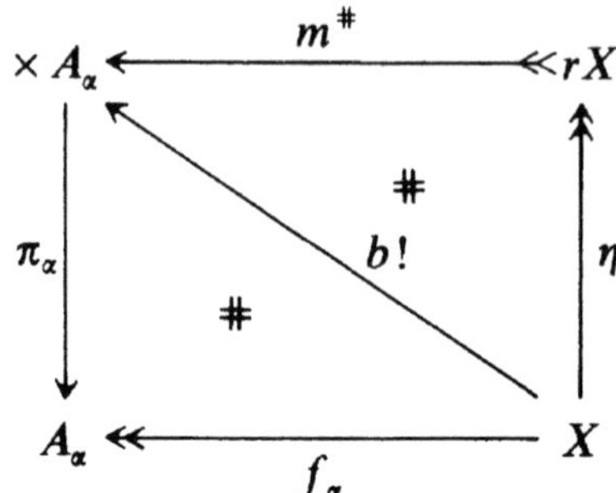

The mapping b exists and is unique from the definition of a product. The maps η and $m^{\#}$ are obtained by factoring b through the closure of its image. Because rX is a closed subspace of a product of objects of $\mathscr{A}$, our hypotheses guarantee that rX is in $\mathscr{A}$.

We will show that rX is the required reflection of X in $\mathscr{A}$. Let $f: X \to A$ be a mapping with A belonging to $\mathscr{A}$. The following diagram illustrates the procedure to construct the factorization of f through rX:

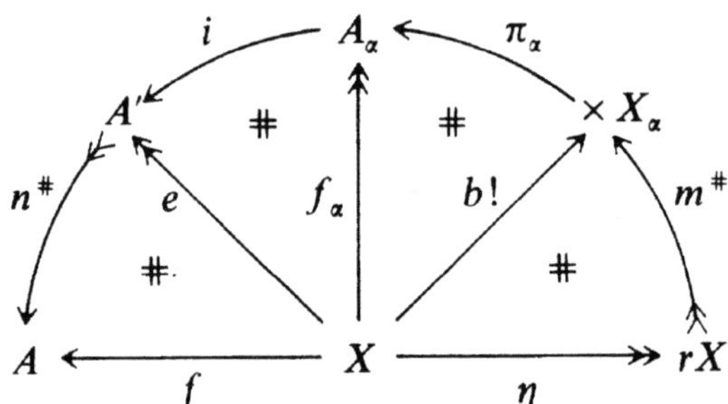

We first factor f through the closure of its image in order to express f as an epimorphism e followed by an extremal monomorphism $n^{\#}$. Because the subcategory $\mathscr{A}$ is closed under extremal subobjects, A' belongs to $\mathscr{A}$. Now because $\{f_\alpha\}$ is a skeleton of epimorphisms, there is an isomorphism i between some A_α and A' such that $e = i \circ f_\alpha$.

It remains to show that $n^{\#} \circ i \circ \pi_\alpha \circ m^{\#} \circ \eta$ is a unique factorization of f. We can write

$$n^{\#} \circ i \circ \pi_\alpha \circ m^{\#} \circ \eta = n^{\#} \circ i \circ \pi_\alpha \circ b$$

by the factorization of b. From the defining condition of a product, we have

$$n^{\#} \circ i \circ \pi_\alpha \circ b = n^{\#} \circ i \circ f_\alpha.$$

The factorization of e through the skeleton member f_α yields

$$n^\# \circ i \circ f_\alpha = n^\# \circ e = f.$$

Finally, the factorization is unique because η is an epimorphism.   ☐

10.22. Since topological products and closed subspaces of compact spaces are compact, the preceding theorem shows that the category of compact spaces is epireflective in the category of Hausdorff spaces. We must reconcile this fact with the fact that we have defined the Stone-Čech compactification only for completely regular spaces.

Observe that since the category of completely regular spaces is productive and hereditary, the theorem also shows that the completely regular spaces form an epi-reflective subcategory of the Hausdorff spaces. In Corollary 1.8, we saw that the space ρX is the completely regular reflection of a space X. Actually, we showed that ρX can be defined for an arbitrary space, a fact which will be placed in categorical context in Exercise 10A.

Now let X be a Hausdorff space and let f map X to a compact space K. Then we have the following diagram:

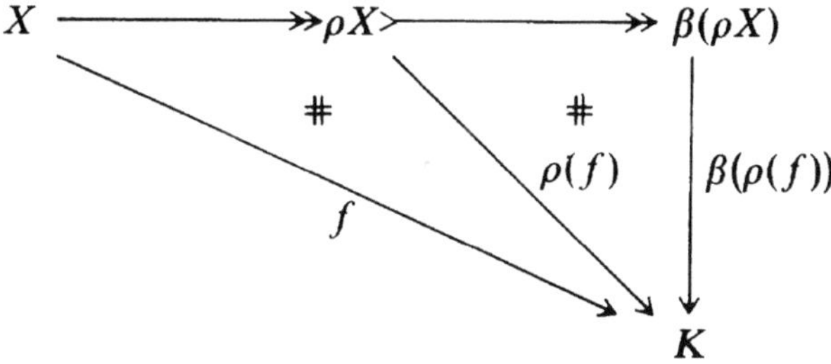

Thus, the reflection of a Hausdorff space X is the Stone-Čech compactification of ρX. Note that if we replace K by a realcompact space, then we can repeat the diagram with β replaced by υ to show that the realcompact spaces are also epi-reflective in the category of Hausdorff spaces.

We can carry the similarity between compact and realcompact spaces further. Recall that the compact spaces are just the closed subspaces of products of copies of the unit interval I. In 1948, E. Hewitt proved the following analogous result for realcompact spaces with I replaced by the real line $\mathbb{R}$.

Theorem (Hewitt). *The realcompact spaces are the closed subspaces of products of copies of* $\mathbb{R}$.

Proof. We have already shown that the realcompact spaces form an epi-reflective subcategory of the category of Hausdorff spaces and that $\mathbb{R}$ is realcompact (Example 1.54). Thus, a closed subspace of a product of real lines must be realcompact.

Now assume that X is realcompact. For each f in $C(X)$, let f^α denote the extension of f to a mapping of βX into the one-point compactification $\alpha\mathbb{R}$ of $\mathbb{R}$. The family of all such extensions separates points and points and closed sets of βX. Hence, the Embedding Lemma 1.5 shows that βX can be embedded as a subspace of a product of copies of $\alpha\mathbb{R}$ indexed by $C(X)$ and because βX is compact, its image under the embedding is closed. Thus, we have the following diagram:

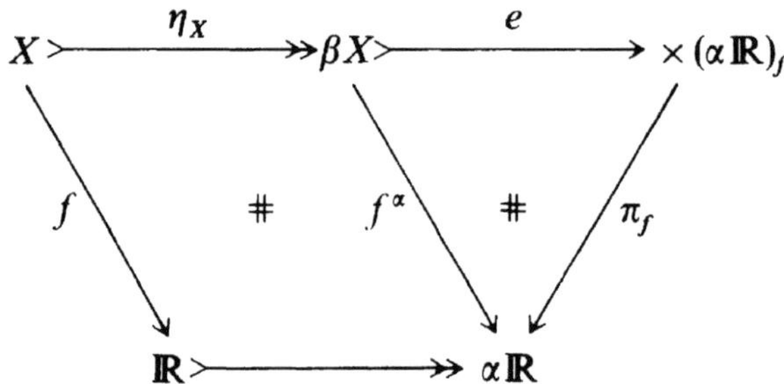

The points of υX correspond to the subspace of $e[\beta X]$ obtained by restricting the coordinates to the points of $\mathbb{R}$ (Section 1.53). Thus, υX is a restriction of the closed subspace $e[\beta X]$ to a product of real lines. Hence, $X = \upsilon X$ is homeomorphic to a closed subspace of a product of real lines. $\square$

10.23. Now assume that $\mathscr{R}$ is an epi-reflective subcategory of the Hausdorff spaces and that I or $\mathbb{R}$ is an object of $\mathscr{R}$. Then Theorem 10.21 shows that $\mathscr{R}$ must contain all compact or all realcompact spaces, respectively. Hence, either space is contained in a smallest epi-reflective subcategory of the Hausdorff spaces which consists of all closed subspaces of products of copies of the space. More generally, if $\mathscr{A}$ is any subcategory of the category of Hausdorff spaces, then the intersection of the productive and closed hereditary subcategories containing $\mathscr{A}$ is also productive and closed hereditary. Note that the Hausdorff spaces form such a category so that the intersection is non-empty. Hence, $\mathscr{A}$ is contained in a smallest epi-reflective subcategory of the Hausdorff spaces. We have verified the

Proposition. *Any subcategory $\mathscr{A}$ of the category of Hausdorff spaces is contained in a smallest epi-reflective subcategory of the Hausdorff spaces*

and the objects of this subcategory are the closed subspaces of products of spaces in $\mathscr{A}$.

The epi-reflective subcategory described in the previous proposition is denoted by **Haus**($\mathscr{A}$) and is called the *epi-reflective hull* of $\mathscr{A}$. **Haus**($\mathscr{A}$) is said to be generated by $\mathscr{A}$ and is said to be *simply generated* if $\mathscr{A}$ has only a single object.

10.24. The systematic study of simply generated epi-reflective categories was begun in the 1958 paper of R. Engelking and S. Mrówka. The 1966 and 1968 papers of Mrówka continue the investigation as does the thesis of R. Blefko. Let E be a topological space. In the terminology introduced by Engelking and Mrówka, *E-completely regular* spaces are defined to be subspaces of products of copies of E and closed subspaces of such products are defined to be *E-compact spaces*. If E is a Hausdorff space, then the E-compact spaces form a simply generated subcategory of the category of Hausdorff spaces. The E-completely regular spaces are precisely those spaces which are densely embedded in their "E-compactifications". Thus, the E-completely regular spaces are related to the E-compact spaces in the same way that the completely regular spaces are related to the compact spaces. The category of E-compact spaces has been extensively investigated for several specific choices of E.

Examples. (a) The I-completely regular spaces are the completely regular spaces and the I-compact spaces are the compact spaces. The reflection of a completely regular space X is its Stone-Čech compactification, βX, which is the set of z-ultrafilters on X with the topology described in Section 1.19.

(b) The $\mathbb{R}$-completely regular spaces are the completely regular spaces and the $\mathbb{R}$-compact spaces are the realcompact spaces. The reflection of a completely regular space is denoted by υX and is the subspace of βX consisting of the z-ultrafilters on X which have the countable intersection property.

(c) Let $\mathbf{2}$ denote the two point discrete space. In his 1955 paper, B. Banaschewski showed that the $\mathbf{2}$-completely regular spaces are the zero-dimensional spaces and that the $\mathbf{2}$-compact spaces are the compact zero-dimensional spaces. The associated reflective functor is denoted by ζ. Banaschewski further showed that for any zero-dimensional space X, ζX is the component space of βX, i.e. the quotient space obtained by identifying two points of βX if they lie in the same component. Thus, we see from Proposition 3.34 that $\zeta X = \beta X$ if and only if X is strongly zero-dimensional. Hence, the space M of Example 3.39 is an example of a zero-dimensional space such that $\zeta M \neq \beta M$.

The space ζX can also be seen to be the Wallman-type compactification of the zero-dimensional space X as described in Section 1.19(a). There the normal base is taken to be the set of clopen subsets. Further, the compact zero-dimensional spaces are characterized as those zero-dimensional spaces in which every ultrafilter of clopen subsets is fixed.

(d) The category of $\mathbb{N}$-compact spaces bears a relation to the category of 2-compact spaces similar to that which the realcompact spaces bear to the compact spaces. In 1967 B, H. Herrlich showed that a zero-dimensional space is $\mathbb{N}$-compact if and only if every ultrafilter of clopen sets having the countable intersection property is fixed. This result also appears in the 1970 paper of K. Chew. Since $\mathbb{N}$ is zero-dimensional, any $\mathbb{N}$-completely regular space is zero-dimensional and therefore is embedded in ζX, its compact zero-dimensional reflection. From (c) above, ζX is the space of ultrafilters of clopen subsets of X. The $\mathbb{N}$-compact reflection of X, vX, is the subspace of ζX consisting of the clopen ultrafilters which have the countable intersection property. Thus, the relation of vX to ζX is analogous to that of υX to βX.

The category of $\mathbb{N}$-compact spaces has been the subject of much recent investigation. Because $\mathbb{N}$ is realcompact, we see that every $\mathbb{N}$-compact space is realcompact. The problem of characterizing those realcompact spaces which are also $\mathbb{N}$-compact is unsolved. The category of $\mathbb{N}$-compact spaces has, however, been shown to contain the category of strongly zero-dimensional realcompact spaces and to be contained in the category of zero-dimensional realcompact spaces. The first of these containments was demonstrated by H. Herrlich in his 1967 B paper. The latter containment is clear since products and subspaces of zero-dimensional spaces are zero-dimensional. Speculation originally focused on the possibility that one of these containments might be reversible. However, both have been shown to be proper. In his thesis, P. Nyikos describes a zero-dimensional realcompact space Δ which fails to be $\mathbb{N}$-compact. A brief account of Δ and its significance is given in his 1971 paper. The space Δ is a highly complicated example originally introduced by P. Roy in his 1962 and 1968 papers to show that small inductive dimension and Lebesgue covering dimension are not the same in metric spaces. In his 1974 paper, S. Mrówka describes an example of a $\mathbb{N}$-compact space which fails to be strongly zero-dimensional, thus showing that the other containment is also proper.

Related questions concerning these two categories remain. Our remarks above and Theorem 10.21 show that the category of zero-dimensional realcompact spaces is epi-reflective. However, it is not known if this category is simply generated. Concerning the category of strongly zero-dimensional realcompact spaces, it is not yet known if even finite products or closed subspaces of strongly zero-dimensional realcompact spaces are strongly zero-dimensional. Other related questions are discussed in the 1972 paper of Mrówka.

(e) The categories of $\mathbb{Q}$-compact spaces and $\mathbb{P}$-compact spaces are both identical with the category of $\mathbb{N}$-compact spaces. We will sketch a proof of this fact, showing the following containments:

$$\mathbf{Haus}(\mathbb{N}) \subseteq \mathbf{Haus}(\mathbb{Q}) \subseteq \mathbf{Haus}(\mathbb{P}) \subseteq \mathbf{Haus}(\mathbb{N})$$

The first containment is clear since $\mathbb{N}$ is a closed subspace of $\mathbb{Q}$.

The proof that $\mathbf{Haus}(\mathbb{Q})$ is contained in $\mathbf{Haus}(\mathbb{P})$ is based on the fact that any countable dense subset of $\mathbb{R}$ is homeomorphic to $\mathbb{Q}$. This fact is then used to embed $\mathbb{Q}$ as the diagonal of a product of copies of $\mathbb{P}$. Let $\mathbb{Q}_D$ denote the dyadic rationals. For each i in $\mathbb{P}$, put $\mathbb{P}_i = \mathbb{R} \setminus (\mathbb{Q}_D \cup \{i\})$. Then $\bigcap \{\mathbb{P}_i : i \in \mathbb{P}\} = \mathbb{Q} \setminus \mathbb{Q}_D$ is homeomorphic to $\mathbb{Q}$ and each $\mathbb{P}_i$ is a copy of $\mathbb{P}$. Hence, the diagonal of $\bigtimes \{P_i : i \in \mathbb{P}\}$ is homeomorphic to $\mathbb{Q}$ and is closed since the diagonal of a product of Hausdorff spaces is closed [D, p. 138]. Hence, $\mathbb{Q}$ is homeomorphic to a closed subspace of a product of copies of $\mathbb{P}$. Thus, $\mathbb{Q}$ is an object of $\mathbf{Haus}(\mathbb{P})$ and $\mathbf{Haus}(\mathbb{Q})$ is contained in $\mathbf{Haus}(\mathbb{P})$.

Finally, $\mathbb{P}$ is homeomorphic to a countable product of copies of $\mathbb{N}$. A proof of this result is outlined in Exercise 24K of Willard, 1970. Thus, $\mathbf{Haus}(\mathbb{P})$ is contained in $\mathbf{Haus}(\mathbb{N})$.

(f) Let k be a cardinal number. A completely regular space is said to be *k-compact* if every z-ultrafilter such that the intersection of every subfamily of less than k members is non-empty is fixed. These categories were introduced by H. Herrlich in his 1967A paper, and were shown to be simply generated by M. Hušek in 1969.

(g) In his 1971 paper, M. Hušek assumes that measurable cardinals exist and describes a metric space S having measurable cardinality. He then relates the category of S-compact spaces to the question of an υ-analog for the Glicksberg Theorem. A family of subsets of a space is said to be *discrete* if every point of the space has a neighborhood meeting at most one of the sets. If $\mathfrak{m}$ is the least measurable cardinal, then a space is said to be *pseudo-$\mathfrak{m}$-compact* if every discrete family of open subspaces has non-measurable cardinal. Hušek shows that S plays a role for pseudo-$\mathfrak{m}$-compactness analogous to that played by $\mathbb{R}$ for pseudocompactness. Denote the S-compact reflection of a completely regular space X by $\beta_S X$. Then he shows that $\beta_S X = \upsilon X$ if and only if X is pseudo-$\mathfrak{m}$-compact. He obtains necessary and/or sufficient conditions relating the pseudo-$\mathfrak{m}$-compactness of $X \times Y$ to the equality $\upsilon(X \times Y) = \upsilon X \times \upsilon Y$. However, Hušek also offers an example to show that the pseudo-$\mathfrak{m}$-compactness of $X \times Y$ does not characterize those pairs for which the equality holds.

Adjunctions in Reflective Subcategories

10.25. Consider the following diagram in a category $\mathscr{C}$.

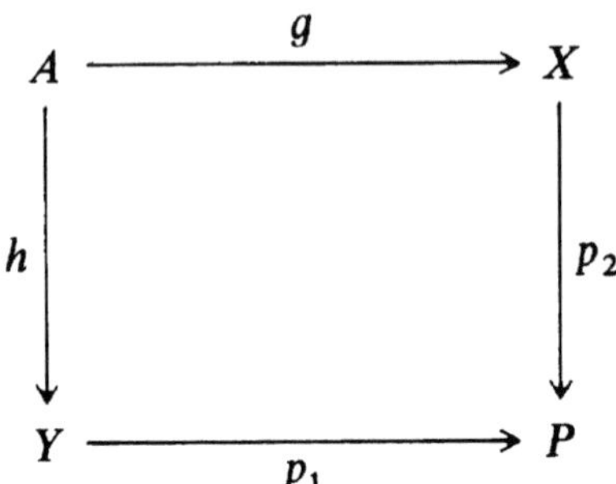

The object P together with the morphisms p_1 and p_2 is said to be the *pushout* of the diagram

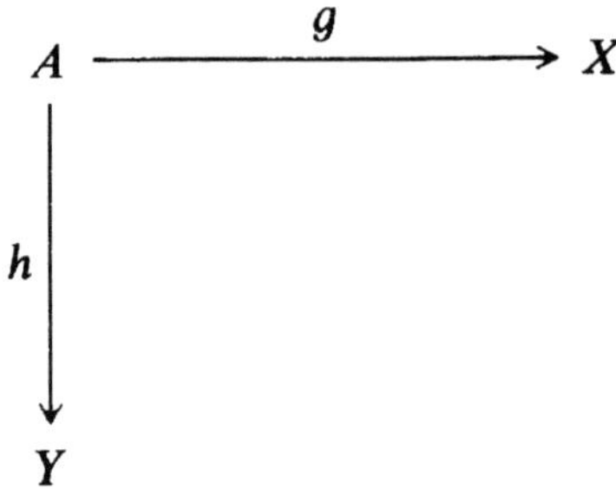

if for every pair of morphisms $l: Y \to Z$ and $k: X \to Z$ such that $l \circ h = k \circ g$, there is a unique morphism φ such that the following diagram commutes:

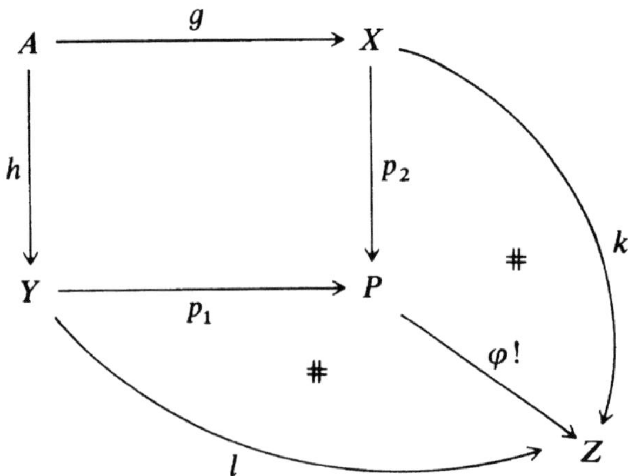

One familiar example of a pushout is the adjunction space in the category of topological spaces. Let A be a closed subspace of a space X and let f map A to Y. Then the *adjunction space* $X \cup_f Y$ formed by "attaching X to Y by f" is the quotient space of $X \oplus Y$ formed by identifying each point a of A with its image $f(a)$ in Y. In the following diagram, q denotes the quotient map and p_1 and p_2 are the compositions of q with the embeddings of Y and X into $X \oplus Y$.

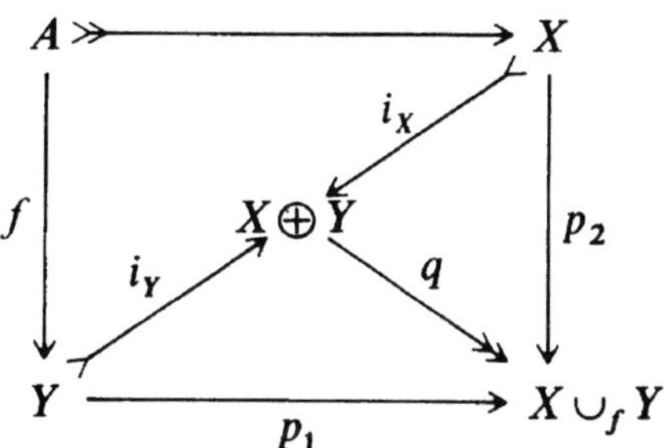

An account of the basic properties of adjunction spaces is given in the text of S. T. Hu.

Proposition. *The adjunction space* $X \cup_f Y$ *is the pushout of the diagram*

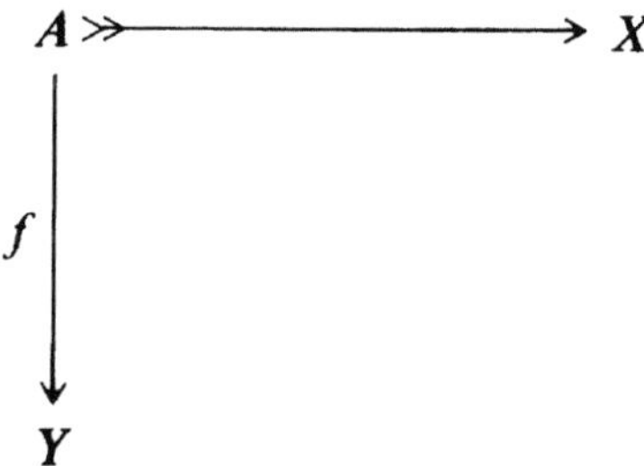

Proof. Let $l: Y \to Z$ and $k: X \to Z$ be maps such that $l \circ f = k|A$, i.e. such that k agrees with $l \circ f$ on A. Then we can define a function φ from $X \cup_f Y$ to Z by

$$\varphi(r) = \begin{cases} k(x) & \text{if } r = q(x) \text{ for } x \in X \setminus A \\ l(y) & \text{if } r = q(y) \text{ for } y \in Y \setminus f[A] \\ k(a) = (l \circ f)(a) & \text{if } r = q(a) = q(f(a)) \text{ for } a \in A. \end{cases}$$

The definition of φ implies that $\varphi \circ p_1 = l$ and $\varphi \circ p_2 = k$. It remains to show that φ is continuous. Let U be open in Z. Then $l^{\leftarrow}(U)$ and $k^{\leftarrow}(U)$ are open in Y and X, respectively. Further, $f^{\leftarrow}(l^{\leftarrow}(U)) = k^{\leftarrow}(U) \cap A$.

Thus, $i_X[l^\leftarrow(U)] \cup i_Y[k^\leftarrow(U)]$ is a saturated open set of $X \oplus Y$. Since $l^\leftarrow(U) = p_1^\leftarrow(\varphi^\leftarrow(U))$ and $k^\leftarrow(U) = p_2^\leftarrow(\varphi^\leftarrow(U))$, we have that

$$\varphi^\leftarrow(U) = q[i_X[l^\leftarrow(U)] \cup i_Y[k^\leftarrow(U)]],$$

which is open in the quotient space $X \cup_f X$. ☐

10.26. The next lemma emphasizes that $X \cup_f Y$ is actually a copy of X joined to a copy of Y along the subspace $f[A]$.

Lemma. *Y and $X \backslash A$ are embedded as subspaces of $X \cup_f Y$ and are closed and open, respectively.*

Proof. $X \backslash A$ is embedded as an open subspace of $X \oplus Y$ and $q|X\backslash A$ is one-to-one. Therefore, $q|X\backslash A$ is an open continuous bijection onto its image, and hence is an embedding. Similarly, Y is embedded as a closed subspace of $X \oplus Y$ and $q|Y$ is a closed continuous bijection onto its image. ☐

10.27. Thus far we have spoken only about pushouts in the category of all topological spaces. By modifying the construction, we can show that pushouts exist in reflective subcategories of the category of topological spaces.

Proposition. *Let $\mathscr{R}$ be a reflective subcategory of the category of topological spaces. Then if A is a closed subspace of X, the following diagram in $\mathscr{R}$ has a pushout in $\mathscr{R}$:*

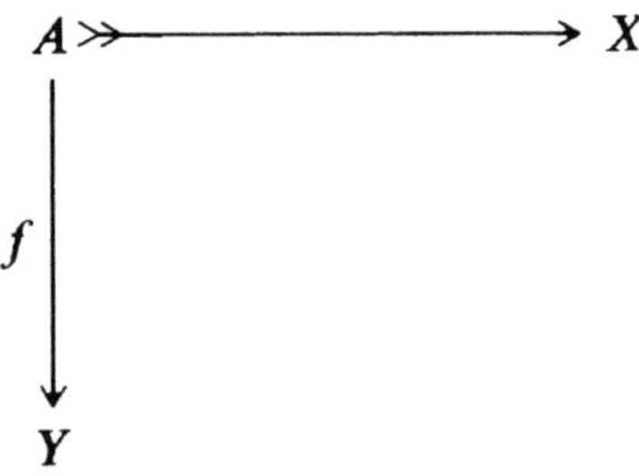

Proof. Let r be the reflective functor associated with $\mathscr{R}$. We will show that the pushout is $r(X \cup_f Y)$. Let Z belong to $\mathscr{R}$ and let $l: Y \to Z$ and

$k: X \to Z$ be such that $l \circ f$ agrees with k on A. Then we have the following diagram:

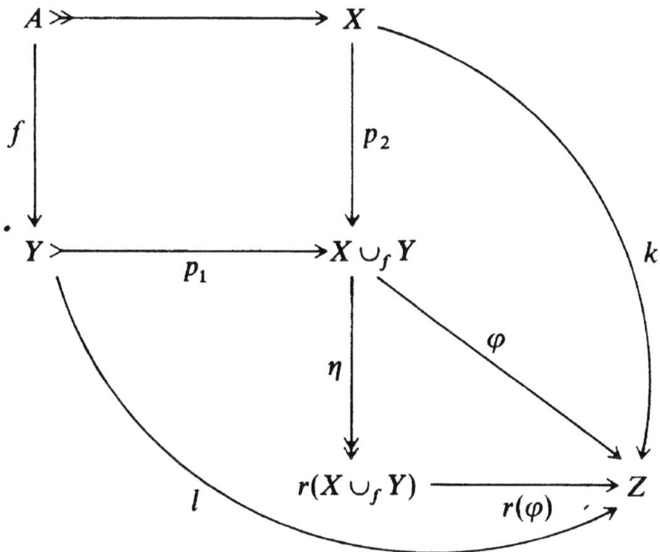

The map φ exists because $X \cup_f Y$ is the pushout in the category of topological spaces. Then because Z is an object of $\mathscr{R}$, φ will factor through η yielding the map $r(\varphi)$ such that $k = r(\varphi) \circ \eta \circ p_2$ and $l = r(\varphi) \circ \eta \circ p_1$. Thus, $r(X \cup_f Y)$ together with the morphisms $\eta \circ p_1$ and $\eta \circ p_2$ is the pushout in $\mathscr{R}$. $\square$

10.28. In particular, we have seen in Section 10.22 that the category of completely regular spaces is reflective in the category of topological spaces. In this case, the preceding proposition yields $\rho(X \cup_f Y)$ as the pushout. This construction can be applied to describe the Stone-Čech compactification of certain spaces. The next lemma is a crucial step in that direction because it deals with C^*-embedded subspaces of $\rho(X \cup_f Y)$.

Lemma. *Consider the following diagram in the category of completely regular spaces. If f is a C^*-embedding of A into Y, then X is C^*-embedded in the pushout $\rho(X \cup_f Y)$.*

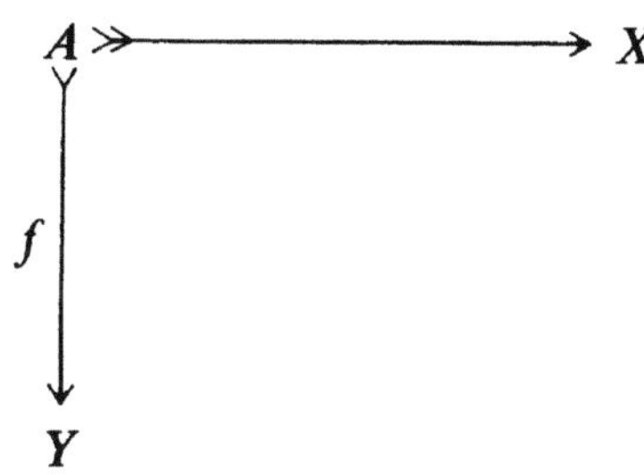

Proof. Let k belong to $C^*(X)$ and let l be the extension to Y of $k|A$. Then we have the following diagram:

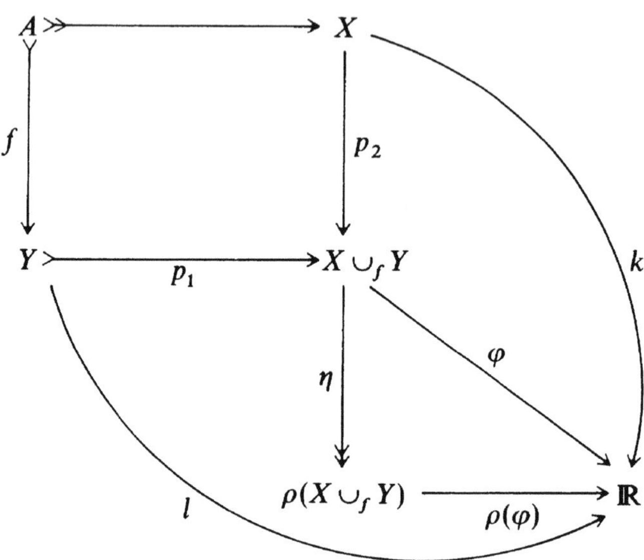

We first show that X is C^*-embedded in $X \cup_f Y$. Lemma 10.26 shows that $X \backslash A$ and Y are both embedded in $X \cup_f Y$. Since we also have here that f is an embedding, $p_2|A$ is just the composition of the embeddings p_1 and f so that p_2 is an embedding of all of X into $X \cup_f Y$. Thus, we have shown that X is C^*-embedded in the adjunction space $X \cup_f Y$ since φ is an extension of the mapping k.

To show that X is C^*-embedded in $\rho(X \cup_f Y)$, we will show that $\eta|p_2[X]$ is a one-to-one open map onto its range and hence is an embedding.

Recall from Theorem 1.6 that for any space S, the map η of S onto ρS identifies those points of S which cannot be separated by members of $C^*(S)$. We will show that every distinct pair of points of $p_2[X]$ is separated so that $\eta|p_2[X]$ is one-to-one. Let $p_2(x)$ and $p_2(z)$ be distinct points of $p_2[X]$. Then there exists a member k of $C^*(X)$ which separates x from z. But then we have the mapping φ such that $k = \varphi \circ p_2$. Therefore, φ separates $p_2(x)$ and $p_2(z)$.

Now we show that $\eta|p_2[X]$ is an open map onto its range. Because $p_2[X]$ is completely regular, the cozero-sets of $p_2[X]$ form a base for the topology and it is sufficient to show that the image of a cozero-set is open. Because $p_2[X]$ is C^*-embedded in $X \cup_f Y$, every cozero-set in $p_2[X]$ is the trace of a cozero-set of $X \cup_f Y$. Let $U = \mathbf{Cz}(g) \cap p_2[X]$

be such a trace. Then we can write $g = \rho(g) \circ \eta$ where $\rho(g)$ belongs to $C^*(\rho(X \cup_f Y))$. But this factorization shows that

$$\eta[U] = \mathbf{Cz}(\rho(g)) \cap (\eta \circ p_2)[X],$$

and hence $\eta[U]$ is open in $(\eta \circ p_2)[X]$.

Hence, $\eta \circ p_2$ is an embedding and X is C^*-embedded in $\rho(X \cup_f Y)$. $\square$
Observe that the analogous result will hold for C-embedding.

10.29. We now use the lemma to describe the Stone-Čech compactification of a completely regular space which contains a closed subspace that meets every non-compact zero-set.

Theorem. *Let A be a closed subspace of a completely regular space X such that every non-compact zero-set of X meets A. Then βX is the pushout of the following diagram in the category of completely regular spaces:*

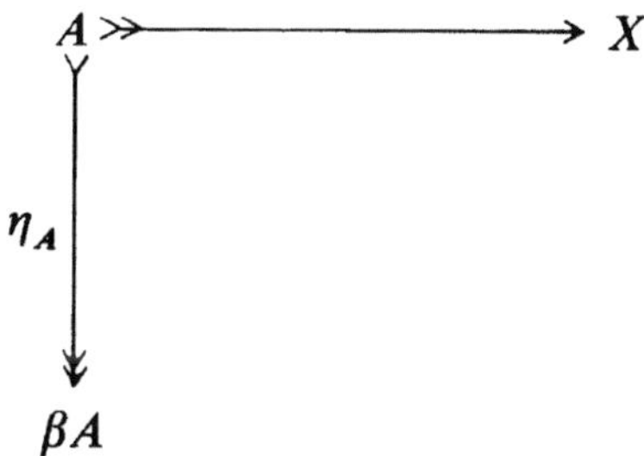

Proof. The lemma shows that X is C^*-embedded in the pushout $\rho(X \cup_{\eta_A} \beta A)$. Because A is dense in βA, X is also dense in the pushout. It remains to show compactness.

Denote the following composition by g:

$$X \oplus \beta A \twoheadrightarrow X \cup_{\eta_A} \beta A \twoheadrightarrow \rho(X \cup_{\eta_A} \beta A).$$

Let $\mathcal{U}$ be a z-ultrafilter on $\rho(X \cup_{\eta_A} \beta A)$. We will show that $\mathcal{U}$ is fixed. If there exists a zero-set Z in $\mathcal{U}$ such that $g^{\leftarrow}(Z)$ misses A, then the zero-set $g^{\leftarrow}(Z)$ is contained in $(\beta A \setminus A) \oplus (X \setminus A)$. Hence, the compactness of βA and our hypothesis imply that $g^{\leftarrow}(Z)$ is compact. Therefore, $Z = g[g^{\leftarrow}(Z)]$ is compact. Since no compact zero-set can belong to a free z-ultrafilter, $\mathcal{U}$ is fixed. If no such Z exists in $\mathcal{U}$, then $g^{\leftarrow}(Z)$ meets A for every Z in $\mathcal{U}$, and the family $\{g^{\leftarrow}(Z) \cap A : Z \in \mathcal{U}\}$ is a z-filter on A. This z-filter on A must cluster in βA, and the image of the cluster point under g is a cluster point of $\mathcal{U}$. Since the z-ultrafilter $\mathcal{U}$ converges to any cluster point (Proposition 1.28), $\mathcal{U}$ is fixed and $\rho(X \cup \eta_A \beta A)$ is compact.

Hence, $\rho(X \cup \eta_A \beta A) = \beta X$.   ☐

10.30. We now consider two applications of the previous theorem to particular spaces. The following example was introduced by J.R. Isbell and is described in [GJ, ex. 5 I].

Example. In Proposition 3.21 we showed that $\mathbb{N}$ admits an almost disjoint family $\mathcal{E}$ of $\mathfrak{c}$ infinite subsets. By enlarging $\mathcal{E}$ if necessary, assume that $\mathcal{E}$ is a maximal almost-disjoint family of infinite subsets. For each E in $\mathcal{E}$, choose a new point ω_E and let $\Psi = \mathbb{N} \cup \{\omega_E : E \in \mathcal{E}\}$. Let the points of $\mathbb{N}$ be isolated and let a basic neighborhood of a point ω_E be any subset containing ω_E and all but finitely many points of E. Ψ is clearly Hausdorff, and since the subspace $E \cup \{\omega_E\}$ is a compact neighborhood of ω_E, Ψ is locally compact. Thus, Ψ is completely regular [D, p. 238].

Put $D = \{\omega_E : E \in \mathcal{E}\}$. Since points of $\mathbb{N}$ are isolated, D is closed. Because the neighborhood $E \cup \{\omega_E\}$ of ω_E meets D in the singleton $\{\omega_E\}$, D is a discrete subspace of cardinality $\mathfrak{c}$. Because Ψ is separable, it can admit only $\mathfrak{c}$ real-valued mappings. Hence, the closed, discrete subspace D cannot be C^*-embedded in Ψ since D admits $2^{\mathfrak{c}}$ real-valued mappings. Hence, Ψ is not normal.

Now we apply the theorem to describe $\beta\Psi$. The maximality of $\mathcal{E}$ shows that any closed set which misses D is a finite subset of $\mathbb{N}$ and is compact. Hence, every non-compact zero-set of Ψ meets the closed subspace D. Thus, the theorem implies that $\beta\Psi$ is $\rho(\Psi \cup \eta_D \beta D)$, the pushout of the following diagram in the category of completely regular spaces:

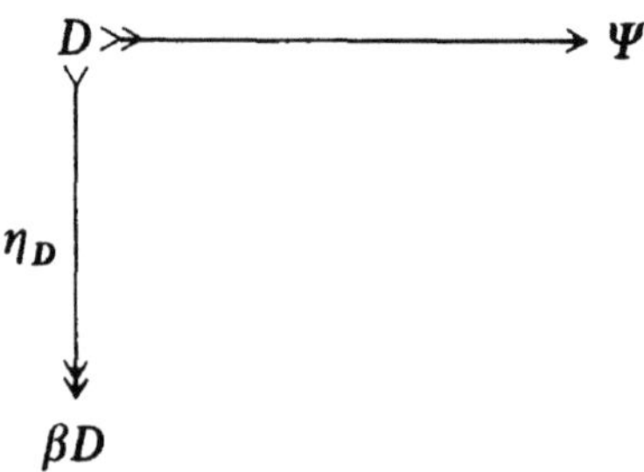

Note also that the pushout of the diagram in the category of topological spaces, i.e. $\Psi \cup \eta_D \beta D$, is not $\beta\Psi$. If the adjunction space were compact and Hausdorff, then the compact subspace βD would be C^*-embedded and as a consequence, so would D. But this would imply that D is C^*-embedded in Ψ, which we have seen is impossible. However, it is possible to show, using the maximality of $\mathscr{E}$, that every open cover of $\Psi \cup \eta_D \beta D$ has a finite subcover. Hence, it must be the case that $\Psi \cup \eta_D \beta D$ fails to be Hausdorff.

10.31. Example. Consider the Tychonoff Plank, $\mathbf{T} = (\omega_1 + 1) \times (\omega_0 + 1)$ as discussed in Sections 4.1 and 4.4 and let X be the subspace obtained from T by omitting the point $\{(\omega_0, 0)\}$ from the "bottom row". Then every non-compact zero-set of X meets the set $N = \{(n,0) : n < \omega_0\}$ and N is homeomorphic with $\mathbb{N}$. Hence, Theorem 10.29 shows that βX is the pushout in the category of completely regular spaces of the following diagram:

$$
\begin{array}{ccc}
N \!\!\!\gg\!\!\!\! & \longrightarrow & X \\
\Big\downarrow{\scriptstyle \eta_N} & & \\
\beta N & &
\end{array}
$$

However, in this case both X and βN are normal and the adjunction space $X \cup \eta_N \beta N$ is therefore Hausdorff [D, p. 145]. One can also show that the adjunction space is compact, so that in this case we have that $\beta X = X \cup \eta_N \beta N$. Thus, βX is obtained by attaching a copy of $\mathbb{N}^*$ and it is not actually necessary to consider the reflection of the adjunction space.

Perfect Mappings

10.32. A mapping $f : X \to Y$ is called a *compact mapping* if its fibers are compact, i.e. if $f^{\leftarrow}(y)$ is compact for every y in Y. The mapping f is said to be *perfect* if it is both closed and compact. In this section, we will investigate the relationships between perfect mappings and the Stone-Čech compactification.

There are numerous examples of perfect maps. Any map from a compact space to a Hausdorff space is perfect as is any closed embedding. A projection parallel to a compact factor is closed [D, p. 227] and has fibers which are homeomorphic to the compact factor. Hence, such a projection is perfect.

If $\{f_\alpha : X_\alpha \to Y_\alpha\}$ is a family of perfect maps, then the product mapping,

$$
\underset{\alpha}{\times} f_\alpha : \underset{\alpha}{\times} X_\alpha \to \underset{\alpha}{\times} Y_\alpha
$$

defined by $(\underset{\alpha}{\times} f_\alpha)((x_\alpha))_\alpha = f_\alpha(x_\alpha)$ is easily seen to be perfect.

10.33. We will find that the dual notion of a pushout will be useful in the description of perfect maps. The following square in a category $\mathscr{C}$

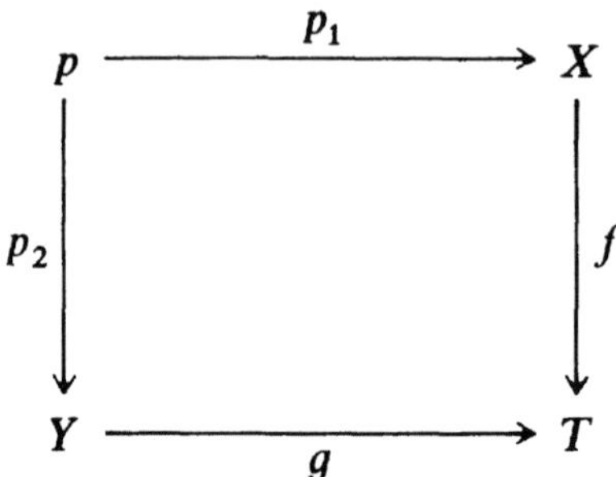

is said to be a *pullback* in $\mathscr{C}$ if for every pair of morphisms $l:Z\to X$ and $k:Z\to Y$ such that $f\circ l=g\circ k$ there exists a unique morphism $h:Z\to P$ such that $p_1\circ h=l$ and $p_2\circ h=k$. Following the example of pushouts, we say that P together with p_1 and p_2 is the pullback of f and g. Pullbacks are easily described in the category of topological spaces.

Proposition. *Pullbacks exist in the category of topological spaces.*

Proof. Let $f:X\to T$ and $g:Y\to T$ be maps and let

$$P = \{(x,y)\in X \times Y : f(x)=g(y)\}\,.$$

Let $p_1:P\to X$ and $p_2:P\to Y$ be the restrictions of the appropriate projections. If $l:Z\to X$ and $k:Z\to Y$ are maps such that $f\circ l=g\circ k$, then the definition $h(z)=(l(z),k(z))$ makes the following diagram commute since the image of l is easily seen to be contained in P:

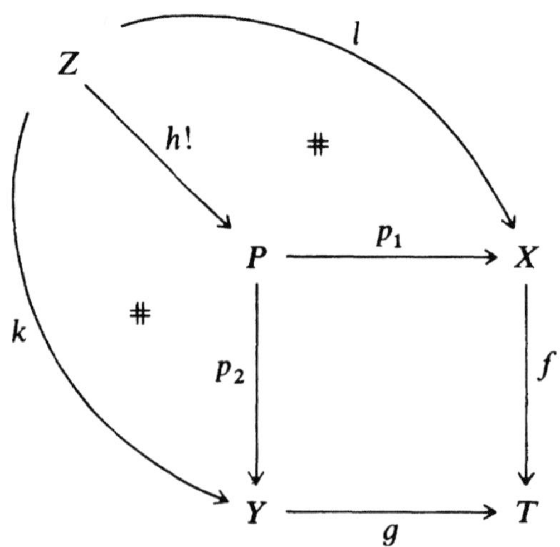

Thus, P together with the maps p_1 and p_2 is the pullback of f and g. □

The structure of P in the preceding proof indicates that pullbacks exist in many subcategories of the category of topological spaces. For instance, pullbacks will exist in any full subcategory which is productive and hereditary.

10.34. We now consider characterizations of perfect maps which relate them to the Stone-Čech compactification.

Proposition. *If X and Y are completely regular spaces, then the following are equivalent for a map $f:X \to Y$:*
  (1) *f is perfect.*
  (2) *If $\mathscr{U}$ is an ultrafilter on X and if $f[\mathscr{U}]$ converges to y in Y, then $\mathscr{U}$ converges (necessarily to some x in $f^{\leftarrow}(y)$).*
  (3) *$\beta(f)$ takes growth to growth, i.e. $\beta(f)[X^*]$ is contained in Y^*.*

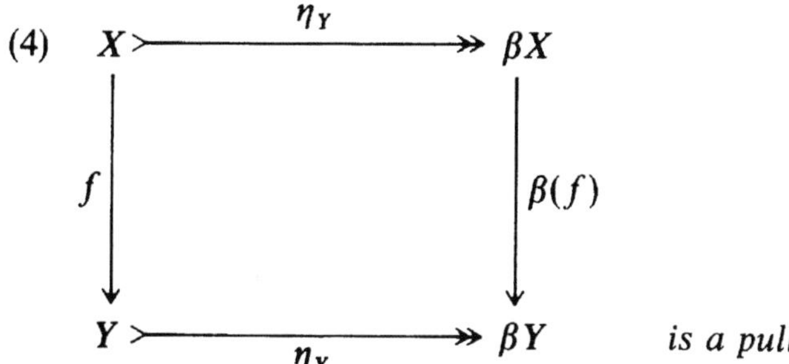

is a pullback.

Condition (2) in the proposition was introduced by N. Bourbaki. Condition (4) was discovered independently by S.P. Franklin and H. Herrlich and perhaps others.

Proof. (1)⇒(2): Let f be perfect and $\mathscr{U}$ be an ultrafilter on X such that $f[\mathscr{U}]$ converges to y in Y. Because f is continuous, if $\mathscr{U}$ converges it must converge to a point of $f^{\leftarrow}(y)$. If $\mathscr{U}$ fails to converge, then for every x in $f^{\leftarrow}(y)$, there is an open neighborhood U_x of x such that U_x is not in $\mathscr{U}$. Since $f^{\leftarrow}(y)$ is compact, it is covered by a finite subfamily $\{U_{x_i}\}$. The open set $V = \bigcup U_{x_i}$ does not belong to $\mathscr{U}$ because $\mathscr{U}$ is an ultrafilter. Thus, $X \setminus V$ does belong to $\mathscr{U}$ so that $f[X \setminus V]$ belongs to $f[\mathscr{U}]$. Because f is a closed map, $Y \setminus f[X \setminus V]$ is a neighborhood of y which fails to belong to $f[\mathscr{U}]$. This contradicts the assumption that $f[\mathscr{U}]$ converges to y. Hence, $\mathscr{U}$ must converge.

(2)⇒(1): We first show that f has compact fibers. Let $\mathscr{V}$ be an ultrafilter on $f^{\leftarrow}(y)$ for some y in Y. Let $\mathscr{U}$ be an ultrafilter on X which contains $\mathscr{V}$. Then $f[\mathscr{U}]$ converges to y in Y. Hence, $\mathscr{U}$ and therefore $\mathscr{V}$ converge to a point in $f^{\leftarrow}(y)$ and $f^{\leftarrow}(y)$ is compact because every ultrafilter on $f^{\leftarrow}(y)$ converges.

Now we show that f is closed. Let F be a closed subset of X and let $\mathscr{V}$ be an ultrafilter on $f[F]$ converging to a point y of Y. For every V in $\mathscr{V}$, $f(f^{\leftarrow}(V) \cap F) = V \cap f[F]$ is non-empty. Hence, the family $\{f^{\leftarrow}(V) \cap F : V \in \mathscr{V}\}$ is contained in an ultrafilter $\mathscr{U}$ on X. Then $f[\mathscr{U}]$ converges to y, and therefore $\mathscr{U}$ converges to a point x in $f^{\leftarrow}(y)$. Since F is closed, x belongs to F. Hence, $y = f(x)$ and $f[F]$ is closed.

$(2) \Rightarrow (3)$: Let p belong to βX. Because X is dense in βX, there is an ultrafilter $\mathscr{U}$ on X which converges to p. Continuity implies that $\beta(f)[\mathscr{U}] = f[\mathscr{U}]$ converges to a point q of βY. If q belongs to Y, then $\mathscr{U}$ converges to a point x in $f^{\leftarrow}(y)$. Because βX is Hausdorff, we must have $p = x$. Thus, the only points of βX which are mapped to points of Y are the points of X.

$(3) \Rightarrow (2)$: Suppose that $\mathscr{U}$ is an ultrafilter on X such that $f[\mathscr{U}]$ converges to y in Y. Because X is dense in βX, $\mathscr{U}$ converges to a point p of βX. Then continuity implies that $\beta(f)[\mathscr{U}] = f[\mathscr{U}]$ converges to y in Y and that p belongs to $\beta(f)^{\leftarrow}(y)$. Since $\beta(f)$ sends X^* into Y^*, p must belong to X.

$(3) \Rightarrow (4)$: Suppose that $h : Z \rightarrow \beta X$ and $g : Z \rightarrow Y$ are mappings such that $\beta(f) \circ h = \eta_Y \circ g$. Since $\eta_Y \circ g[Z]$ is contained in βY and $\beta(f)$ sends X^* into Y^*, we have that $h[Z]$ is contained in X. Hence, defining $l : Z \rightarrow X$ by $l(z) = h(z)$ shows that the square is a pullback.

$(4) \Rightarrow (3)$: Choose p in βX and assume that $\beta(f)(p) = y$ belongs to Y. Then let h be the map which embeds $\{p\}$ into βX and g be the map from the subspace $\{p\}$ which sends p to $\beta(f)(p)$. Then $\beta(f) \circ h = \eta_Y \circ g$ so that there exists a map $l : \{p\} \rightarrow X$ such that $h = \eta_X \circ l$. Hence, p belongs to X.   ☐

10.35. Since the extension of a perfect map sends growth to growth, it is clear that the inverse image of a compact space under a perfect mapping is compact. Thus, the fibers of a composition of perfect maps will be compact and we have verified the

Corollary. *The inverse image of a compact space under a perfect mapping is compact. Hence, the composition of perfect maps is perfect.*

10.36. A subcategory $\mathscr{A}$ of the completely regular spaces is said to be *left-fitting* if whenever Y is an object of $\mathscr{A}$ and $f : X \rightarrow Y$ is a perfect mapping, then X must also be an object of $\mathscr{A}$. *Right-fitting subcategories* are defined analogously.

Examples. We have just seen that the subcategory of compact spaces is left-fitting as a consequence of the fact that the extension of a perfect map sends growth to growth. By using this same condition on the extension we can show that the subcategories of locally compact, real-compact, and σ-compact spaces are also left-fitting.

Since the growth of a space is closed if and only if the space is locally compact, it is clear from this condition that the subcategory of locally compact spaces is both left and right-fitting.

Recall that a space X is realcompact if and only if every point of X^* is contained in a zero-set which misses X (Theorem 1.53). Then the condition on the growths shows that the category of realcompact spaces is left-fitting since the inverse image of a zero-set is a zero-set. In 1958 B, S. Mrówka gave an example to show that a perfect image of a realcompact space need not be realcompact. This example has also been discussed by M. Weir in his thesis.

The same condition also shows that the category of σ-compact spaces is left-fitting. (See Exercise 1B.) Since the continuous image of a compact space is compact, this subcategory is also right-fitting.

10.37. Now consider any subcategory $\mathscr{A}$ of completely regular spaces. Let $\mathscr{B}$ be the subcategory of completely regular spaces whose objects are the family of spaces which map perfectly onto a space belonging to $\mathscr{A}$. Then $\mathscr{B}$ contains $\mathscr{A}$ since an identity map is perfect. Further, $\mathscr{B}$ is left-fitting since perfect maps compose. Thus, every subcategory of completely regular spaces is contained in a left-fitting subcategory. Left-fitting subcategories are easily seen to be closed under intersections, and therefore there is a smallest left-fitting subcategory containing any subcategory $\mathscr{A}$. We will call this smallest left-fitting subcategory the *left-fitting hull* of $\mathscr{A}$.

A more complete description of the spaces in the left-fitting hull of is provided by the following result.

Proposition. *Let $\mathscr{A}$ be a subcategory of the category of completely regular spaces. Then the following are equivalent for any completely regular space X:*

    (1) *X belongs to the left-fitting hull of $\mathscr{A}$.*

    (2) *There exists a space A in $\mathscr{A}$ and a perfect map $f: X \to A$.*

    (3) *X can be embedded as a closed subspace of a product of a compact space and a space in $\mathscr{A}$.*

If $\mathscr{A}$ is countably productive, so is its left-fitting hull.

Proof. Let $\mathscr{L}$ be the left-fitting hull of $\mathscr{A}$.

    (1)$\Rightarrow$(2): The subcategory of $\mathscr{L}$ whose objects satisfy (2) is easily seen to contain $\mathscr{A}$ and is left-fitting because the composition of perfect maps is again perfect. Because $\mathscr{L}$ is the smallest left-fitting subcategory containing $\mathscr{A}$, the subcategory of $\mathscr{L}$ just described must be all of $\mathscr{L}$.

(2)$\Rightarrow$(1): If a space X maps perfectly to a space of $\mathscr{A}$, then X must belong to the smallest left-fitting subcategory containing $\mathscr{A}$. Hence, X belongs to $\mathscr{L}$.

(2)$\Rightarrow$(3): Let $f:X\to A$ be perfect with A belonging to $\mathscr{A}$ and let $\eta_X:X\to\beta X$ be the usual embedding. Then the Embedding Lemma, 1.5, implies that the evaluation map

$$e:X\to X\times\beta X$$

defined by $e(x)=(x,\eta_X(x))$ is an embedding. Further, $e[X]$ is the graph of η_X and is therefore closed in the product space because βX is Hausdorff [D, p. 140]. The mapping

$$f\times 1_{\beta X}:X\times\beta X\to A\times\beta X$$

is perfect because it is the product of two perfect maps. Also, $f\times 1_{\beta X}$ is clearly one-to-one. Hence, $(f\times 1_{\beta X})\circ e$ is a closed embedding since it is a continuous, closed bijection onto its range.

(3)$\Rightarrow$(2): Let $e:X\to A\times K$ be a closed embedding where A is an object of $\mathscr{A}$ and K is compact. The closed embedding e is perfect as is the projection π_A of $A\times K$ onto A. Hence, $\pi_A\circ e$ is the required perfect map from X to an object of $\mathscr{A}$. $\quad\square$

10.38. We now see that condition (3) of the previous proposition allows us to characterize left-fitting subcategories in a result similar to Theorem 10.21 for epi-reflective subcategories.

Proposition. *The left-fitting subcategories of the category of completely regular spaces are the closed hereditary subcategories which are closed under products with compact spaces.*

Proof. Let $\mathscr{A}$ be a left-fitting subcategory. Then $\mathscr{A}$ is closed hereditary because a closed embedding is a perfect map. The product of a space in $\mathscr{A}$ and a compact space is again in $\mathscr{A}$ because the projection parallel to a compact factor is a perfect map [D, p. 227].

Conversely, let $\mathscr{A}$ be a closed hereditary subcategory which is closed under products with compact spaces. Let Y be an object of $\mathscr{A}$ and $f:X\to Y$ be a perfect map. Then Proposition 10.37 implies that X is homeomorphic to a closed subspace of $A\times K$ where K is a compact space and A is an object of $\mathscr{A}$. The hypotheses on $\mathscr{A}$ then imply that X is an object of $\mathscr{A}$, so that $\mathscr{A}$ is left-fitting. $\quad\square$

10.39. Examples. (a) In Proposition 8.36, we saw that the product of a paracompact space and a compact space is paracompact. It is easy to see that a closed subspace of a paracompact space is paracompact, so that the subcategory of paracompact spaces is left-fitting.

(b) In a similar way, one can show that the Lindelöf spaces, the countably compact spaces, the metacompact spaces, and the countably paracompact spaces all form left-fitting subcategories.

(c) In 1964, K. Morita introduced the category of M-spaces in connection with the problem of characterizing those spaces whose products with metric spaces are normal. He showed that this subcategory is left-fitting. A survey of the theory of M-spaces is given in the 1971 paper of Morita.

(d) In his 1961 paper, Z. Frolík showed that the left-fitting hull of the category of completely metrizable spaces is the category of paracompact, topologically complete spaces.

(e) Frolík also showed that the left-fitting hull of the category whose objects are the open subspaces of $\mathbb{R}$ is the category of σ-compact, locally compact spaces.

(f) It is easy to see that the left-fitting hull of the category whose only object is the singleton space is the category of all compact spaces.

(g) In 1964, K. Morita showed that the left-fitting hull of the category of metric spaces is the category of paracompact M-spaces.

10.40. Earlier, in Section 10.24, we considered the epi-reflective hull of a single space. Now we will describe the spaces which belong to the epi-reflective hull of a left-fitting subcategory of completely regular spaces. We first need a preliminary result.

Proposition. *The intersection of a family of subspaces belonging to an epi-reflective subcategory $\mathscr{R}$ of the Hausdorff spaces also belongs to $\mathscr{R}$.*

Proof. Let $\{R_\alpha\}$ be a family of subspaces of a space X with each R_α in $\mathscr{R}$. Then the map $(e(y))_\alpha = y$ embeds $\bigcap R_\alpha$ as the diagonal of $\bigtimes R_\alpha$. Because the diagonal of a product of Hausdorff spaces is closed, Theorem 10.21 shows that $\bigcap R_\alpha$ belongs to $\mathscr{R}$. □

10.41. Following the terminology of Section 10.24, if $\mathscr{A}$ is a subcategory of completely regular spaces, then the $\mathscr{A}$-*completely regular spaces* are those spaces which are subspaces of products of spaces of $\mathscr{A}$. The $\mathscr{A}$-*compact spaces* are the $\mathscr{A}$-completely regular spaces which are closed subspaces of such products. In Chapter 1 we saw that the reflection βX of an I-completely regular space X in the category of I-compact spaces can be characterized as that compactification of X to which every mapping of X into I will extend. We now mimic the proof of Corollary 1.11 to derive the analogous result for the $\mathscr{A}$-compact reflection of an $\mathscr{A}$-completely regular space. Call a subspace S of X $\mathscr{A}$-*embedded* if every map from S to an object of $\mathscr{A}$ will extend to X.

Proposition. *The reflection aX of an $\mathscr{A}$-completely regular space X in the category of $\mathscr{A}$-compact spaces is the unique $\mathscr{A}$-compact space in which X is dense and $\mathscr{A}$-embedded.*

Proof. Since any object of $\mathscr{A}$ is $\mathscr{A}$-compact, aX must clearly have this property. Now suppose that X is dense and $\mathscr{A}$-embedded in an $\mathscr{A}$-compact space Y. We will show that Y is aX. Let f map X to an $\mathscr{A}$-compact space K. We must show that f extends to Y. Since a reflection must be unique (Proposition 10.11), this will show that Y is aX. Because K is $\mathscr{A}$-compact, there is a closed embedding e of K into a product $\bigtimes A_\alpha$ of spaces of $\mathscr{A}$. Then the composition $\pi_\alpha \circ e \circ f$ maps X to A_α and therefore extends to a mapping g_α of Y into A_α. If h from Y to $\bigtimes A_\alpha$ is defined by $h(y)_\alpha = g_\alpha(y)$, we have the following diagram:

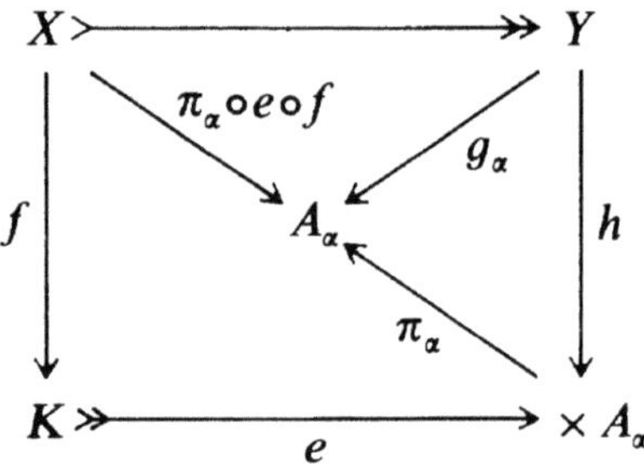

Since K is closed in $\bigtimes A_\alpha$, $h[Y] \subset e[K]$ as in Theorem 1.11 and $e^\leftarrow \circ h$ is the required extension of f. □

10.42. If $\mathscr{A}$ is a left-fitting subcategory of completely regular spaces, then the following result which appears in the 1971 paper of S. P. Franklin shows that aX is a subspace of βX.

Theorem (Franklin). *If $\mathscr{A}$ is a left-fitting subcategory of completely regular spaces, then the reflection aX of a completely regular space X in **Haus**($\mathscr{A}$) is the intersection of the subspaces of βX which contain X and belong to $\mathscr{A}$.*

The key to the proof will lie in the fact that any mapping $f : X \to Y$ can be extended to a subspace of βX so that the extended map is perfect and retains the same codomain as f. This observation follows from the fact that a map is perfect if its Stone-Čech extension sends growth to growth. Thus, $\beta(f)|\beta(f)^\leftarrow(Y)$ is the required perfect extension of f.

Proof. Let Y be the intersection of all subspaces of βX which contain X and belong to $\mathscr{A}$. Such subspaces must exist since it follows easily from Proposition 10.38 that any non-trivial left-fitting subcategory must contain all compact spaces and hence βX must be such a subspace.

Then X is dense in Y and Proposition 10.40 shows that Y is an object of **Haus**($\mathscr{A}$). By Proposition 10.41, we need only show that every map f of X into a space A of $\mathscr{A}$ will extend to Y. But $\beta(f)|\beta(f)^\leftarrow(A)$ is a perfect map into A and hence $\beta(f)^\leftarrow(A)$ is an object of $\mathscr{A}$ since $\mathscr{A}$ is left-fitting. Then by definition, Y is a subspace of $\beta(f)^\leftarrow(A)$ so that $\beta(f)|Y$ is the required extension. $\Box$

10.43. Note that we have shown in the previous result that if $\mathscr{A}$ is left-fitting, then every completely regular space is $\mathscr{A}$-completely regular. Since any $\mathscr{A}$-compact space is homeomorphic to its $\mathscr{A}$-compact reflection, the following corollary is immediate.

Corollary. *If $\mathscr{A}$ is a left-fitting subcategory of completely regular spaces, then every completely regular space is $\mathscr{A}$-completely regular. A space is $\mathscr{A}$-compact if and only if it is the intersection of a family of subspaces of βX which belong to $\mathscr{A}$.*

10.44. The definition of realcompact given in Section 1.53 shows that υX is the intersection of perfect inverse images of the real line $\mathbb{R}$. Many of the examples of left-fitting subcategories are themselves subcategories of the category of realcompact spaces and contain $\mathbb{R}$. Thus, they have the realcompact spaces as their epi-reflective hull. Examples of such subcategories are the Lindelöf spaces, the σ-compact spaces, and the σ-compact locally compact spaces. In addition, one can show that the epi-reflective hull of the paracompact spaces and the paracompact M-spaces will be the category of realcompact spaces if and only if no measurable cardinal exists.

10.45. In Theorem 10.41, we saw that the reflection of a space X in **Haus**($\mathscr{A}$) can be characterized by the existence of extensions of mappings of X into spaces belonging to $\mathscr{A}$. We now show that if we know $\mathscr{A}$ to be a left-fitting subcategory and that $\mathscr{A}$ is the left-fitting hull of a subcategory $\mathscr{B}$, then there is an alternate characterization of aX in terms of extensions of maps of X into spaces belonging to the smaller subcategory $\mathscr{B}$.

Theorem. *If $\mathscr{A}$ is the left-fitting hull of $\mathscr{B}$, then aX is characterized as a space Y with the following properties:*
   (a) *Y is an object of **Haus**($\mathscr{A}$) and $X \subset Y \subset \beta X$.*
   (b) *Any mapping of X into a space in $\mathscr{B}$ extends to Y.*
Proof. It is clear from Theorem 10.42 that aX satisfies (a) and (b) and also that

$$aX = \bigcap\{\beta(f)^\leftarrow(A) : f : X \to A \text{ with } A \in_o \mathscr{A}\}.$$

Since the embedding of X into an $\mathscr{A}$-compact subspace of βX must extend to aX, if Y satisfies (a) we must have aX contained in Y. We will exhibit a subspace Z of βX satisfying (a) and (b) such that any subspace Y of βX that also satisfies (a) and (b) will be contained in Z. Then we will have $aX \subset Y \subset Z$ and the proof will be completed by showing that $aX = Z$. Define Z by

$$Z = \bigcap \{\beta(f)^\leftarrow(B) : f : X \to B \text{ with } B \in_o \mathcal{B}\}\,.$$

It is clear that Z satisfies (a) and we can see that Z satisfies (b) by considering restriction of extensions to βX. The definition of Z shows that Z contains any subspace Y satisfying (a) and (b).

To show that $aX = Z$, we will show that the two families of subspaces of βX whose intersections define aX and Z are actually the same. By Proposition 10.37, if A is an object of $\mathcal{A}$, then there exists a perfect map $g : A \to B$ where g is onto and B is an object of $\mathcal{B}$. Now if $f : X \to A$, we have two extensions to βX: $\beta(f)$ and $\beta(g \circ f)$. Further, the extension $\beta(g) : \beta A \to \beta B$ satisfies $\beta(g)[A^*] \subset B^*$. Thus, we have

$$\beta(g \circ f)^\leftarrow(B) = \beta(f)^\leftarrow(\beta(g)^\leftarrow(B)) = \beta(f)^\leftarrow(A)\,.$$

Thus, every subspace of βX which appears in the intersection aX is also a subspace involved in the intersection Z. Hence, $Y = aX$. $\quad\square$

Observe that condition (a) in the theorem plays an essential role. If $\mathcal{A}$ is the category of all compact spaces, then we must have $aX = \beta X$. But $\mathcal{A}$ is the left-fitting hull of the singleton space and every constant map of X will of course extend to any compactification of X. Hence, it is necessary to require that Y be a subspace of βX.

Projectives

10.46. In the remainder of the chapter, we will adapt the definition of a projective object from homological algebra and module theory to investigate an analogous definition in a topological context. We will see that the adaptation is a fruitful one in that it provides additional information about the relationship between extremally disconnected spaces and complete Boolean algebras described in Proposition 2.5. There we saw that the Boolean algebra of clopen sets of a zero-dimensional space is complete precisely when the space is extremally disconnected.

Consider the following diagram in a category $\mathscr{C}$ of topological spaces and continuous mappings. We will say that the space P is a *projective object* in $\mathscr{C}$ if there exists a mapping $\psi : P \to X$ such that $f \circ \psi = g$ whenever f is a perfect onto map.

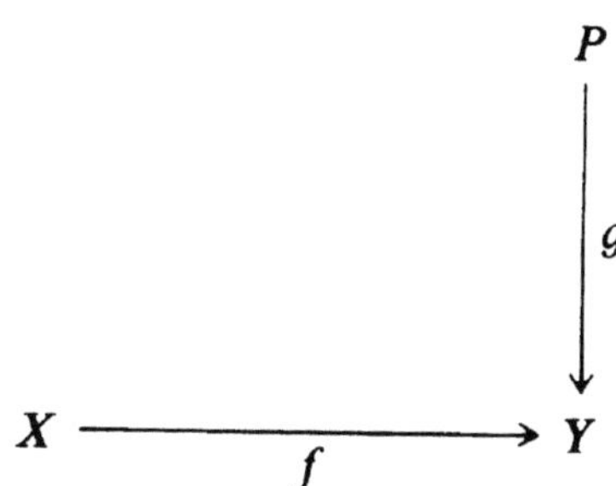

Recall from Section 2.5 that an *extremally disconnected space* is one in which the closure of every open subset is open. We will show that all projective objects in certain subcategories are extremally disconnected. Call a full subcategory $\mathscr{C}$ of Hausdorff spaces *acceptable* if:

(a) Whenever Y is an object of $\mathscr{C}$ and $\{0,1\}$ is the two point discrete space, then $A \times \{0,1\}$ is an object of $\mathscr{C}$.

(b) A closed subspace of an object of $\mathscr{C}$ is also an object of $\mathscr{C}$.

Observe that the categories of Hausdorff spaces, completely regular spaces, and compact Hausdorff spaces are all acceptable. In his 1958 paper, A. M. Gleason proved the following result which shows that a projective object in any of these three categories is extremally disconnected.

Proposition (Gleason). *Any projective object in an acceptable category is extremally disconnected.*

Proof. Let P be a projective object in an acceptable category $\mathscr{C}$ and let U be an open subspace of P. We must show that cl U is open. Let X be the closed subspace of $P \times \{0,1\}$ defined by

$$X = ((P \setminus U) \times \{0\}) \cup ((\mathrm{cl}\ U) \times \{1\}).$$

Because $\mathscr{C}$ is acceptable, X is an object of $\mathscr{C}$. Let π_P be the projection of $P \times \{0,1\}$ onto P and let e be the embedding of X into $P \times \{0,1\}$. Both π_P and e are perfect so that the composition $\pi_P \circ e$ is perfect. Since $P \subseteq (P \setminus U) \cup \mathrm{cl}\ U$, $\pi_P \circ e$ is also onto. Since P is a projective object of $\mathscr{C}$, there exists a mapping ψ such that the following diagram commutes:

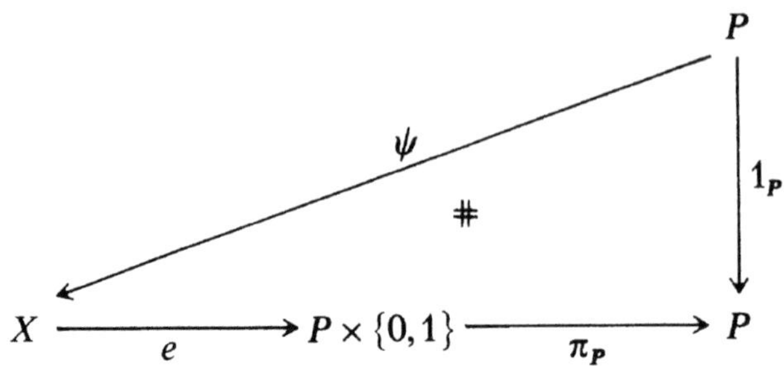

Because $\pi_P \circ e$ is one-to-one on $U \times \{1\}$, we must have $\psi(p) = (p, 1)$ for p in U. Hence, by continuity, $\psi(p) = (p, 1)$ for p in cl U. Similarly, if p is not in cl U, we have $\psi(p) = (p, 0)$. Thus, cl $U = \psi^{\leftarrow}(\text{cl } U \times \{1\})$ and because cl $U \times \{1\}$ is clopen in X, cl U is open in P. $\square$

10.47. The proposition indicates that the category of extremally disconnected spaces will play an important role in the investigation of projectives. The next sequence of results describe the properties of extremally disconnected spaces which we will need later. The following proposition is based on [GJ, ex. 1 H, 6 M].

Proposition. *The following are equivalent:*
   (1) *X is extremally disconnected.*
   (2) *Disjoint open subsets of X have disjoint closures.*
   (3) *βX is extremally disconnected.*
   (4) *Every dense subspace of X is C^*-embedded.*

Proof. (1)$\Rightarrow$(2): Let U and V be disjoint open subsets of X. Then cl $V \cap U = \emptyset$ because U is open. Similarly, cl $V \cap$ cl $U = \emptyset$ because cl V is open.

(2)$\Rightarrow$(1): Let U be open in X. Then U and $X \backslash$cl U have disjoint closures whose union is X. Hence, cl U is open.

(2)$\Rightarrow$(3): Let U be an open subset of βX. Then $U \cap X$ and $X \backslash \text{cl}(U \cap X)$ have disjoint closures whose union is X. Thus, $\text{cl}(U \cap X)$ is clopen in X and therefore, $\text{cl}_{\beta X} U = \text{cl}_{\beta X}(\text{cl}_X(U \cap X))$ is clopen in βX. Hence, βX is extremally disconnected.

(3)$\Rightarrow$(1): Let U be open in X. Then $U = X \cap V$ for some open set V of βX. Because X is dense in βX, $\text{cl}_X U = X \cap \text{cl}_{\beta X} V$, and $\text{cl}_X U$ is thus clopen in X.

(1) and (2)$\Rightarrow$(4): Let Y be dense in X. We will show that completely separated subsets of Y are completely separated in X. Thus, Urysohn's Embedding Lemma, 1.2, will show that Y is C^*-embedded. Completely

separated subsets of Y are contained in disjoint open subsets of Y. Disjoint open subsets of Y are easily seen to be the traces on Y of disjoint open subsets of X. Then (1) and (2) imply that completely separated subsets of Y are contained in disjoint clopen subsets of X and are thus completely separated in X.

(4)$\Rightarrow$(1): Let U be open in X. Then $Y = U \cup (X \backslash \text{cl}\, U)$ is dense in X, and hence is C^*-embedded. Since U and $X \backslash \text{cl}\, U$ are completely separated in Y, they are completely separated in X. But then their closures in X are disjoint closed sets whose union is X. Hence, $\text{cl}\, U$ is open and X is extremally disconnected. $\square$

10.48. A mapping f of X onto Y is said to be *irreducible* if Y is not the image under f of any proper closed subspace of X. The following results on irreducible mappings are taken from the 1967 paper of D. P. Strauss. Gleason proved similar lemmas under more restrictive assumptions. First we show that every compact map has an irreducible restriction.

Lemma. *If f is a compact mapping of a space X onto a space Y, then there is a closed subspace F of X such that $f|F$ is an irreducible mapping onto Y.*

Proof. Let $\mathscr{F}$ be the family of closed subspaces of X which are mapped onto Y by f. Let $\{H_\alpha : \alpha < \beta\}$ be a descending family of members of $\mathscr{F}$. For a point y of Y, $\{H_\alpha \cap f^\leftarrow(y)\}$ is a descending family of non-empty compact subsets and hence, $\bigcap (H_\alpha \cap f^\leftarrow(y))$ is not empty [D, p. 225]. Therefore, $f[\bigcap H_\alpha] = Y$ and Zorn's Lemma shows the existence of the subspace F. $\square$

10.49. Lemma. *Let $f : X \to Y$ be an irredicible map. If U is an open subset of X, then*

$$F[U] \subset \text{cl}(Y \backslash f[X \backslash U]).$$

Proof. We can assume that U is non-empty, since otherwise there is nothing to prove. Let y belong to $f[U]$ and let V be a neighborhood of y. It will suffice to show that V meets $Y \backslash f[X \backslash U]$. Because $U \cap f^\leftarrow(V)$ is a non-empty open set and f is irreducible, there is a point z belonging to $Y \backslash f[X \backslash (U \cap f^\leftarrow(V))]$. Choose x such that $f(x) = z$. Then $z = f(x)$ is in $V = f[f^\leftarrow(V)]$ so that we have $z \in V \cap (Y \backslash f[X \backslash U])$. $\square$

10.50. Now we relate irreducible maps to extremally disconnected spaces.

Lemma. *A closed irreducible mapping of a Hausdorff space onto an extremally disconnected space is a homeomorphism.*

Proof. Let $f : X \to Y$ be closed and irreducible, let Y be extremally disconnected, and let X be Hausdorff. We show that distinct points x and y of X have distinct images. Let U and V be disjoint open neighborhoods of x and y, respectively. Since U and V are disjoint, every point of Y belongs to either $f[X \backslash U]$ or $f[X \backslash V]$, so that we have $Y = f[X \backslash U] \cup f[X \backslash V]$. Hence, $Y \backslash f[X \backslash U]$ and $Y \backslash f[X \backslash V]$ are disjoint open sets, and since Y is extremally disconnected, they have disjoint closures. But $f(x)$ belongs to $\mathrm{cl}(Y \backslash f[X \backslash U])$ and $f(y)$ belongs to $\mathrm{cl}(Y \backslash f[X \backslash V])$ by Lemma 10.49 so that $f(x) \neq f(y)$. Hence, f is a closed, continuous bijection and is therefore a homeomorphism. $\square$

10.51. We are now able to characterize the projective objects in the category of compact spaces.

Theorem (Gleason). *The projective objects in the category of compact spaces are precisely the extremally disconnected spaces.*

Proof. Proposition 10.46 shows that the projective objects are all extremally disconnected. We must show that every extremally disconnected space is projective.

Let P be an extremally disconnected compact space. Let f map a compact space X onto another compact space Y. Then f must be perfect. Let g map P to Y. The pullback $L = \{(x, p) : f(x) = g(p)\}$ is a closed subspace of $X \times P$ and is therefore compact. Because f is onto, the projection π_P carries L onto P. Lemma 10.48 yields a closed subspace F of L such that $\pi_P | F$ is an irreducible mapping onto P. Hence, we have the following diagram:

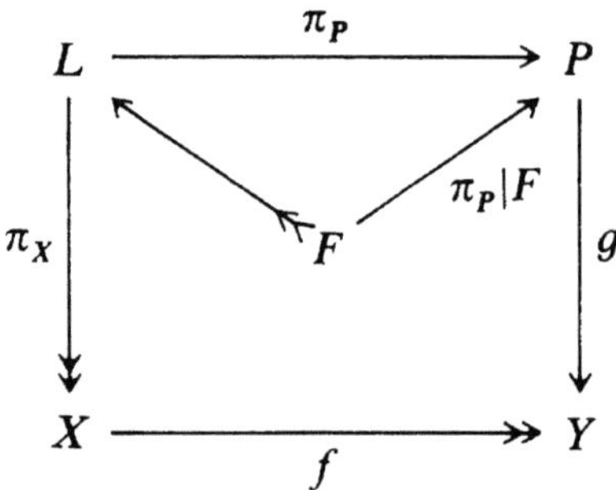

Since P is extremally disconnected, Lemma 10.50 shows that $\pi_P | F$ is a homeomorphism. Then $h = \pi_X \circ (\pi_P | F)^{\leftarrow}$ is the required map: If p is in P, then $f \circ h(p) = g(p)$ because L is the pullback. Hence, P is a projective object. $\square$

10.52. The previous result for compact spaces can be applied, together with a characterization of perfect maps, to obtain the analogous result for completely regular spaces.

Corollary. *The projective objects in the category of completely regular spaces are precisely the extremally disconnected spaces.*

Proof. It follows from Proposition 10.46 that a projective space is extremally disconnected.

Now we show that an extremally disconnected space is projective. Let P be extremally disconnected and let $f: X \to Y$ be a perfect-onto mapping between two completely regular spaces. Let $g: P \to Y$ be any mapping. Then by taking Stone-Čech compactifications and using the preceding theorem, there exists a mapping ψ making the following diagram commute since βP is extremally disconnected (Proposition 10.47):

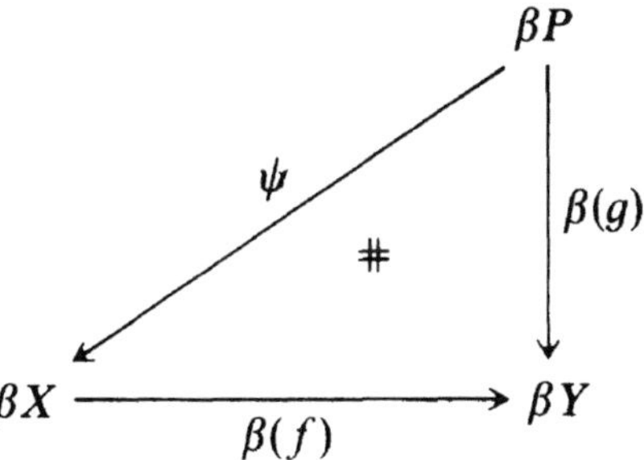

Because f is perfect, $\beta(f)$ sends X^* to Y^*. Hence, $f \circ (\psi | P) = g$ showing that P is projective. $\square$

10.53. In the algebraic setting of projectives, every module is the image of a projective module. To establish the analogous result in our topological setting, we will associate with each completely regular space a projective space $E(X)$ and a mapping g of $E(X)$ onto X in such a way that the topological structure of X is expressed in $E(X)$ and g. In particular, we would like $E(X)$ to be homeomorphic to X whenever X is already projective, i.e. when X is extremally disconnected. Lemma 10.50 indicates that this would be the case if we require g to be a closed, irreducible mapping of $E(X)$ onto X. We will be able to accomplish this and also make the restriction that the fibers of g be compact, i.e. that g will also be a compact mapping. We will call the projective space $E(X)$ together with the perfect irreducible mapping g the *projective cover* of X. Before considering the general question of the existence of projective covers, we look at a specific case.

Example. Denote by S_1 the space consisting of a convergent sequence $\{x_n : n \geqslant 1\}$ and its limit x_0. Define a map $g: \beta \mathbb{N} \to S_1$ by sending n to x_n and points of the growth to x_0. We have seen that $\beta \mathbb{N}$ is extremally disconnected and g is easily seen to be perfect and irreducible. Thus, $E(S_1) = \beta \mathbb{N}$.

10.54. Our procedure to show the existence of projective covers will follow the lines of the characterization of projectives. We will show the compact case first and then establish the completely regular case as a corollary. To show that every compact space has a projective cover, we must associate to each compact space X an extremally disconnected compact space $E(X)$ and an irreducible mapping of $E(X)$ onto X. The extremally disconnected space can be obtained in a natural way. The Boolean algebra $R(X)$ of regular closed subsets of X is complete (Proposition 2.3) and therefore its Stone space $S(R(X))$ is extremally disconnected (Proposition 2.5). Moreover, because the points of $S(R(X))$ are the maximal filters of $R(X)$ and X is compact, there is a natural way to define a mapping of $S(R(X))$ onto X. Each maximal filter $\mathscr{F}$ of regular closed sets is a family of closed subsets of X having the finite intersection property and therefore has non-empty intersection. Further, X has a base of regular closed sets which makes it easy to see that the intersection of each maximal filter $\mathscr{F}$ is a single point of X. Thus, to each maximal filter $\mathscr{F}$ in $R(X)$ we can associate the unique point of X belonging to $\bigcap \mathscr{F}$. To show that $S(R(X))$ is the projective cover of X, it remains only to show that the function just described is an irreducible mapping onto X and that any other compact extremally disconnected space which maps irreducibly onto X is homeomorphic to $S(R(X))$.

Theorem (Gleason). *The projective cover of a compact space X is $S(R(X))$ together with the mapping which assigns to each maximal filter of $R(X)$ its limit in X. Further, any compact extremally disconnected space which maps irreducibly onto X is homeomorphic to $S(R(X))$.*

The proof will require the use of the Boolean algebra isomorphism between $R(X)$ and the clopen subsets of $S(R(X))$. Recall from Theorem 2.10 that this isomorphism associates to each regular closed subset of X the family of maximal filters which contain it. We will use this assignment to show that the function described above is both continuous and irreducible.

Proof. Let $g: S(R(X)) \to X$ be the function described above. Note that if U is any open set containing the point $g(\mathscr{F})$, then cl U belongs to $\mathscr{F}$. This follows from the maximality of $\mathscr{F}$ since (cl $U)'$ does not contain $g(\mathscr{F})$ and hence cannot belong to $\mathscr{F}$.

We now show that g is continuous. Let $\mathscr{F}$ belong to $S(R(X))$ and let U be a neighborhood of $g(\mathscr{F})$. Because X is regular, there exists an open set V such that

$$g(\mathscr{F}) \subset V \subset \text{cl } V \subset U .$$

Now cl V is a regular closed subset of X and hence determines a clopen set W of $S(R(X))$ via the Stone isomorphism h of $R(X)$ with the clopen sets of $S(R(X))$. Thus, we have

$$W = h(\operatorname{cl} V) = \{\mathscr{G} \in S(R(X)): \operatorname{cl} V \in \mathscr{G}\} \,.$$

But then the definition of g implies that if $\mathscr{G}$ belongs to W,

$$g(\mathscr{G}) \in \bigcap \mathscr{G} \subset \operatorname{cl} V \subset U \,.$$

Thus, $g[W]$ is contained in U and g is continuous.

To show that g maps $S(R(X))$ onto X, we must show that every point x of X is the limit of a maximal filter of regular closed sets. But this is clear since the set of regular closed sets containing x in their interior is closed under finite meets and hence is contained in a maximal filter.

Finally, we show that g is irreducible. The complement of any proper closed subset F of $S(R(X))$ contains a non-empty basic clopen set W. The Stone Representation Theorem implies that there exists a non-empty regular closed subset K of X such that

$$W = h(K) = \{\mathscr{F} \in S(R(X)): K \in \mathscr{F}\} \,.$$

Now if $g(\mathscr{F})$ is in $\operatorname{int} K$, we have seen that K must belong to $\mathscr{F}$ and that $\mathscr{F}$ is in W. Hence, $g[F]$ misses $\operatorname{int} K$, and g is irreducible.

Hence, $S(R(X))$ and g form a projective cover of X.

Now suppose that Y is a compact extremally disconnected space which is mapped irreducibly onto X by f. Then we have a map ψ such that the following diagram commutes:

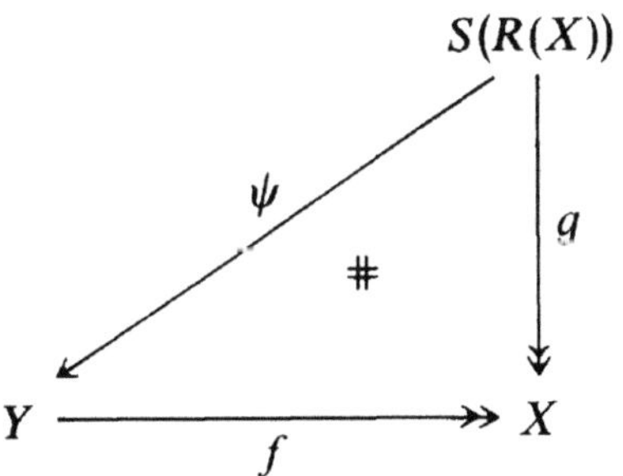

Because $S(R(X))$ is compact, ψ is closed. Because $f \circ \psi[S(R(X))] = X$ and f is irreducible, the closed subspace $\psi[S(R(X))]$ of Y must be all of Y. On the other hand, if F is a proper closed subspace of $S(R(X))$, $g[F] = f \circ \psi[F]$ cannot be all of X because g is irreducible. Hence, $\psi[F]$ is a proper subspace of Y and ψ is therefore irreducible. Therefore,

Lemma 10.50 implies that ψ is a homeomorphism because Y is extremally disconnected. ☐

Note that the preceding theorem together with Example 10.53 shows that that the Stone space of $R(S_1)$ is $\beta\mathbb{N}$.

10.55. As was the case for the characterization of projective objects, we now use the existence of a projective cover of a compact space to obtain a projective cover for an arbitrary completely regular space.

Corollary. *Every completely regular space has a projective cover which is unique up to homeomorphism.*

Proof. Let X be completely regular and let $g:S(R(\beta X))\to\beta X$ be the projective cover of βX. Put $E(X)=g^{\leftarrow}(X)$. The restriction $g|E(X)$ is easily seen to be perfect and irreducible because it is the restriction to a saturated set. If U is a non-empty open subset of $S(R(\beta X))$ missing $E(X)$, then $g[S(R(\beta X))\backslash U]$ is a closed subspace of βX containing X, and hence, is all of βX. Thus, $S(R(\beta X))\backslash U = S(R(\beta X))$ since g is irreducible, and U therefore is empty. Hence, $E(X)$ is dense in $S(R(\beta X))$. Proposition 10.47 implies that $E(X)$ is C^*-embedded in the extremally disconnected space $S(R(\beta X))$. Hence, $\beta(E(X))=S(R(\beta X))$ so that Proposition 10.47 now implies that $E(X)$ is extremally disconnected. Hence, $g|E(X):E(X)\to X$ is the projective cover of X. The uniqueness of $E(X)$ follows from that of $S(R(\beta X))$. ☐

10.56. Note that in the preceding proof we showed that $\beta(E(X))=S(R(\beta X))$. Since $S(R(\beta X))$ is $E(\beta X)$, we have established that E and β commute.

Proposition. *If X is any completely regular space, then $\beta(E(X))=E(\beta X)$.*

10.57. The 1967 paper of Strauss contains a construction of projective covers using filters of open sets. The 1959 paper of J. Rainwater describes an additional construction. The 1971 paper of B. Banaschewski describes projective covers in more general categorical terms. The 1971 paper of A. W. Hager provides a fourth proof of the existence of a projective cover for a compact space.

10.58. Now we will borrow again from module theory and briefly consider injective Boolean algebras. Consider the following diagram in the category of Boolean algebras and Boolean algebra homomorphisms:

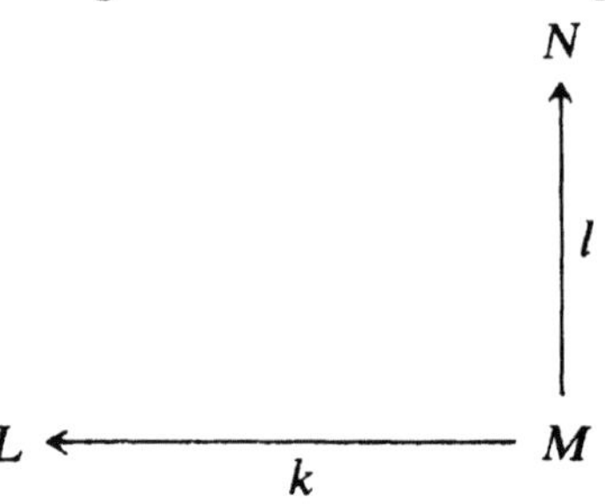

The Boolean algebra N is said to be *injective* if there exists a homomorphism $\varphi: L \to N$ such that $\varphi \circ k = l$ whenever k is a monomorphism, i.e. when k is one-to-one.

We will use the duality between Boolean algebras and compact totally disconnected spaces to characterize the injective Boolean algebras. If we express the Boolean algebras as algebras of clopen sets of compact totally disconnected spaces, then we have the following two diagrams:

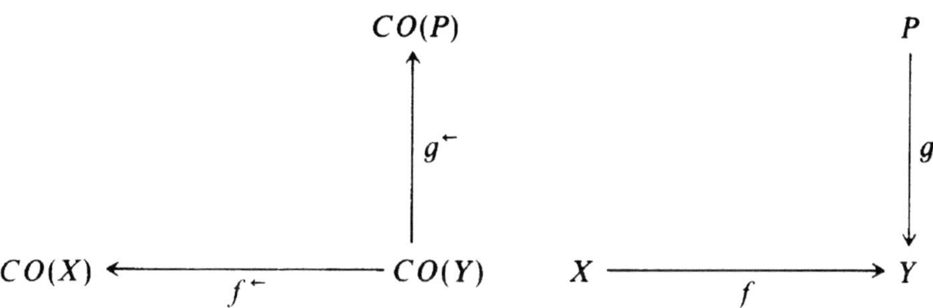

Further, we can easily see from Proposition 2.8 that $f^{\leftarrow}$ is a monomorphism exactly when f is onto. Since X is compact and Y is Hausdorff, f will always be perfect.

Since a mapping $\psi: P \to X$ such that $f \circ \psi = g$ will always exist precisely when P is extremally disconnected, we see that the homomorphism $\varphi = \psi^{\leftarrow}$ will exist precisely when $CO(P)$ is complete (Proposition 2.5). Hence, we have established the following result from the 1948 paper of R. Sikorski.

Theorem (Sikorski). *The injective Boolean algebras are precisely the complete ones.*

Exercises

10A. *Epi-reflective subcategories*

The category of all topological spaces and all mappings has the following properties:
1. The epimorphisms are the onto mappings.
2. The extremal monomorphisms are the embeddings.
3. Every mapping factors into an epimorphism followed by an extremal monomorphism.

4. The epi-reflective subcategories are the productive and hereditary subcategories.
5. The completely regular spaces form an epi-reflective subcategory.
6. A space is *functionally Hausdorff* if distinct points are completely separated. The functionally Hausdorff spaces form an epi-reflective subcategory. The reflection is the quotient space obtained by identifying points which are not completely separated.

10 B. *Coreflective subcategories*

A functor $s:\mathscr{C}\to\mathscr{A}$ is a *coreflective functor* if $\mathscr{A}$ is a subcategory of $\mathscr{C}$ and there is a morphism $\varepsilon_C:sC\to C$ such that any morphism $f:A\to C$ with A an object of $\mathscr{A}$ factors uniquely through ε_C. $\mathscr{A}$ is called a *coreflective subcategory* of $\mathscr{C}$ and sC is called the *coreflection* of C in $\mathscr{A}$. We will give three examples of coreflective subcategories of the category of all topological spaces by describing the coreflection.

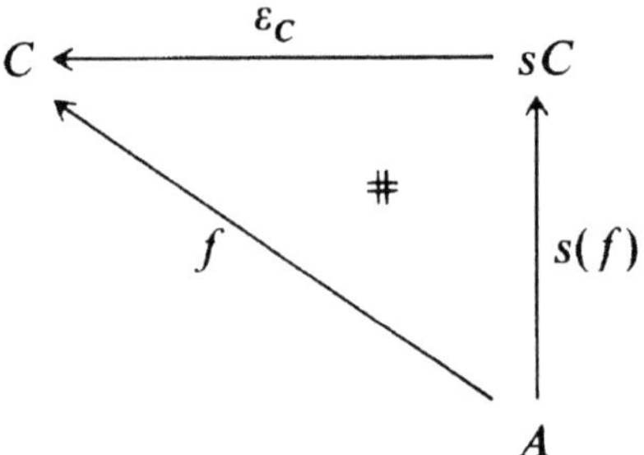

1. The category of P-spaces (Section 1.65) is a coreflective subcategory and the coreflection is obtained by enlarging the topology on the space to include all G_δ sets as open sets.
2. The P-space coreflection of a completely regular space can be obtained by enlarging the topology to include the zero-sets.
3. The category of k-spaces is a coreflective subcategory. [The coreflection is obtained by enlarging the topology in a suitable way. See Exercise 8 G.]
4. A subset of a space is called *sequentially open* if every sequence converging to a point in the subset is eventually in the subset. Any open set is sequentially open but not conversely. $[\omega_1+1]$
5. A *sequential space* is one in which every sequentially open set is open. The sequential spaces form a coreflective subcategory. The coreflection is obtained by enlarging the topology to include the sequentially open sets.

References: The notion of topological coreflections are discussed in Herrlich, 1968, and in the papers of Herrlich and Strecker.

10C. *Characterization of coreflective subcategories*

A coreflective subcategory is called *mono-coreflective* if the mapping $\varepsilon_C : sC \to C$ is a monomorphism. An epimorphism $e^{\#}$ is an *extremal epimorphism* if whenever $e^{\#}$ can be factored as illustrated

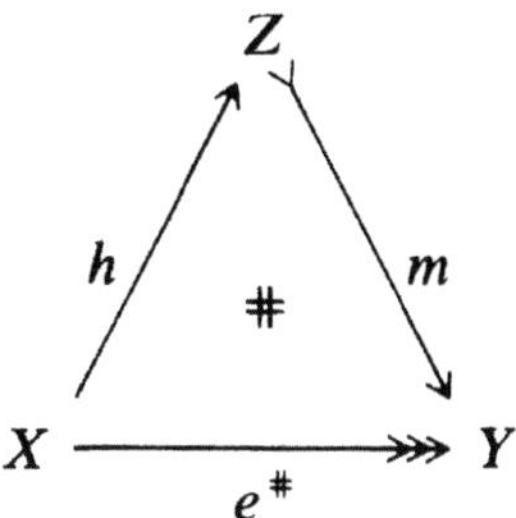

so that m is a monomorphism, then m is an isomorphism. Extremal epimorphisms will be indicated by "triple-headed arrows" and a "$\#$" on the name of the morphism. The category of all topological spaces has the following properties:

1. The monomorphisms are the one-to-one mappings.
2. The extremal epimorphisms are the quotient mappings.
3. Every mapping factors into an extremal epimorphism followed by a monomorphism. [Identify fibers of the mapping.]
4. There is only a set of pairwise non-homeomorphic spaces which can be mapped to a given space by a monomorphism.
5. The monocoreflective subcategories are those which are closed under quotients and disjoint topological sums. [Dualize Theorem 10.21.]

10D. *The categories of k-spaces and sequential spaces*

As in Example 10.53, let S_1 denote a convergent sequence and its limit point.

1. A subspace of a sequential space is closed if and only if its intersection with every copy of S_1 in the space is closed in S_1.
2. Every first countable space is sequential.
3. Not every sequential space is first countable. [Let $\mathbb{Z}$ denote the subspace of the integers in $\mathbb{R}$. Consider the quotient $\mathbb{R}/\mathbb{Z}$ obtained by identifying the points of Z.]
4. Every sequential space is a k-space. [Exercise 8G and S_1 is compact.]
5. Not every k-space is sequential. [$\omega_1 + 1$.]

References: For sequential spaces, S. P. Franklin, 1965, 1967A, 1967B, and S. Baron, 1968. For k-spaces, Kelley, 1955, and Franklin, 1967B. N. E. Steenrod, 1967, discusses the importance of the category of k-spaces in the context of algebraic topology. Arhangel'skiĭ and Franklin, 1968, introduce ordinal invariants based on these two categories of spaces.

10E. *Locally connected spaces*

A space is *locally connected* if each point has a base of connected neighborhoods.

1. A space is locally connected if and only if the components of its open sets are open.
2. A quotient of a locally connected space is locally connected.
3. A disjoint topological sum of a family of locally connected spaces is locally connected.
4. The locally connected spaces form a mono-coreflective subcategory of the category of all topological spaces.

References: A.M. Gleason, 1963. One might except from Exercise 10B that the coreflection is obtained by enlarging the topology to include components of open sets. The coreflection is actually obtained by iterating this process.

10F. *The category of pseudocompact spaces*

1. The product of a pseudocompact space and a compact space is pseudocompact. [Proposition 8.21.]
2. A regular closed subspace of a pseudocompact space is pseudo-compact. [Proposition 8.5.]
3. A closed subspace of a pseudocompact space need not be pseudo-compact. Hence, the category of pseudocompact spaces is not left-fitting. [Example 6.8 and Proposition 10.38.]
4. Consider the space Ψ of Example 10.30. Let X be the space obtained by attaching to Ψ a sequence S_E of isolated points for each E in $\mathscr{E}$ such that S_E converges to ω_E. X is not pseudocompact.
5. Define a mapping of X onto Ψ by sending points of Ψ to themselves and points of S_E to ω_E. Such a mapping is perfect, showing again that the category of pseudocompact spaces is not left-fitting.
6. Let $\mathscr{P}$ be the class of spaces such that if Y is in $\mathscr{P}$, then $X \times Y$ is pseudocompact for any pseudocompact space X. If Y is a pseudo-compact space such that every point of Y has a neighborhood be-longing to $\mathscr{P}$, then Y is in $\mathscr{P}$. [Modify Proposition 8.21.]
7. $\mathscr{P}$ is finitely productive.

Reference: The example described in (4) and (5) was suggested by S.P. Franklin and B.V.S. Thomas. (6) is from Frolík, 1960B.

10G. *One-point compactification as functor*

1. Let X and Y be non-compact, locally compact spaces. A mapping

$f: X \to Y$ is perfect if and only if the extension $\alpha(f): \alpha X \to \alpha Y$ defined by

$$\alpha(f)(x) = \begin{cases} f(x) & \text{for } x \in X \\ \infty & \text{for } x = \infty \end{cases}$$

is continuous.

2. Let $\mathscr{C}$ be the (non-full) category of all non-compact, locally compact spaces and perfect mappings. The one-point compactification defines a functor from $\mathscr{C}$ to the category of compact spaces.

10 H. *Left-fitting and epi-reflective*

1. An epi-reflective subcategory of the category of Hausdorff spaces is left-fitting if and only if it contains all the compact spaces. [Theorem 10.21 and Proposition 10.38.]
2. υX is the smallest realcompact subspace of βX which contains X.

10 I. *Another definition of projective*

Define an object P of a category $\mathscr{C}$ of topological spaces and mappings to be projective if whenever $f: X \to Y$ is an onto mapping and $g: P \to Y$ is any mapping, then there exists a map $\psi: P \to X$ such that $f \circ \psi = g$. With this definition, the projective objects in any full subcategory of topological spaces and mappings which contains the discrete spaces are just the discrete spaces.

10 J. *An equivalent of the Axiom of Choice*

In Lemma 10.48, we used a form of the Axiom of Choice, Zorn's Lemma, to show that any compact onto mapping has an irreducible restriction. Show that the existence of an irreducible restriction for any compact onto mapping implies the Axiom of Choice. [Let $\{F_\alpha : \alpha \in \mathscr{A}\}$ be any family of non-empty sets. Define a topology on each F_α by letting the closed subsets be the finite subsets. Map $\oplus F_\alpha$ onto the index set $\mathscr{A}$ with $\mathscr{A}$ having the discrete topology.]
Reference: S. P. Franklin and B. V. S. Thomas, 1971.

10 K. *Coproducts of projective objects*

If $\mathscr{C}$ is any full subcategory in which coproducts exist, then a coproduct of a family of objects is projective if and only if each object in the family is projective.

10 L. *Categories closed under pullbacks*

A subcategory of the category of Hausdorff spaces is closed under pullbacks if and only if it is closed hereditary and closed under finite products.

10M. *A mapping property of projective covers*

Let X and Y be completely regular and let $g_X: E(X) \to X$ and $g_Y: E(Y) \to Y$ be their projective covers.
1. If f is a mapping of Y onto X, then there is a mapping h of $E(Y)$ into $E(X)$ such that $f \circ g_Y = g_X \circ h$.
2. If X and Y are compact, then h is onto.
Reference: Henriksen and Jerison, 1965, considers the question of when the mapping h is unique in (2).

10N. *βD and extremally disconnected spaces*

1. If the composition $f \circ g$ of two maps is an embedding, then g is an embedding.
2. If X an extremally disconnected space having density $\mathfrak{n}$ and D is the discrete space of cardinality $\mathfrak{n}$, then X is a subspace of βD. [X is projective and βX is an image of βD.]
3. Every separable extremally disconnected space is a subspace of $\beta \mathbb{N}$.
Reference: Suggested by G. Naber.

10O *Indecomposable continua as growths*

Apply Theorem 10.29 to give another proof of sufficiency in Theorem 9.30: It X is a non-compact locally connected generalized continuum which has the strong complementation property, then X^* is homeomorphic to the growth of the ray $[1, \infty)$ and thus is an indecomposable continuum. [Lemma 9.29.]

Bibliography

Alexandroff, P.:
1939   Bikompakte Erweiterungen topologischer Räume. Mat. Sbornik 5, 403–423 (1939). (Russian. German summary.) MR 1, p. 318.

Alexandroff, P. and Hopf, H.:
1935   Topologie, Berlin: Springer 1935.

Alexandroff, P., and Urysohn, P.:
1929   Mémoire sur les espaces topologiques compacts. Verh. Akad. Wetensch. Amsterdam **14**, 1–96 (1929).

Alò, R.A.:
1969   Uniformities and embeddings. Proc. Int. Sympos. on Topology and Its Appls. (Herceg-Novi, 1968), pp. 45–59. Belgrad: Savez Drustava Mat. Fiz, i Astronom. 1969. MR 42 # 3737.

1972   Some Tietze type extension theorems. General Topology and Its Relations to Modern Analysis and Algebra ,III (Proc. Third Prague Topological Sympos., 1971), pp. 23–27, Prague: Acàdemia 1972.

Alò, R. A., Imler, L., and Shapiro, H. L.:
1970   P- and z-embedded subspaces. Math. Ann. **188**, 13–22 (1970). MR 42 # 1062.

Alò, R.A. and Sennot, L.:
1971   Extending linear space-valued functions. Math. Ann. **191**, 79–86 (1971). MR 43 # 6877.

1972   Collectionwise normality and the extension of functions on product spaces. Fund. Math. **76**, 231–243 (1972).

Alò, R.A. and Shapiro, H.L.:
1968A   A note on compactifications and semi-normal spaces. J. Austral. Math. Soc **8**, 102–108 (1968). MR 37 # 3527.

1968B   Normal bases and compactifications. Math. Ann. **175**, 337–340 (1968). MR 36 # 3312.

1969A   Wallman compact and realcompact spaces. Contributions to Extension Theory of Topological Structures (Proc. Sympos., Berlin, 1967), pp. 9–14. Berlin: Deutscher Verlag Wissensch. 1969. MR 40 # 872.

1969B $\mathscr{Z}$-realcompactifications and normal bases. J. Austral. Math. Soc. **9**, 489–495 (1969). MR 39 # 3455.

1970   Continuous uniformities. Math. Ann. **185**, 322–328 (1970). MR 41 # 4484.

1974   Normal Topological Spaces. Cambridge Tracts in Mathematics No. 65, Cambridge and New York: Cambridge University Press 1974.

Aquaro, G.:
1961   Completamenti di spazii uniformi. Ann. Mat. Pura Appl. **56**, 87–98 (1961).
        MR 25 # 3505.
1962A   Convergenza localmente quasi-uniforme ed estensione di Hewitt di uno spazio
        completamente regolare. Annali della Scuola Normale Superiore di Pisa **16**,
        207–212 (1962). MR 26 # 5536.
1962B   Completions of uniform spaces. General Topology and Its Relations to Modern
        Analysis and Algebra (Proc. Prague Topological Sympos., 1961), pp. 69–71,
        Prague: Academia 1962.

Arhangel'skiĭ, A. V.:
1967   An extremally disconnected bicompactum of weight c is inhomogeneous. Dokl.
        Akad. Nauk SSSR **175**, 751–754 (1967). (Russian.) English Translation: Soviet
        Math. Dokl. **8**, 897–900 (1967). MR 36 # 2122.

Arhangel'skiĭ, A. V. and Franklin, S. P.:
1968   Ordinal invariants for topological spaces. Michigan Math. J. **15**, 313–320 (1968).
        MR 39 # 2112.

Aull, C. E.:
1971   Properties of side points of sequences. General Topology and Appls. **1**, 201–208
        (1971). MR 44 # 4705.

Bagley, R. W. and McKnight, J. D., Jr.:
1959   On Q-spaces and collections of closed sets with the countable intersection property.
        Quart. J. Math. Oxford Ser. (2), **10**, 233–235 (1959). MR 22 # 964.

Bagley, R. W., Connell, E. H., and McKnight, J. D.:
1958   On properties characterizing pseudo-compact spaces. Proc. Amer. Math. Soc. **9**,
        500–506 (1958). MR 20 # 3523.

Banach, S.:
1932   Théorie des opérations linéaires, Warsaw, 1932; New York: Chelsea Publishing Co.
        1955.

Banaschewski, B.:
1955   Über nulldimensionale Räume. Math. Nachr. **13**, 129–140 (1955). MR 19, p. 157.
1956A   Überlagerungen von Erweiterungsräumen. Arch. Math. **7**, 107–115 (1956). MR 18,
        p. 224.
1956B   Local connectedness of extension spaces. Canad. J. Math. **8**, 395–398 (1956).
        MR 17, p. 1229.
1959   On the Katětov and Stone-Čech extensions. Canad. Math. Bull. **2**, 1–4 (1959),
        MR 21 # 3823.

1960   On homeomorphisms between extension spaces. Canad. J. Math. **12**, 252–262
        (1960). MR 22 # 4046.
1963   On Wallman's method of compactification. Math. Nachr. **27**, 105–114 (1963).
        MR 28 # 3400.

1964   Extensions of topological spaces. Canad. Math. Bull. **7**, 1–22 (1964). MR 28 # 4501.
1971   Projective covers in categories of topological spaces and topological algebras.
        General Topology and Its Relations to Modern Analysis and Algebra, III (Proc.
        Conf., Kanpur, 1968), pp. 63–91. Prague: Academia, 1971.

Baron, S.:
1968   The coreflective subcategory of sequential spaces. Canad. Math. Bull. **11**, 603–604 (1968). MR 38 # 5158.
1969   Reflectors as composition of epireflectors. Trans. Amer. Math. Soc. **136**, 499–508 (1969). MR 38 # 4535.

Bellamy, D. P.:
1968   Topological properties of compactifications of a half-open interval, Thesis, Michigan State Univ., East Lansing, Mich., 1968.
1971   A non-metric indecomposable continuum. Duke Math. J. **38**, 15–20 (1971). MR 42 # 6792.

1971A   Aposyndesis in the remainder of Stone-Čech compactifications. Bull. Acad. Polon. Sci. Sér. Sic. Math. Astronom. Phys. **19**, 941–944 (1971).

Bellamy, D. P. and Rubin, L. R.:
1973   Indecomposable continua in Stone-Čech compactifications. Proc. Amer. Math. Soc. **39**, 427–432 (1973). Mr 47 # 4219.

Bernstein, A. R.:
1970   A new kind of compactness for topological spaces. Fund. Math. **66**, 185–193 (1970). MR 40 # 4924.

Biles, C. M.:
1970   Wallman-type compactifications. Proc. Amer. Math. Soc. **25**, 363–368 (1970). MR 41 # 7634.

Birkhoff, G.:
1948   Lattice Theory. Amer. Math. Soc. Colloquium Publications, Vol. 25 (revised edition) 1948. MR 10, p. 673.

Blair, R. L.:
1974   On v-embedded sets in topological spaces. In: Proceedings of the Second Pittsburgh International Conference on General Topology and Its Applications, Lecture Notes in Mathematics, Vol. **378**, pp. 46–79. Berlin-Heidelberg-New York: Springer 1974.

Blair, R. L. and Hager, A. W.:
        Extensions of zero-sets and of real-valued functions. Math. Z. **136**, 41–52 (1974).
        Z-embedding in $\beta X \times \beta Y$. (To appear.)
        Notes on the Hewitt realcompactification of a product space. (To appear.)

Blass, A. R.:
1970   Orderings of ultrafilters, Thesis, Harvard University, 1970.
1973   The Rudin-Keisler ordering of P-points. Trans. Amer. Math. Soc. **179**, 145–166 (1973).

Blefko, R.:
1965   On E-compact spaces, Thesis, Pennsylvania State University, 1965.

Blefko, R. and Mrówka, S.:
1966   On the extensions of continuous functions from dense subspaces. Proc. Amer. Math. Soc. **17**, 1396–1400 (1966). MR 34 # 1989.

Booth, D. D.:
1969   Countably indexed ultrafilters, Thesis, University of Wisconsin, 1969.
1970   Ultrafilters on a countable set. Annals of Math. Logic **2**, 1–24 (1970). MR 43 # 3104.

Bourbaki, N.:
1966   General Topology, Reading, Mass.: Addison-Wesley 1966. MR 34 # 5044.

Brooks, R. M.:
1967   On Wallman compactifications. Fund. Math. **60**, 157–173 (1967). MR 35 # 964.

Buchwalter, H.:
1971   Sur le théorème de Glicksburg-Frolík. C. R. Acad. Sci. Paris Ser. A–B. **273**, A11–A14 (1971). MR 45 # 9291.

Buddenhagen, J. R.:
1971   Subsets of a countable set. Amer. Math. Monthly **78**, 536–537 (1971).

Calder, A.:
1972   The cohomotopy groups of Stone-Čech increments. Indag. Math. **34**, 37–44 (1972). MR 46 # 2673.

Cartan, H.:
1937A Théorie des filtres. C.R. Acad. Sci. Paris Ser. A–B. **205**, 595–598 (1937). Z 17, p. 243.
1937B Filtres et ultrafiltres. C.R. Acad. Sci. Paris Ser. A–B. **205**, 777–779 (1937). Z 18, p. 3.

Čech, E.:
1937   On bicompact spaces. Ann. of Math. **38**, 823–844 (1937). Z 17, p. 428. Russian Translation: Uspehi Mat. Nauk **26**, 165–186 (1971). MR 45 # 2663.

Čech, E. and Novák, J.:
1947   On regular and combinatorial embedding. Časopis Pěst. Mat. **72**, 7–16 (1947). (English. Czech. summary.) MR 9, p. 98.

Chandler, R. E.:
1972   New compactifications from old. Amer. Math. Monthly **79**, 501–503 (1972). MR 45 # 9292.

Chew, Kim-Peu:
1970   A characterization of $\mathbb{N}$-compact spaces. Proc. Amer. Math. Soc. **26**, 679–682 (1970). MR 42 # 2436.

Choquet, G.:
1968A Construction d'ultrafilters sur $\mathbb{N}$. Bull. Sci. Math. **92**, 41–48 (1968). MR 38 # 2722.
1968B Deux classes remarquables d'ultrafilters sur $\mathbb{N}$. Bull. Sci. Math. **92**, 143–153 (1968). MR 38 # 5154.

Chou, Ching:
1969   Minimal sets and ergodic measures for $\beta N \backslash N$. Illinois J. Math. **13**, 777–788 (1969). MR 40 # 2814.

Cohen, H. B.:
1964   The k-extremally disconnected spaces as projectives. Canad. J. Math. **16**, 253–260 (1964). MR 28 # 4502.

Comfort, W. W.:
1963   An example in density character. Arch. Math. **14**, 422–423 (1963). MR 28 # 4503.
1965   Retractions and other continuous maps from βX onto $\beta X - X$. Trans. Amer. Math. Soc. **114**, 1–9 (1965). MR 32 # 3035.
1967A A non-pseudocompact product space whose finite products are pseudocompact. Math. Ann. **170**, 41–44 (1967). MR 35 # 965.
1967B Locally compact realcompactifications. General Topology and Its Relations to

Modern Analysis and Algebra, II (Proc. Second Prague Topological Sympos., 1966), pp. 95–100, Prague: Academia 1967.

1968A A theorem of Stone-Čech type, and a theorem of Tychonoff type, without the axiom of choice; and their realcompact analogues. Fund. Math. **63**, 97–110 (1968). MR 38 # 5174.

1968B On the Hewitt realcompactification of a product space. Trans. Amer. Math. Soc. **131**, 107–118 (1968). MR 36 # 5896.

1969 A short proof of Marczewski's separability theorem. Amer. Math. Monthly **76**, 1041–1042 (1969). MR 40 # 1993.

1970 Closed Baire sets are (sometimes) zero-sets. Proc. Amer. Math. Soc. **25**, 870–875 (1970). MR 41 # 3695.

1971 A survey of cardinal invariants. General Topology and Appls. **1**, 163–199 (1971). MR 44 # 7510.

Comfort, W. W. and Gordan, H.:
1964 Disjoint open subsets of $\beta X - X$. Trans. Amer. Math. Soc. **111**, 513–520 (1964). MR 29 # 583.

Comfort, W. W. and Hager, A. W.:
1970 Estimates for the number of real-valued continuous functions. Trans. Amer. Math. Soc. **150**, 619–631 (1970). MR 41 # 7621.

1971 The projection mapping and other continuous functions on a product space. Math. Scand. **28**, 77–90 (1971). MR 47 # 4206.

1972 Cardinality of t-complete Boolean algebras. Pacific J. Math. **40**, 541–545 (1972).

Comfort, W. W., Hindman, N., and Negrepontis, S.:
1969 F'-spaces and their product with P-spaces. Pacific J. Math. **28**, 489–502 (1969). MR 39 # 3440.

Comfort, W. W. and Negrepontis, S.:
1965 The ring $C(X)$ determines the category of X. Proc. Amer. Math. Soc. **16**, 1041–1045 (1965). MR 31 # 6209.

1966 Extending continuous functions on $X \times Y$ to subsets of $\beta X \times \beta Y$. Fund. Math. **59**, 1–12 (1966). MR 34 # 782.

1968 Homeomorphs of three subspaces of $\beta \mathbb{N} - \mathbb{N}$. Math. Z. **107**, 53–58 (1968). MR 38 # 2739.

1972 On families of large oscillation. Fund. Math. **75**, 275–290 (1972). MR 46 # 4473.

1974 The Theory of Ultrafilters, Grundlehren der mathematischen Wissenschaften, Bd. 211, Berlin-Heidelberg-New York: Springer 1974.

Comfort, W. W. and Ross, K.:
1963 On the infinite product of topological spaces. Arch. Math. **14**, 62–64 (1963). MR 26 # 6926.

1966 Pseudocompactness and uniform continuity in topological groups. Pacific J. Math. **16**, 483–496 (1966). MR 34 # 7699.

Corson, H. H.:
1959 Normality in subsets of product spaces. Amer. J. Math. **81**, 785–796 (1959). MR 21 # 5947.

Curtis, P. C., Jr.:
1960 A note concerning certain product spaces. Arch. Math. **11**, 50–52 (1960). MR 22 # 1876a.

Day, M. M.:
1962 Normed Linear Spaces, Ergebnisse der Mathematik und ihrer Grenzgebiete, Heft 21, Berlin-Göttingen-Heidelberg: Springer 1962. MR 26 # 2847.

Dickman, R. F., Jr.:
1967   Unicoherence and related properties. Duke Math. J. **34**, 343–351 (1967). MR 35 # 3632.
1969   On closed extensions of functions. Proc. Nat. Acad. Sci. U.S.A. **62**, 326–332 (1969). MR 42 # 5233.
1972   A necessary and sufficient condition for $\beta X \backslash X$ to be an indecomposable continuum. Proc. Amer. Math. Soc. **33**, 191–194 (1972). MR 45 # 4364.

Doss, R.:
1949   On uniform spaces with a unique structure. Amer. J. Math. **71**, 19–23 (1949). MR 10, p. 557.

Dowker, C. H.:
1955   Local dimension of normal spaces. Quart. J. Math. Oxford **6**, 101–120 (1955). MR 19, p. 157.

Duda, E.:
1963   Brouwer property spaces. Duke Math. J. **30**, 647–660 (1963). MR 27 # 5236.

Dugundji, J.:
1966   Topology, Boston: Allyn and Bacon 1966. MR 33 # 1824.

Dunford, N. and Schwartz, J.:
1966   Linear Operators, I, New York: Interscience Publishers 1966. MR 22 # 8302.

Dwinger, P. H.:
1961   Introduction to Boolean Algebras, Würzburg: Physica-Verlag 1961. MR 26 # 6093.
1968   On a class of reflective subcategories. Indag. Math. **30**, 36–45 (1968). MR 37 # 1431.

Dykes, N.:
1969   Mappings and realcompact spaces. Pacific J. Math. **31**, 347–358 (1969). MR 41 # 7644.

Efremovič, V.A.:
1951   Infinitesimal spaces. Dokl. Akad. Nauk. SSSR **76**, 341–343 (1951). (Russian.) MR 12, p. 744.
1952   Proximity geometry, I. Mat. Sb. **31**, 189–200 (1952). (Russian.) MR 14, p. 1106.

Engelking, R.:
1964   Remarks on real-compact spaces. Fund. Math. **55**, 303–308 (1964). MR 31 # 4000.
1968   Outline of General Topology, New York: John Wiley and Sons, Inc. 1968. MR 36 # 4508.

Engelking, R. and Mrówka, S.:
1958   On E-compact spaces. Bull. Acad. Polon. Sci. Sér, Sci. Math. Astronom. Phys. **6**, 429–436 (1958). MR 20 # 3522.

Fan, K. and Gottesman, N.:
1952   On compactifications of Freudenthal and Wallman. Indag. Math. **14**, 504–510 (1952). MR 14, p. 669.

Fine, N. J. and Gillman, L.:
1960   Extension of continuous functions in βN. Bull. Amer. Math. Soc. **66**, 376–381 (1960). MR 23 # A619.
1962   Remote points in βR. Proc. Amer. Math. Soc. **13**, 29–36 (1962). MR 26 # 732.

Firby, P.A.:
1971   Finiteness at infinity. Proc. Edinburgh Math. Soc. **17**, 299–304 (1970/71). MR 46 # 2641.

Flachsmeyer, J.:
1961 Zur Spektralentwicklung topologischer Räume. Math. Ann. **144**, 253–274 (1961). MR 26 # 735.

Fleischer, I. and Franklin, S. P.:
1969 On compactness and projections. Contributions to Extension Theory of Topological Structures (Proc. Sympos., Berlin, 1967), pp. 77–79, Berlin: Deutscher Verlag Wissensch. 1969.

Franklin, S. P.:
1965 Spaces in which sequences suffice. Fund. Math. **57**, 107–115 (1965). MR 31 # 5184.
1967A Spaces in which sequences suffice II. Fund. Math. **61**, 51–56 (1967). MR 36 # 5882.
1967B The categories of k-spaces and sequential spaces. Class Notes, Carnegie Institute of Technology, 1967.
1970 Topics in categorical topology. Class Notes, Carnegie-Mellon University, 1970.
1971 On epi-reflective hulls. General Topology and Appls. **1**, 29–31 (1971). MR 44 # 3287.

Franklin, S. P. and Rajagopalan, M.:
1971 Some examples in topology. Trans. Amer. Math. Soc. **155**, 305–314 (1971). MR 44 # 972.

Franklin, S. P. and Thomas, B. V. S.:
1971 Another topological equivalent of the axiom of choice. Amer. Math. Monthly **78**, 1109–1110 (1971). MR 45 # 3195.

Franklin, S. P. and Walker, R. C.:
1972 Normality of powers implies compactness. Proc. Amer. Math. Soc. **36**, 295–296 (1972).

Freudenthal, H.:
1951 Kompaktisierungen und Bikompaktisierungen. Indag. Math. **13**, 184–192 (1951). MR 12, p. 728.

Freyd, P.:
1964 Abelian Categories: An Introduction to the Theory of Functors, New York: Harper and Row 1964. MR 29 # 3517.

Frink, O.:
1964 Compactifications and semi-normal spaces. Amer. J. Math. **86**, 602–607 (1964). MR 29 # 4028.

Frolík, Z.:
1959 Generalizations of compact and Lindelöf spaces. Czechoslovak Math. J. **9**, 172–217 (1959). (Russian. English summary.) MR 21 # 3821.
1960A The topological product of countably compact spaces. Czechoslovak Math. J. **10**, 329–338 (1960). (English. Russian summary.) MR 22 # 8480.
1960B The topological product of two pseudocompact spaces. Czechoslovak Math. J. **10**, 339–349 (1960). (English. Russian summary.) MR 22 # 7099.
1961 On approximation and uniform approximation of spaces. Proc. Japan Acad. **37**, 530–532 (1961). MR 25 # 1533.
1967A Types of ultrafilters on countable sets. General Topology and Its Relations to Modern Analysis and Algebra, II (Proc. Second Prague Topological Sympos., 1966), pp. 142–143, Prague: Academia 1967. MR 38 # 654.
1967B Sums of ultrafilters. Bull. Amer. Math. Soc. **73**, 87–91 (1967). MR 34 # 3525.
1967C Non-homogeneity of $\beta P \backslash P$. Comment. Math. Univ. Carolinae **8**, 705–709 (1967). MR 42 # 1068.

1967D Homogeneity problems for extremally disconnected spaces. Comment. Math. Univ. Carolinae **8**, 757–763 (1967). MR 41 # 9176.

1967E On two problems of W. W. Comfort. Comment. Math. Univ. Carolinae **8**, 139–144 (1967). MR 35 # 966.

1968A Fixed points of maps of βN. Bull. Amer. Math. Soc. **74**, 187–191 (1968). MR 36 # 5897, Errata MR 37, p. 1470.

1968B Fixed points of maps of extremally disconnected spaces and complete Boolean algebras. Bull. Acad. Polon. Sci. Sér. Sci. Math. Astronom. Phys. **16**, 269–275 (1968). MR 38 # 1665.

1971 Maps of extremally disconnected spaces, theory of types, and applications. General Topology and Its Relations to Modern Analysis and Algebra, III (Proc. Conf., Kanpur, 1968), pp. 131–142, Prague: Academia 1971. MR 45 # 4373.

Frolík, Z. and Mrówka, S.:
1971 Perfect images of $\mathscr{R}$- and $\mathscr{N}$-compact spaces. Bull. Acad. Polon. Sci. Sér. Sci. Math. Astronom. Phys. **19**, 369–371 (1971). (Russian summary.) MR 46 # 6288.

Gagrat, M. S. and Naimpally, S. A.:
1966 Proximity approach to extension problems. Fund. Math. **71**, 63–76 (1966).

1973 Wallman compactifications and Wallman realcompactifications. J. Austral. Math. Soc. **15**, 417–427 (1973).

Gait, J.:
1970 Spaces having no large dyadic subspace. Bull. Austral. Math. Soc. **2**, 261–265 (1970). MR 42 # 2437.

Gelfand, I. and Kolmogoroff, A.:
1939 On rings of continuous functions on topological spaces. Dokl. Akad. Nauk SSSR **22**, 11–15 (1939).

Gelfand, I. and Shilov, G.:
1941 Über verschiedene Methoden der Einführung der Topologie in die Menge der maximalen Ideale eines normierten Ringes. Rec. Mat. (Mat. Sbornik) N.S. **9**, 25–38 (1941).

Gillman, L.:
1960 A P-space and an extremally disconnected space whose product is not an F-space. Arch. Math. **11**, 53–55 (1960). MR 22 # 1876 b.

1961 A note on F-spaces. Arch. Math. **12**, 67–68 (1961). MR 23 # A2187.

1967 The space βN and the continuum hypothesis. General Topology and Its Relations to Modern Analysis and Algebra, II (Proc. Second Prague Topological Sympos., 1966), pp. 144–146, Prague: Academia 1967. MR 38 # 1656.

Gillman, L. and Henriksen, M.:
1954 Concerning rings of continuous functions. Trans. Amer. Math. Soc. **77**, 340–362 (1954). MR 16, p. 156.

1956A Some remarks about elementary divisor rings. Trans. Amer. Math. Soc. **82**, 362–365 (1956). MR 18, p. 9.

1956B Rings of continuous functions in which every finitely generated ideal is principal. Trans. Amer. Math. Soc. **82**, 366–391 (1956). MR 18, p. 9.

Gillman, L., Henriksen, M., and Jerison, M.:
1954 On a theorem of Gelfand and Kolmogoroff concerning maximal ideals in rings of continuous functions. Proc. Amer. Math. Soc. **5**, 447–455 (1954). MR 16, p. 607.

Gillman, L. and Jerison, M.:
1959   Stone-Čech compactifications of a product. Arch. Math. **10**, 443–446 (1959). MR 22 # 223.
1960   Rings of Continuous Functions, Princeton: Van Nostrand 1960. MR 22 # 6994.

Ginsburg, J. and Saks, V.:
       Some applications of ultrafilters in topology. Pacific J. Math. (to appear).

Gleason, A. M.:
1958   Projective topological spaces. Illinois J. Math. **2**, 482–489 (1958). MR 22 # 12509.
1963   Universal locally connected refinements. Illinois J. Math. **7**, 521–531 (1963). MR 29 # 1612.

Glicksburg, I.:
1952   The representation of functionals by integrals. Duke Math. J. **19**, 253–261 (1952). MR 14, p. 288.
1959   Stone-Čech compactifications of products. Trans. Amer. Math. Soc. **90**, 369–382 (1959). MR 21 # 4405.

Gould, G. G.:
1964   A Stone-Čech-Alexandroff-type compactification and its application to measure theory. Proc. London Math. Soc. **14**, 221–244 (1964). MR 30 # 442.

Green, J. W.:
1972   Filter characterizations of C- and C^*-embedding. Proc. Amer. Math. Soc. **35**, 574–580 (1972). MR 46 # 2634.

de Groot, J. and McDowell, R. H.:
1967   Locally connected spaces and their compactifications. Illinois J. Math. **11**, 353–364 (1967). MR 35 # 6129.

Hager, A. W.:
1966   Some remarks on the tensor product of function rings. Math. Z. **92**, 210–224 (1966). MR 33 # 1831.
1969A Projections of zero-sets (and the fine uniformity on a product space). Trans. Amer. Math. Soc. **140**, 87–94 (1969). MR 39 # 3448.
1969B On inverse closed subalgebras of $C(X)$. Proc. London Math. Soc. **19**, 233–257 (1969). MR 39 # 6261.
1971   The projective resolution of a compact space. Proc. Amer. Math. Soc. **28**, 262–266 (1971). MR 42 # 6788.
1972   Uniformities on a product. Canad. J. Math. **24**, 379–389 (1972). MR 45 # 5948.
1973   Compactification and completion as absolute closure. Proc. Amer. Math. Soc. **40**, 635–638 (1973).

Hahn, H.:
1914   Mengentheorctische Charakterisierung der stetigen Kurven. Sitzungsberichte, Akad. der Wissenschaften **123**, 2433 (1914).

Halmos, P. R.:
1963   Lectures on Boolean Algebras, New York: Van Nostrand 1963. MR 29 # 4713.

Hamburger, P.:
1972   On k-compactifications and realcompactifications. Acta Math. Acad. Sci. Hungar. **23**, 255–262 (1972). MR 47 # 1025.

Hanf, W.:
1957   On some fundamental problems concerning isomorphism of Boolean algebras. Math. Scand. **5**, 205–217 (1957).

Harris, D.:
1971   The Wallman compactification as a functor. General Topology and Appls. **1**,
        273–281 (1971). MR 45 # 1122.
1972   The Wallman compactification is an epireflection. Proc. Amer. Math. Soc. **31**,
        265–267 (1972). MR 44 # 5927.

Hausdorff, F.:
1936   Über zwei Sätze von G. Fichtenholz und L. Kantorovitch. Studia Math. **6**, 18–19
        (1936).
1957   Set Theory, New York: Chelsea 1957. MR 19, p. 111.

Hechler, S. H.:
1972   Short complete nested sequences in $\beta N \backslash N$ and small maximal almost-disjoint
        families. General Topology and Appls. **2**, 139–149 (1972).

Heider, L. J.:
1959   Compactifications of dimension zero. Proc. Amer. Math. Soc. **10**, 377–384 (1959).
        MR 21 # 3822.

Henriksen, M.:
1956   On the equivalence of the ring, lattice, and semigroup of continuous functions.
        Proc. Amer. Math. Soc. **7**, 959–960 (1956). MR 18, p. 559.
1959   Multiplicative summability methods and the Stone-Čech compactification. Math.
        Z. **71**, 427–435 (1959). MR 21 # 7434.

Henriksen, M. and Isbell, J. R.:
1957A   Local connectedness in the Stone-Čech compactification. Illinois J. Math. **1**,
        574–582 (1957). MR 20 # 2688.
1957B   On the Stone-Čech compactification of a product of two spaces. Bull. Amer.
        Math. Soc. **63**, 145–146 (1957).
1958   Some properties of compactifications. Duke Math. J. **25**, 83–105 (1958). MR
        20 # 2689.

Henriksen, M. and Jerison, M.:
1957   A non-normal subspace of βN. Bull. Amer. Math. Soc. **63**, 146 (1957).
1965   Minimal projective extensions of compact spaces. Duke Math. J. **32**, 291–295
        (1965). MR 32 # 1670.

Herrlich, H.:
1967A   Fortsetzbarkeit stetiger Abbildungen und Kompaktheitsgrad topologischer Räu-
        me. Math. Z. **96**, 64–72 (1967). MR 34 # 8370.
1967B   *E*-kompakte Räume. Math. Z. **96**, 228–255 (1967). MR 34 # 5051.
1968   Topologische Reflexionen und Coreflexionen, Lecture Notes in Mathematics,
        Vol. 78, Berlin-Heidelberg-New York: Springer 1968. MR 41 # 988.
1969   On the concept of reflections in general topology. Contributions to Extension
        Theory of Topological Structures (Proc. Sympos., Berlin, 1967) pp. 105–114.
        Berlin: Deutscher Verlag Wissensch. 1969. MR 44 # 2210.
1971   Categorical topology. General Topology and Appls. **1**, 1–15 (1971). MR 44 # 974.

Herrlich, H. and Strecker, G. E.:
1968   *H*-closed spaces and reflective subcategories. Math. Ann., **177**, 302–309 (1968).
1971A   Algebra $\cap$ Topology = Compactness. General Topology and Appl. **1**, 283–287
        (1971).
1971B   Coreflective subcategories. Trans. Amer. Math. Soc. **157**, 205–226 (1971). MR
        43 # 6281.

1972 Coreflective subcategories in general topology. Fund. Math. **73**, 199–218 (1972).

Hewitt, E.:
1943 A problem in set-theoretic topology. Duke Math. J. **10**, 309–333 (1943). MR 5, p. 46.
1946 A remark on density characters. Bull. Amer. Math. Soc. **52**, 641–643 (1946). MR 8, p. 139.
1948 Rings of real-valued continuous functions I. Trans. Amer. Math. Soc. **64**, 45–99 (1948). MR 10, p. 126.

Hindman, N.:
1969A On P-like spaces and their product with P-spaces, Thesis, Wesleyan University, Middletown, Conn., 1969.
1969B On the existence of c-points in $\beta\mathbb{N} - \mathbb{N}$. Proc. Amer. Math. Soc. **21**, 277–280 (1969). MR 39 # 922.
1973 Preimages of points under the natural map from $\beta(\mathbb{N} \times \mathbb{N})$ to $\beta\mathbb{N} \times \beta\mathbb{N}$. Proc. Amer. Math. Soc. **37**, 603–608 (1973).

Hocking, J.G. and Young, G.S.:
1961 Topology, Reading, Mass.: Addison-Wesley 1961. MR 23 # A2857.

Hu, S.T.:
1964 Elements of General Topology, San Francisco-London-Amsterdam: Holden-Day, Inc. 1964. MR 31 # 1643.

Hung, H.H. and Negrepontis, S.:
1973 Spaces homeomorphic to $(2^\alpha)_\alpha$. Bull. Amer. Math. Soc. **79**, 143–146 (1973).

Hušek, M.:
1969 The class of k-compact spaces is simple. Math. Z. **110**, 123–126 (1969). MR 39 # 6260.
1970 The Hewitt realcompactification of a product. Comment. Math. Univ. Carolinae **11**, 393–395 (1970). MR 42 # 2438.
1971 Pseudo-m-compactness and $\upsilon(P \times Q)$. Nederl. Akad. Wetensch. Proc. Ser. A74 = Indag. Math. **33**, 320–326 (1971). MR 45 # 5957.
1972A Realcompactness of function spaces and $\upsilon(P \times Q)$. General Topology and Appls. **2**, 165–179 (1972). MR 46 # 6301.
1972B Perfect images of E-compact spaces. Bull. Acad. Polon. Sci. Sér. Sci. Math. Astronom. Phys. **20**, 41–45 (1972). (Russian summary.) MR 46 # 6289.

Isbell, J.R.:
1955 Zero-dimensional spaces. Tôhoku Math. J. **7**, 1–8 (1955). MR 19, p. 156.
1964A Uniform Spaces, Mathematical Surveys, No. 12, Providence: American Mathematical Society 1964. MR 30 # 561.
1964B Subobjects, adequacy, completeness and categories of algebras. Rozprawy Mat. **36**, 3–33 (1964). MR 29 # 1238.

Iséki, K.:
1957 A characterization of pseudo-compact spaces. Proc. Japan Acad. **33**, 320–322 (1957). MR 19, p. 757.

Isiwata, T.:
1952 On uniform space with complete structure. Sugaku Kenkyuroku **1**, 68–74 (1952). (Japanese.)
1957A Some classes of completely regular T_1-spaces. Sci. Rep. Tokyo Kyoiku Daigaku. Sect. A**5**, 287–292 (1957). MR 20 # 1963.
1957B On the ring of all bounded continuous functions. Sci. Rep. Tokyo Kyoiku Daigaku. Sect. A**5**, 293–294 (1957). MR 20 # 1964.

1957C A generalization of Rudin's theorem for the homogeneity problem. Sci. Rep. Tokyo Kyoiku Daigaku. Sect. A5, 300–303 (1957). MR 20 # 1965.

1957D On subspaces of Čech compactification space. Sci. Rep. Tokyo Kyoiku Daigaku Sect. A5, 304–309 (1957). MR 20 # 1966.

1958   On Stonian spaces. Sci. Rep. Tokyo Kyoiku Daigaku Sect. A6, 147–176 (1958). MR 21 # 3824.

1959   Some properties of F-spaces. Proc. Japan Acad. 35, 71–76 (1959). MR 21 # 3825.

1960   On the duality concerning Stone-Čech compactifications. Sûgaku 11, 226–228 (1959/60). MR 25 # 553. (Japanese.)

1963   Normality and perfect mappings. Proc. Japan Acad. 39, 95–97 (1963). MR 27 # 718.

1964   Some classes of countably compact spaces. Czechoslovak Math. J. 14, 22–26 (1964). MR 29 # 2766.

1967   Mappings and spaces. Pacific J. Math. 20, 455–480 (1967). MR 36 # 2127.

1969 Z-mappings and C^*-embeddings. Proc. Japan Acad. 45, 889–893 (1969). MR 42 # 1078.

1974   Some properties of the remainder of Stone-Čech compactifications. Fund. Math. 83, 129–142 (1974).

Jacobson, N.:
1945   A topology for the set of primitive ideals in an arbitrary ring. Proc. Nat. Acad. Sci. U.S.A. 31, 333–338 (1945), MR 7, p. 110.

Jerison, M.:
1950   The space of bounded maps into a Banach space. Ann. of Math. 52, 309–327 (1950). MR 12, p. 188.

1965   Sur l'anneau des germes des fonctions continues. C.R. Acad. Sci. Paris Sér. A–B 260, 6507–6509 (1965). MR 31 # 3884.

1967   Rings of germs of continuous functions, Functional Analysis (Proc. Conf., Irvine, Calif., 1966), pp. 168–174, London: Academic Press; Washington, D.C.: Thompson Book Co. 1967. MR 36 # 4342.

Jerison, M., Siegel, J. and Weingram, S.:
1969   Distinctive properties of Stone-Čech compactifications. Topology 8, 195–201 (1969). MR 39 # 4607.

Johnson, D. G. and Mandelker, M.:
1973   Functions with pseudocompact support. General Topology and Appls. 3, 331–338 (1973).

Johnson, D.R.:
1971   The construction of elementary extensions and the Stone-Čech compactification, Thesis, Yale University, 1971. '

Juhász, I.:
1969   On the character of points in $\beta\mathbb{N}_m$, Contributions to Extension Theory of Topological Structures (Proc. Sympos., Berlin, 1967), pp. 138–139, Berlin: Deutscher Verlag Wissensch. 1969.

Kakutani, S.:
1941   Concrete representations of abstract (M)-spaces. Ann. of Math. 42, 994–1024 (1941).

Kamke, E.:
1950   Theory of Sets, New York: Dover Publications, Inc. 1950.

Katětov, M.:
1951   Remarks on Boolean algebras. Colloq. Math. 2, 229–235 (1951). MR 14, p. 237.

1953 On real-valued functions in topological spaces. Fund. Math. **38**, 85–91 (1951); **40**, 203–205 (1953). MR 14, p. 304; MR 15, p. 640.

1960 Über die Berührungsräume. Wiss. Z. Humboldt-Univ. Berlin Math.-Natur. Reihe **9**, 685–691 (1959/60). (Russian, English, and French summaries.) MR 32 # 1672.

1967 A theorem on mappings. Comment. Math. Univ. Carolinae **8**, 431–433 (1967). MR 37 # 4802.

1968 Products of filters. Comment. Math. Univ. Carolinae **9**, 173–189 (1968). MR 40 # 3496.

Keesling, J.:
1969A Open and closed mappings and compactification. Fund. Math. **65**, 73–81 (1969). MR 40 # 2003.

1969B Compactification and the continuum hypothesis. Fund. Math. **66**, 53–54 (1969/70). MR 41 # 1008.

Keisler, H.J.:
1966 Universal homogeneous Boolean algebras. Michigan Math. J. **13**, 129–132 (1966). MR 33 # 3968.

Kelley, J.:
1955 General Topology, Princeton: Van Nostrand 1955. MR 16, p. 1136.

Kennison, J.F.:
1962 m-pseudocompactness. Trans. Amer. Math. Soc. **104**, 436–442 (1962). MR 26 # 3009.

1965 Reflective functors in general topology and elsewhere. Trans. Amer. Math. Soc. **118**, 303–315 (1965). MR 30 # 4812.

1967 A note on reflection maps. Illinois J. Math. **11**, 404–409 (1967). MR 35 # 1649.

1968 Full reflective subcategories and generalized covering spaces. Illinois J. Math. **12**, 353–365 (1968). MR 37 # 2832.

Kim, J.:
1972 Sequentially complete spaces. J. Korean Math. Soc. **9**, 39–43 (1972). MR 46 # 2645.

Kroonenberg, N.S.:
1971 A topological compact Hausdorff space with countably many isolated points in which sets of isolated points cannot be left out. Bull. Acad. Polon. Sci. Sér. Sci. Math. Astronom. Phys. **19**, 501–503 (1971). (Russian summary.) MR 46 # 2642.

Kunen, K.:
1970 On the compactification of the integers. Notices Amer. Math. Soc. **17**, 299 (1970). Abstract # 70 T–G 7.

Liu, C.T. and Strecker, G.E.:
1972 Concerning almost realcompactifications. Czechoslovak Math. J. **22**, 181–190 (1972).

Loeb, P.A.:
1969 Compactifications of Hausdorff spaces. Proc. Amer. Math. Soc. **22**, 627–634 (1969). MR 39 # 6263.

Lozier, F.W.:
1969 A class of compact rigid 0-dimensional spaces. Canad. J. Math. **21**, 817–821 (1969). MR 39 # 6243.

Lynn, F.:
1970 A theory of generalized filters, Masters Thesis, Naval Postgraduate School, Monterey, California, 1970.

Mack, J., Rayburn, M. and Woods, R. G.:
1974   Lattices of topological extensions. Trans. Amer. Math. Soc. **189**, 163–174 (1974).

MacLane, S.:
1971   Categories for the Working Mathematician, Graduate Texts in Mathematics, Vol. 5, Berlin-Heidelberg-New York: Springer 1971.

Magill, K. D., Jr.:
1965 N-point compactifications. Amer. Math. Monthly **72**, 1075–1081 (1965). MR 32 # 3036.
1966A   Countable compactifications. Canad. J. Math. **18**, 616–620 (1966). MR 33 # 6578.
1966B   A note on compactifications. Math. Z. **94**, 322–325 (1966). MR 34 # 3530.
1968   The lattice of compactifications of a locally compact space. Proc. London Math. Soc. **18**, 231–244 (1968). MR 37 # 4783.
1970   More on remainders of spaces in compactifications. Bull. Acad. Polon. Sci. Sér. Sci. Math. Astronom. Phys. **18**, 449–451 (1970). MR 42 # 3751.

Mandelker, M.:
1968A   Prime z-ideal structure of $C(\mathbb{R})$. Fund. Math. **63**, 145–166 (1968). MR 38 # 2589.
1968B   Prime z-ideal structure in rings of bounded continuous functions. Proc. Amer. Math. Soc. **19**, 1432–1438 (1968). MR 37 # 6761.
1969   Round z-filters and round subsets of βX. Israel J. Math. **7**, 1–8 (1969). MR 39 # 6264.
1971   Supports of continuous functions. Trans. Amer. Math. Soc. **156**, 73–83 (1971). MR 43 # 1124.

Marczewski, E.:
1947   Séparabilitié et multiplication cartésienne des espaces topologiques. Fund. Math. **34**, 127–143 (1947). MR 9, p. 98.

Marjanović, M. M.:
1971   A pseudocompact space having no dense countably compact subspace. Glasnik Mat. Ser. III **6**, 149–151 (1971). (Serbo-Croatian summary.) MR 46 # 856.

Martin, D. A. and Solovay, R. M.:
1970   Internal Cohen extensions. Ann. of Math. Logic **2**, 143–178 (1970). MR 42 # 5787.

Mazurkiewicz, S.:
1920   Sur les lignes de Jordan. Fund. Math. **1**, 166–209 (1920).

McArthur, W. G.:
1970   Hewitt realcompactification of products. Canad. J. Math. **22**, 645–656 (1970). MR 42 # 1069.
1973   Realcompactifications of products of ordered spaces. Proc. Amer. Math. Soc. **38**, 186–192 (1973). MR 47 # 1027.

McAuley, L. F.:
1956   Paracompactness and an example due to F. B. Jones. Proc. Amer. Math. Soc. **7**, 1155–1156 (1956). MR 18, p. 496.

McDowell, R. H.:
1958   Extension of functions from dense subspaces. Duke Math. J. **25**, 297–304 (1958). MR 20 # 4251.

Michael, E.:
1971   A theorem on perfect maps. Proc. Amer. Math. Soc. **28**, 633–634 (1971). MR 43 # 1144.

Misra, A. K.:
1970   Spaces in which G-delta sets are open, Thesis, Indian Institute of Technology, Kanpur, 1970.

1972   A topological view of P-spaces. General Topology and Appls. **2**, 349–362 (1972). MR 47 # 5851.

Mitchell, B.:
1965   Theory of Categories, New York: Academic Press 1965. MR 34 # 2647

Morita, K.:
1961   A note on paracompactness. Proc. Japan Acad. **37**, 1–3 (1961). MR 25 # 3502.
1964   Products of normal spaces with metric spaces. Math. Ann. **154**, 365–382 (1964). MR 29 # 2773.
1970A   Topological completions and M-spaces. Sci. Rep. Tokyo Kyoiku Daigaku Sect. A **10**, 271–288 (1970). MR 42 # 6785.
1970B   Paracompactifications of M-spaces. Proc. Japan Acad. **46**, 511–513 (1970). MR 43 # 8050.
1971   A survey of the theory of M-spaces. General Topology and Appls. **1**, 49–55 (1971). MR 44 # 3276.

Mrówka, S.:
1954   On completely regular spaces. Fund. Math. **41**, 105–106 (1954). MR 16, p. 157.
1956   On quasi-compact spaces. Bull. Acad. Polon. Sci. Sér. Sci. Math. Astronom. Phys. **4**, 483–484 (1956). MR 18, p. 590.
1957   Some properties of Q-spaces. Bull. Acad. Polon. Sci Sér. Sci. Math. Astronom. Phys. **5**, 947–950 (1957). (Russian summary.) MR 20 # 1967.
1958A   A property of Hewitt-extension vX of topological spaces. Bull. Acad. Polon Sci. Sér. Sci. Math. Astronom. Phys. **6**, 95–96 (1958). MR 20 # 3521.
1958B   On the unions of Q-spaces. Bull. Acad. Polon. Sci. Sér. Sci. Math. Astronom. Phys. **6**, 365–368 (1958). MR 21 # 1572.
1959   On the potency of subsets of $\beta\mathbb{N}$. Colloq. Math. **7**, 23–25 (1959). MR 22 # 5018.
1966   On E-compact spaces II. Bull. Acad. Polon. Sci. Sér. Sci. Math. Astronom. Phys. **14**, 597–605 (1966). MR 34 # 6712.
1968   Further results on E-compact spaces. Acta. Math. **120**, 161–185 (1968). MR 37 # 2165.
1970   Some comments on the author's example of a non-$\mathscr{R}$-compact space. Bull. Acad. Polon. Sci. Sér. Sci. Math. Astronom. Phys. **18**, 443–448 (1970). MR 42 # 3749.
1971A E-complete regularity and E-compactness. General Topology and its Relations to Modern Analysis and Algebra, III (Proc. Conf., Kanpur, 1968), pp. 207–214. Prague: Academia 1971.
1971B β-like compactifications. General Topology and its Relations to Modern Analysis and Algebra, III (Proc. Conf., Kanpur, 1968), pp. 215–217. Prague: Academia 1971.
1971C   Some consequences of Archangelskii's theorem. Bull. Acad. Polon. Sci. Sér. Sci. Math. Astronom. Phys. **19**, 373–376 (1971). MR 46 # 2622.
1972   Recent results on E-compact spaces and structures of continuous functions. Proceedings of the University of Oklahoma Topology Conference, 1972, pp. 168–221.
1973 $\mathbb{N}$-compactness and strong 0-dimensionality. Notices Amer. Math. Soc. **20**, A-534 (1973). Abstract # 706-54-7.
1974   Recent results on E-compact spaces. In: Proceedings of the Second Pittsburgh International Conference on General Topology and Its Applications, 1972, Lecture Notes in Mathematics, Vol. 378, pp. 298–301. Berlin-Heidelberg-New York: Springer 1974.

Nachbin, L..
1954   Topological vector spaces of continuous functions. Proc. Nat. Acad. Sci. U.S.A. **40**, 471–474 (1954). MR 16, p. 156.

1970   Sur les espaces vectoriels topologiques d'applications continues. C.R. Acad. Sci. Paris Sér. A–B **271**, A596–A598 (1970). MR 42 # 6593.

Nagata, J.:
1968   Modern General Topology, Amsterdam: North-Holland Publishing Company 1968. MR 41 # 9171.
1969   Some aspects of extension theory in general topology. Contributions to Extension Theory of Topological Structures (Proc. Sympos., Berlin, 1967), pp. 157–161. Berlin: Deutscher Verlag Wissensch. 1969.

Nakamura, M. and Kakutani, S.:
1943   Banach limits and the Čech compactification of a countable discrete set. Proc. Imp. Acad. Tokyo **19**, 224–229 (1943). MR 7, p. 306.

Negrepontis, S.:
1967A   Absolute Baire sets. Proc. Amer. Math. Soc. **18**, 691–694 (1967). MR 35 # 4883.
1967B   A note on the r-compactifications of some discrete sets. Arch. Math. **18**, 264–266 (1967). MR 36 # 852.
1967C   Baire sets in topological spaces. Arch. Math. **18**, 603–608 (1967). MR 36 # 3314.
1969A   Extension of continuous functions in βD. Contributions to Extension Theory of Topological Structures (Proc. Sympos., Berlin, 1967), pp. 163–169. Berlin: Deutscher Verlag Wissensch. 1969. MR 40 # 6509.
1969B   On the product of F-spaces. Trans. Amer. Math. Soc. **136**, 339–346 (1969). MR 38 # 2724.
1969C   An example on realcompactifications. Arch. Math. (Basel) **20**, 162–164 (1969). MR 39 # 6265.
1969D   The Stone space of the saturated Boolean algebras. Trans. Amer. Math. Soc. **141**, 515–527 (1969). MR 40 # 1311.

Njastad, O.:
1966   On Wallman-type compactifications. Math. Z. **91**, 267–276 (1966). MR 32 # 6404.

Noble, N.:
1967   A generalization of a theorem of A. H. Stone. Arch. Math. **18**, 394–395 (1967). MR 36 # 5883.
1969A   A note on z-closed projections. Proc. Amer. Math. Soc. **23**, 73–76 (1969). MR 39 # 7575.
1969B   Products with closed projections. Trans. Amer. Math. Soc. **140**, 381–391 (1969). MR 40 # 3500.
1969C   Countably compact and pseudo-compact products. Czechoslovak Math. J. **19**, 390–397 (1969). MR 40 # 1968.
1970   The continuity of functions on Cartesian products. Trans. Amer. Math. Soc. **149**, 187–198 (1970). MR 41 # 2636.
1971   Products with closed projections II. Trans. Amer. Math. Soc. **160**, 169–183 (1971). MR 44 # 979.
1972 C-embedded subsets of products. Proc. Amer. Math. Soc. **31**, 613–614 (1972). MR 44 # 2202.

Noble, N. and Ulmer, M.:
1972   Factoring functions on Cartesian products. Trans. Amer. Math. Soc. **163**, 329–339 (1972). MR 44 # 5917.

Novák, J.:
1953A   On some problems of Luzin ·concerning the subsets of natural numbers. Czechoslovak Math. J. **3**, 385–395 (1953). (Russian. English summary.) MR 16, p. 20.
1953B   On the Cartesian product of two compact spaces. Fund. Math. **40**, 106–112 (1953). MR 15, p. 640.

1955   On a problem concerning completely regular sets. Fund. Math. **41**, 103–104 (1955). MR 16, p. 157.

Nyikos, P.:
1971A N-compact spaces, Thesis, Carnegie-Mellon University, 1971.
1971B   Not every 0-dimensional realcompact space is N-compact. Bull. Amer. Math. Soc. **77**, 392–396 (1971). MR 43 # 8048.
1973   Prabır Roy's space Δ is not N-compact. General Topology and Appls. **3**, 197—210 (1973).

Onuchic, N.:
1957 P-spaces and the Stone-Čech compactification. An. Acad. Brasil. Ci. **29**, 43–45 (1957).
1960   On the Nachbin uniform structure. Proc. Amer. Math. Soc. **11**, 177–179 (1960). MR 26 # 6930.

Parovičenko, I. I.:
1963   On a universal bicompactum of weight $\aleph$. Dokl. Akad. Nauk. SSSR **150**, 36–39 (1963). (Russian.) English translation: Soviet Mathematics **4**, 592–595 (1963). MR 27 # 719.

Pfeifer, G. L.:
1969   The Stone-Čech compactification of an irreducibly connected space. Proc. Amer. Math. Soc. **20**, 531–532 (1969). MR 38 # 3828.

Pierce, R. S.:
1958   A note on complete Boolean algebras. Proc. Amer. Math. Soc. **9**, 892–896 (1958). MR 21 # 1280.
1967   Modules over commutative regular rings. Memoirs of the American Mathematical Society, No. 70, Providence: American Mathematical Society 1967. MR 36 # 151.

Plank, D.:
1966   On a class of subalgebras of $C(X)$ with applications to $\beta X \setminus X$, Thesis, University of Rochester, 1966.
1969   On a class of subalgebras of $C(X)$ with applications to $\beta X - X$. Fund. Math. **64**, 41–54 (1969). MR 39 # 6266.

Pondiczery, E. S.:
1944   Power problems in abstract spaces. Duke Math. J. **11**, 835–837 (1944).

Pospíšil, B.:
1937   Remark on bicompact spaces. Ann. of Math. **38**, 845–846 (1937).

Purisch, S.:
1973A   On the suborderability of metric spaces and Stone-Čech compactifications. Notices Amer. Math. Soc. **20**, A 182 (1973). Abstract # 701-54-49.
1973B   The orderability, and suborderability of topological spaces, Thesis, Carnegie-Mellon University, Pittsburgh, 1973.
        On the orderability of Stone-Čech compactifications, Proc. Amer. Math. Soc. (to appear).

Raimi, R.:
1966   Homeomorphisms and invariant measures for $\beta N - N$. Duke Math. J. **33**, 1–12 (1966). MR 33 # 6608.
1968   Translation properties of finite partitions of the positive integers. Fund. Math. **61**, 253–256 (1967/68). MR 36 # 5924.

Rainwater, J.:
1959   A note on projective resolutions. Proc. Amer. Math. Soc. **10**, 734–735 (1959). MR 23 # 618.

Rajagopalan, M.:
1972 $\beta\mathbb{N}\setminus\mathbb{N}\setminus\{p\}$ is not normal. Journal of the Indian Math. Soc. **36**, 173–176 (1972).

Richardson, G. D.:
1970 A Stone-Čech compactification for limit spaces. Proc. Amer. Math. Soc. **25**, 403–404 (1970). MR 41 # 992.

Robinson, S. M.:
1966 The intersection of the free maximal ideals in a complete space. Proc. Amer. Math. Soc. **17**, 468–469 (1966). MR 32 # 6401.
1969A A note on the intersection of free maximal ideals. J. Austral. Math. Soc. **10**, 204–206 (1969). MR 39 # 7438.
1969B Some properties of $\beta X - X$ for complete spaces. Fund. Math. **64**, 335–340 (1969). MR 39 # 6267.

Rogers, J. W., Jr.:
1969 On compactifications with continua as remainders. Notices Amer. Math. Soc. **16**, 1045 (1969). Abstract # 669–9.

Roy, P.:
1962 Failure of equivalence of dimension concepts for metric spaces. Bull. Amer. Math. Soc. **68**, 609–613 (1962). MR 25 # 5495.
1968 Nonequality of dimensions for metric spaces. Trans. Amer. Math. Soc. **134**, 117–132 (1968). MR 37 # 3544.

Rudin, M. E.:
1956 A separable normal nonparacompact space. Proc. Amer. Math. Soc. **7**, 940–941 (1956). MR 18, p. 429.
1966 Types of ultrafilters. Topology Seminar (Wisconsin, 1965), pp. 147–151, Ann. of Math. Studies, No. 60, Princeton: Princeton Univ. Press 1966. MR 35 # 7284.
1970 Composants and $\beta\mathbb{N}$. Proc. Wash. State Univ. Conf. on Gen. Topology (Pullman, Wash., 1970), pp. 117–119. MR 42 # 1070.
1971 Partial orders on the types in $\beta\mathbb{N}$. Trans. Amer. Math. Soc. **155**, 353–362 (1971). MR 42 # 8459.

Rudin, W.:
1956 Homogeneity problems in the theory of Čech compactifications. Duke Math. J. **23**, 409–419, 633 (1956). MR 18, p. 324.

Ryll-Nardzewski, C.:
1968 Remark on Raimi's theorem on translations. Fund. Math. **61**, 257–258 (1967/68). MR 36 # 5925.

Ryll-Nardzewski, C. and Telgarsky, R.:
1970 On the scattered compactification. Bull. Acad. Polon. Sci. Sér. Sci. Math. Astronom. Phys. **18**, 233–234 (1970). MR 41 # 7635.

Saks, S. and Stephenson, R. M., Jr.:
1971 Products of m-compact spaces. Proc. Amer. Math. Soc. **28**, 279–288 (1971). MR 42 # 8448.

Samuel, P.:
1948 Ultrafilters and compactifications of uniform spaces. Trans. Amer. Math. Soc. **64**, 100–132 (1948). MR 10, p. 54.

Shirota, T.:
1952 A class of topological spaces. Osaka J. Math. **4**, 23–40 (1952). MR 14, p. 395.

Sierpiński, W.:
1928 Sur une decomposition d'ensembles. Monatsh. Math. **35**, 239–248 (1928).

1965   Cardinal and Ordinal Numbers, Second Revised Edition, Warsaw: Polish Scientific
         Publishers 1965. MR 20 # 2288, MR 33 # 2549.

Sikorski, R.:
1948   A theorem on extensions of homomorphisms. Ann. Soc. Polon. Math. **21**, 332–335
         (1948). MR 11, p. 76.
1964   Boolean Algebras, Ergebnisse der Mathematik und ihrer Grenzgebiete, Band 25,
         Berlin-Göttingen-Heidelberg-New York: Springer 1964. MR 31 # 2178.

Simmons, G.:
1963   Introduction to Topology and Modern Analysis, New York: McGraw-Hill 1963.
         MR 26 # 4145.

Simon, B.:
1969   Some pictorial compactifications of the real line. Amer. Math. Monthly **76**, 536–538
         (1969). MR 39 # 7565.

Smirnov, Yu. M.:
1952   On proximity spaces. Mat. Sb. **31**, 543–574 (1952). (Russian.) MR 14, p. 1107.
1966A   On the dimension of remainders in bicompact extensions of proximity and topolo-
         gical spaces. Mat. Sb. **69 (111)**, 141–160 (1966). (Russian.) MR 33 # 6579.
1966B   Dimension of increments of proximity spaces and of topological spaces. Dokl.
         Akad. Nauk SSSR **168**, 528–531 (1966). (Russian.) English translation: Soviet
         Math. Dokl. 7, 688–692 (1966). MR 35 # 4886.
1966C   On the dimension of remainders in bicompact extensions of proximity and topolo-
         gical spaces II. Mat Sb. **71 (113)**, 454–482 (1966). (Russian.) MR 35 # 4887.
1967   Proximity and construction of compactifications with given properties. General
         Topology and its Relations to Modern Analysis and Algebra, II (Proc. Sympos.,
         Prague, 1966), pp. 332–340, Prague: Academia 1967.

Snyder, A. K..
1969   The Čech compactification and regular matrix summability. Duke. Math. J. **36**,
         245–252 (1969). MR 40 # 2005.

Solomon, R. C.:
1973   A type of $\beta\mathbb{N}$ with $\aleph_0$ relative types. Fund. Math. **69**, 209–212 (1973).

Solovay, R. M. and Tennenbaum, S.:
1971   Iterated Cohen extensions and Souslin's problem. Ann. of Math. **94**, 201–245
         (1971). MR 45 # 3212.

Steen, L. A. and Seebach, J. A., Jr.:
1970   Counterexamples in Topology, New York: Holt, Rinehart, and Winston 1970.
         MR 42 # 1040.

Steenrod, N. E.:
1967   A convenient category of topological spaces. Michigan Math. J. **14**, 133–152
         (1967). MR 35 # 970.

Steiner, A. K.:
1971   On the topological completion of M-space products. Proc. Amer. Math. Soc. **29**,
         617–620 (1971). MR 43 # 8051.

Steiner, A. K. and Steiner, E. F.:
1968A   Compactifications as closures of graphs. Fund. Math. **63**, 221–223 (1968).
         MR 38 # 6546.
1968B   Precompact uniformities and Wallman compactifications. Indag. Math. **30**,
         117–118 (1968). MR 37 # 3525.

1968C   Wallman and Z-compactifications. Duke Math. J. **35**, 269–275 (1968). MR 37#3526.
1969   On countable multiple point compactifications. Fund. Math. **65**, 133–137 (1969). MR 40#868.
1970A   Graph closures and metric compactifications of $\mathbb{N}$. Proc. Amer. Math. Soc. **25**, 593–597 (1970). MR 41#9206.
1970B   Nest generated intersection rings in Tychonoff spaces. Trans. Amer. Math. Soc. **148**, 589–601 (1970). MR 41#7637.
1971   Relative types of points in $\beta\mathbb{N}\setminus\mathbb{N}$. Trans. Amer. Math. Soc. **160**, 279–286 (1971).
1972   Binding spaces: A unified completion and extension theory. Fund. Math. **76**, 43–61 (1972).

Steiner, E. F.:
1966   Normal families and completely regular spaces. Duke Math. J. **33**, 743–745 (1966). MR 33#7975.
1968   Wallman spaces and compactifications. Fund. Math. **61**, 295–304 (1967/68). MR 36#5899.

Stephenson, R. M., Jr.:
1968   Pseudocompact spaces. Trans. Amer. Math. Soc. **134**, 437–448 (1968). MR 38#674.
1969   Product spaces for which the Stone-Weierstrass theorem holds. Proc. Amer. Math. Soc. **21**, 284–288 (1969). MR 40#3499.

Stone, A. H.:
1948   Paracompactness and product spaces. Bull. Amer. Math. Soc. **54**, 977–982 (1948). MR 10, p. 204.

Stone, M. H.:
1936   The theory of representation for Boolean algebras. Trans. Amer. Math. Soc. **40**, 37–111 (1936).
1937   Applications of the theory of Boolean rings to general topology. Trans. Amer. Math. Soc. **41**, 375–481 (1937). Z 17, p. 135.
1937A   Algebraic characterizations of special Boolean rings. Fund. Math. **29**, 223–303 (1937).
1948   On the compactification of topological spaces. Ann. Soc. Polon. Math. **21**, 153–160 (1948). MR 10, p. 137.

Strauss, D. P.:
1967   Extremally disconnected spaces. Proc. Amer. Math. Soc. **18**, 305–309 (1967). MR 35#961.

Sundaresan, K.:
1971   Spaces of continuous functions into $\mathbb{R}^2$, Research Report 71-7, Department of Mathematics, Carnegie-Mellon University, 1971.
1973   Spaces of continuous functions into a Banach space. Studia. Math. **48**, 15–23 (1973).

Taĭmanov, A. D.:
1952   On extension of continuous mappings of topological spaces. Mat. Sb. **31**, 459–463 (1952). (Russian.) MR 14, p. 395.

Tamano, H.:
1960A   A note on the pseudocompactness of the product of two spaces. Mem. Coll. Sci. Univ. of Kyoto, Ser. A, Math, **33**, 225–230 (1960). MR 22#11369.
1960B   Properties of the Stone-Čech compactification. J. Math. Soc. Japan **12**, 104–117 (1960). MR 26#6929.
1960C   On paracompactness. Pacific J. Math. **10**, 1043–1047 (1960). MR 23#A2186.
1962   On compactifications. J. Math. Kyoto Univ. **1**, 161–193 (1961/62). MR 25#5489.

1969   The role of compactifications in the theory of Tychonoff spaces. Contributions to Extension Theory of Topological Structures (Proc. Sympos., Berlin, 1967), pp. 219–220, Berlin: Deutscher Verlag Wissensch. 1969.

Tarski, A.:
1928   Sur la decomposition des ensembles en sous-ensembles presque disjoints. Fund. Math. **12**, 188–205 (1928).

Terasaka, H.:
1952   On the cartesian product of compact spaces. Osaka J. Math. **4**, 11–15 (1952). MR 14, p. 489.

Thrivikraman, T.:
1972A   On the lattices of compactifications. J. London Math. Soc. **4**, 711–717 (1972). MR 45 # 5961.
1972B   On compactifications of Tychonoff spaces. Yokohama Math. J. **20**, 99–105 (1972). MR 46 # 4481.

Tong, H.:
1949   On some problems of Čech. Ann. of Math. **50**, 154–157 (1949). MR 10, p. 315.
1970   Solutions of problems of P.S. Alexandroff on extensions of topological spaces. Ann. Mat. Pura Appl. **86**, 47–51 (1970). MR 43 # 1127.

Tychonoff, A.:
1930   Über die topologische Erweiterung von Räumen. Math. Ann. **102**, 544–561 (1930).

Urysohn, P.:
1925A   Zum Metrisationsproblem. Math. Ann. **94**, 309–315 (1925).
1925B   Über die Mächtigkeit der zusammenhängenden Mengen. Math. Ann. **94**, 262–295 (1925).

van der Slot, J.:
1969   An elementary proof of the Hewitt-Shirota theorem. Compositio Math. **21**, 182—184 (1969). MR 40 # 1970.
1971   A note on perfect irreducible mappings. Bull. Acad. Polon. Sci. Sér. Sci. Math. Astronom. Phys. **19**, 377—382 (1971). (Russian summary.) MR 46 # 6272.

Vedenisov, N. B.:
1948   Bicompact spaces. Uspehi Mat. Nauk. **3**, 67—79 (1948). (Russian.) MR 10, p. 137.

Venkataraman, M., Rajagopalan, M., and Soundararajan, T.:
1972   Orderable topological spaces. General Topology and Appls. **2**, 1—10 (1972). MR 45 # 7683.

Wagner, F. J.:
1957.   Notes on compactification. I, II. Indag. Math. **19**, 171–176, 177–181 (1957). MR 19, p. 436.
1964   Normal base compactifications. Indag. Math. **26**, 78–83 (1964). MR 28 # 2522.

Wallace, A. D.:
1951   Extensional invariance. Trans. Amer. Math. Soc. **70**, 97—102 (1951). MR 12, p. 845.

Wallman, H.:
1938   Lattices and topological spaces. Ann. of Math. **39**, 112—126 (1938).

Warner, S.:
1958   The topology of compact convergence on continuous function spaces. Duke Math. J. **25**, 265—282 (1958). MR 21 # 1521.

Warren, N. M.:
1970   Extending continuous functions in Stone-Čech compactifications of discrete

spaces and in zero-dimensional spaces, Thesis, University of Wisconsin, Madison, Wisconsin, 1970.
1972   Properties of Stone-Čech compactifications of discrete spaces. Proc. Amer. Math. Soc. **33**, 599—606 (1972). MR 45 # 1123.

Weil, A.:
1937   Sur les espaces à structure uniforme et sur la topologie générale, Actualités Scientifiques et Industrielles, No. 551, Paris: Hermann 1937.

Weir, M.:
1970   A treatise on realcompactness, Thesis, Carnegie-Mellon University, 1970.

Willard, S.:
1966   Absolute Borel sets in their Stone-Čech compactifications. Fund. Math. **58**, 323–333 (1966). MR 33 # 4892.
1970   General Topology, Reading: Addison-Wesley 1970. MR 41 # 9173.

Woods, R. G.:
1968   Certain properties of $\beta X \setminus X$ for σ-compact X, Thesis, McGill University, Montreal, 1968.
1971A   A Boolean algebra of regular closed subsets of $\beta X - X$. Trans. Amer. Math. Soc. **154**, 23—36 (1971). MR 42 # 5230.
1971B   Co-absolutes of remainders of Stone-Čech compactifications. Pacific J. Math. **37**, 545–560 (1971); **39**, 827 (1971). MR 46 # 6300.
1971C   Homeomorphic sets of remote points. Canad. J. Math. **23**, 495–502 (1971). MR 43 # 6880.
1971D   Some $\aleph_0$-bounded subsets of Stone-Čech compactifications. Israel J. Math. **9**, 250–256 (1971). MR 43 # 3997.
1972A   Ideals of pseudocompact regular closed sets and absolutes of Hewitt realcompactifications. General Topology and Appls. **2**, 315–331 (1972).
1972B   On the local connectedness of $\beta X - X$. Canad. Math. Bull. **15**, 591–594 (1972). MR 47 # 2556.
1973   Maps that characterize normality properties and pseudocompactness. J. London Math. Soc. **7**, 453–461 (1973).
1974   A Tychonoff almost realcompactification. Proc. Amer. Math. Soc. **43**, 200–208 (1974).

Wulbert, D. E.:
1969A   Locally connected Stone-Čech compactifications. Contributions to Extension Theory of Topological Structures (Proc. Sympos., Berlin, 1967), pp. 247–248, Berlin: Deutscher Verlag. Wissensch. 1969. MR 39 # 7566.
1969B   A characterization of $C(X)$ for locally connected X. Proc. Amer. Math. Soc. **21**, 269–272 (1969). MR 39 # 1951.

Zenor, P.:
1970   Extending completely regular spaces with inverse limits. Glasnik Mat. Ser III **5**, 157–162 (1970). MR 43 # 1128.

List of Symbols